SELF-ORGANIZING NETWORKS

SELF-ORGANIZING NETWORKS

SELF-PLANNING, SELF-OPTIMIZATION AND SELF-HEALING FOR GSM, UMTS AND LTE

Editors

Juan Ramiro
Ericsson, Málaga, Spain

Khalid Hamied
Ericsson, Atlanta, GA, USA

WILEY

A John Wiley & Sons, Ltd., Publication

Registered Office
John Wiley & Sons Ltd, The Atrium, Southern Gate, Chichester, West Sussex, PO19 8SQ, United Kingdom

For details of our global editorial offices, for customer services and for information about how to apply for permission to reuse the copyright material in this book please see our website at www.wiley.com.

Library of Congress Cataloging-in-Publication Data

Self-organizing networks : self-planning, self-optimization and self-healing for GSM, UMTS and LTE / editors, Juan Ramiro, Khalid Hamied.
    p.   cm.
  Includes index.
  Summary: "The book offers a multi-technology approach as it will consider the implications of the different SON requirements for 2G and 3G networks, as well as 4G wireless technologies"– Provided by publisher.
  ISBN 978-0-470-97352-3 (hardback)
  1. Wireless communication systems–Automatic control.   2. Self-organizing systems.   I. Ramiro, Juan.
II. Hamied, Khalid.
  TK5103.2.S46 2011
  003'.7–dc23
                                        2011025918

A catalogue record for this book is available from the British Library.

Print ISBN: 9780470973523 (H/B)
ePDF ISBN: 9781119954217
oBook ISBN: 9781119954224
ePub ISBN: 9781119960928
mobi ISBN: 9781119960935

Set in 10/12pt Times Roman by SPi Publisher Services, Pondicherry, India
Printed and bound by CPI Group (UK) Ltd, Croydon, CR0 4YY

Contents

Foreword

Consumer uptake of mobile broadband represents the fastest adoption of any technology that our society has ever experienced. Faster than the Internet and earlier generations of mobile communications, the widespread use and acceptance of smartphone technology has been phenomenal. Tablets, Android devices, iPhones, application stores, social media and the data exchanges between end-users and clouds are all growing at near breakneck speed, and operators (our customers) are challenged to keep pace. Together as an industry, we are charged with providing the necessary bandwidth and capacity that people everywhere are coming to expect and anticipate.

The 4G LTE networks provide an incredible customer experience. I live in Stockholm, Sweden, and have had access to the LTE/4G network since TeliaSonera first launched it. It was the first commercial launch of an LTE network anywhere, and I was privileged to be a part of it. As a 4G subscriber, I have come to depend on the best possible connections wherever I find myself in the world. The same is true for subscribers of 3G/2G networks. Best possible connectivity plays an important role in my experience as a customer, and, like most customers, I expect it to work seamlessly.

Another important consideration of the 2G/3G evolution to LTE is the affordable operation of these overlaid multi-standard networks, during a time when operators are required to reduce operational expenditures. Most operators need to keep their OPEX at a constant level or even squeeze out additional operational efficiencies while they are enhancing their service capabilities and data capacity. Many opportunities including machine to machine (m2m) services are extremely price sensitive and cannot take off in full force until certain price thresholds are attained. The forecasted demand is clear, and therefore we must plan accordingly.

Densification of the network, smaller cells and more heterogeneous networks (HetNets) are coming rapidly. We anticipate that this ongoing change will bring more multi-standard demands and multi-vendor challenges. Efficient and effective operations must overcome such complexity, which is growing not linearly, but by factors. The only way these challenges can be cost-effectively, efficiently and humanely overcome is through the use of more automated and autonomous systems – such as Self-Organizing Networks (SON), and that is what this book is all about.

The opportunities we have are tremendous. We are enabling innovation while empowering people, businesses and society. This book is part of the story, and provides a good place to begin this ongoing discussion. There is certainly a lot more to come. You can count on it.

Ulf Ewaldsson
Vice President, Head of Ericsson Radio
Stockholm, Sweden

Preface

This book provides an in-depth description of multi-technology Self-Organizing Networks (SON) for 2G, 3G and Long Term Evolution (LTE), and presents critical business aspects of the proposed SON functionalities. Multi-technology SON allows operators to completely transform and streamline their operations, and extends the automation-related operational savings to all radio access technologies. The availability of a multi-technology SON solution results in more comprehensive, holistic and powerful optimization strategies that deal with several radio access technologies simultaneously.

The book is primarily aimed at engineers who manage or optimize mobile networks, though major network operators, business leaders and professionals of academia will also derive value from the included chapters. With the deployment of LTE, most operators will have three simultaneous radio access technologies to manage, which will add extra pressure to their already tight cost structures. Deploying and operating cellular networks is a complex task that comprises many activities, such as planning, dimensioning, installation, testing, prelaunch optimization, postlaunch optimization, comprehensive performance monitoring, failure mitigation, failure correction and general maintenance. Today, such critical activities are extremely labor-intensive and, hence, costly and prone to errors, which may result in customer dissatisfaction and increased churn.

In order to alleviate this situation, clear requirements have been stated by the Next Generation Mobile Networks (NGMN) Alliance to enable a set of functionalities for automated Self-Organization of LTE networks, so that human intervention is minimized in the planning, deployment, optimization and maintenance activities of these new networks. Subsequently, the support for this new network management paradigm is being translated into concrete functionalities, interfaces and procedures during the standardization of E-UTRAN in the Third Generation Partnership Project (3GPP).

The objective of this book is to discuss the state-of-the-art engineering and automation practices realizing the SON paradigm for multi-technology and multi-vendor wireless infrastructure. Each chapter has been carefully organized to give the reader a comprehensive, layered understanding of SON development and deployment with a glimpse into the industry's natural trending to automated engineering functions that optimize cellular network performance and maximize efficiency.

The layout of the book is structured as follows. Chapter 1 discusses the challenges associated with the explosive growth in mobile broadband and analyzes potential solutions. This chapter also introduces the unique solution that SON techniques provide in contemporary 2G/3G networks. Chapter 2 provides a high-level overview of SON by covering NGMN objectives and 3GPP activities. Advanced readers desiring a more comprehensive view of the 3GPP activities are encouraged to go directly to the 3GPP references provided within the chapter for the updated status of SON support in the standards. Chapter 3 mainly describes an architecture for multi-technology SON and provides a conceptual framework to coordinate SON functions.

Chapters 4, 5 and 6 cover the multi-vendor and multi-technology aspects of the Self-Planning, Self-Optimization and Self-Healing of wireless networks, respectively, covering processes, algorithms and enabling technologies for these activities. Engineers who manage or optimize mobile networks (operators' engineers and consultants) may be very interested in the material contained in these chapters.

Chapter 7 provides a model for Return-on-Investment (ROI) of the proposed SON functionalities and Use Cases. The included model can be applied to build business cases and provide ROI analyses for optimization projects. Chapter 7 would be of interest to managers, executives and sales professionals at operators and vendors.

Appendix A discusses geo-location technology for UMTS and describes the use and application of observed time differences (OTDs) for geo-location. Finally, Appendix B provides an overview of the X-map estimation for LTE and detailed simulation results of two different approaches for a given scenario.

Acknowledgements

First of all, the editors would like to thank all those who contributed to this book: Mehdi Amirijoo, Mark Austin, Rubén Cruz, Juan Carlos del Río, Patricia Delgado Quijada, Andreas Eisenblätter, Nizar Faour, Rafael Ángel García, Juanjo Guerrero, Gustavo Hylander, Thomas Kürner, Frank Lehser, Remco Litjens, Andreas Lobinger, Raúl Moya, Javier Muñoz, Christos Neophytou, Michaela Neuland, José Outes, Salvador Pedraza, Gabriel Ramos, Miguel A. Regueira Caumel, Philippe Renaut, Javier Romero, Lars Christoph Schmelz, Octavian Stan, Szymon Stefański, Ken Stewart, John Turk, Carlos Úbeda and Josko Zec. Moreover, we would like to express our gratitude to Neil Scully for facilitating all contributions from the SOCRATES project.

We also thank our colleagues at Ericsson for their continued support and encouragement and for providing suggestions and comments on the content. Special thanks to Josko Zec, Alejandro Gil, Miguel A. Regueira Caumel, Paul Cowling and Kai Heikkinen.

Additionally, the great assistance provided by Miguel A. Regueira Caumel, Marina Cañete and Jennifer Johnson during the editing process is greatly appreciated. During these months, the guidance and support provided by the great team at John Wiley & Sons, Ltd. has made a difference. Special thanks to Mark Hammond, Sophia Travis, Mariam Cheok, Lynette James and Suvesh Subramanian for all the help they provided.

Finally, we are immensely grateful to our families for their patience and support during the weekend and holiday editing work.

Juan Ramiro and Khalid Hamied

List of Contributors

Mehdi Amirijoo
Ericsson, Linköping, Sweden

Mark David Austin
AT&T, Atlanta, GA, USA

Rubén Cruz
Ericsson, Málaga, Spain

Juan Carlos del Río Romero
Ericsson, Madrid, Spain

Patricia Delgado Quijada
Ericsson, Málaga, Spain

Andreas Eisenblätter
atesio, Berlin, Germany

Nizar Faour
Ericsson, Atlanta, GA, USA

Rafael Ángel García Garaluz
Ericsson, Málaga, Spain

Juan José Guerrero García
Ericsson, Málaga, Spain

Khalid Hamied
Ericsson, Atlanta, GA, USA

Gustavo Hylander Aguilera
Ericsson, Stockholm, Sweden

Thomas Kürner
TU Braunschweig, Braunschweig,
Germany

Frank Lehser
Deutsche Telekom AG, Bonn, Germany

Remco Litjens
TNO, Delft, The Netherlands

Andreas Lobinger
Nokia Siemens Networks, Munich,
Germany

Manuel Raúl Moya de la Rubia
Ericsson, Málaga, Spain

Javier Muñoz
Ericsson, Málaga, Spain

Christos Neophytou
Ericsson, Atlanta, GA, USA

Michaela Neuland
TU Braunschweig, Braunschweig,
Germany

José Outes Carnero
Ericsson, Málaga, Spain

Salvador Pedraza
Ericsson, Málaga, Spain

Juan Ramiro Moreno
Ericsson, Málaga, Spain

Gabriel Ramos Escaño
Ericsson, Málaga, Spain

Miguel Angel Regueira Caumel
Ericsson, Málaga, Spain

Philippe Renaut
Ericsson, Málaga, Spain

Javier Romero
Ericsson, Málaga, Spain

Lars Christoph Schmelz
Nokia Siemens Networks, Munich, Germany

Octavian Stan
Ericsson, Atlanta, GA, USA

Szymon Stefański
Nokia Siemens Networks, Wrocław, Poland

Kenneth Stewart
Motorola, Libertyville, IL, USA

John Turk
Vodafone, Newbury, United Kingdom

Carlos Úbeda Castellanos
Ericsson, Madrid, Spain

Josko Zec
Ericsson, Atlanta, GA, USA

List of Abbreviations

2G	2nd Generation
3G	3rd Generation
3GPP	Third Generation Partnership Project
3GPP2	Third Generation Partnership Project 2
4G	4th Generation
A/C	Air Conditioning
AAC	Advance Audio Coding
AAC-ELD	AAC Enhanced Low Delay
AAC-LD	AAC Low Delay
AC	Admission Control
ACR	Adjacent Channel Rejection
AFP	Automatic Frequency Planning
aGW	access Gateway
AINI	ATM Inter-Network Interface
AMR	Adaptive Multi-Rate
AMR-NB	AMR Narrowband
AMR-WB	AMR Wideband
ANR	Automatic Neighbor Relation
AP	Access Point
API	Application Programming Interface
ARFCN	Absolute Radio Frequency Channel Number
ARP	Address Resolution Protocol
ARPU	Average Revenue Per User
AS	Active Set
ATC	Ancillary Terrestrial Component
ATM	Asynchronous Transfer Mode
AVC	Advance Video Coding
AWS	Advanced Wireless Services
BAL	BCCH Allocation List
BCC	Base Station Color Code
BCCH	Broadcast Control Channel
BCH	Broadcast Channel
BER	Bit Error Rate
BLER	Block Error Rate
BSC	Base Station Controller

BSIC	Base Station Identity Code
BTS	Base Transceiver Station
BWA	Broadband Wireless Access
C/I	Carrier-to-Interference
CAGR	Compounded Annual Growth Rate
CAPEX	Capital Expenditure
CBS	Carrier Branded Services
CCO	Capacity and Coverage Optimization
CDF	Cumulative Distribution Function
CDMA	Code Division Multiple Access
CDR	Charging Data Record
CE	Channel Element
CF	Cash Flow
CGI	Cell Global Identifier
CLPC	Fast Close-Loop Power Control
CM	Configuration Management
CN	Core Network
COC	Cell Outage Compensation
COD	Cell Outage Detection
COM	Cell Outage Management
CPC	Continuous Packet Connectivity
CPICH	Common Pilot Channel
CPU	Central Processing Unit
CQI	Channel Quality Indicator
CRC	Cyclic Redundancy Check
CS	Circuit Switched
CSFR	Call Setup Failure Rate
CSG	Closed Subscriber Group
CT	Call Traces
CT	Core network and Terminals
CW	Continuous Wave
DCR	Dropped Call Rate
DHCP	Dynamic Host Configuration Protocol
DL	Downlink
DM	Domain Manager
DNS	Domain Name System
DO-B	EV-DO Revision B
DO-C	EV-DO Revision C
DRX	Discontinuous Reception
DSP	Digital Signal Processing
DSS	Direct Sequence Spread Spectrum
EBITDA	Earnings Before Interest, Taxes, Depreciation and Amortization
ECGI	Evolved Cell Global Identifier
E-DCH	Enhanced Dedicated Channel
EDGE	Enhanced Data rates for GSM Evolution
EFL	Effective Frequency Load

EGPRS	Enhanced General Packet Radio Service
EIGRP	Enhanced Interior Gateway Routing Protocol
EM	Element Manager
eNodeB	enhanced NodeB
ES	Energy Saving
E-SMLC	Evolved Serving Mobile Location Centre
E-UTRA	Evolved Universal Terrestrial Radio Access
E-UTRAN	Evolved Universal Terrestrial Radio Access Network
EV-DO	Evolution Data Optimized
EVRC	Enhanced Variable Rate Codec
EVS	Enhance Voice Service
FACH	Forward Access Channel
FCC	Federal Communications Commission
FDD	Frequency Division Duplex
FER	Frame Erasure Rate
FFR	Full Frequency Reuse
FLV	Flash Video
FM	Fault Management
FPC	Fractional Power Control
GB	Gigabyte
GERAN	GSM EDGE Radio Access Network
GGSN	Gateway GPRS Serving Node
GNSS	Global Navigation Satellite System
GoS	Grade of Service
GPRS	General Packet Radio Service
GPS	Global Positioning System
GSM	Global System for Mobile Communications
GUI	Graphical User Interface
HCL	Hierarchical Cell Layout
HDMI	High-Definition Multimedia Interface
HFR	Hard Frequency Reuse
HI	Health Indicator
HII	High Interference Indicator
HLS	HTTP Live Steaming
HO	HandOver
HR	Half Rate
HRPD	High Rate Packet Data
HSDPA	High Speed Downlink Packet Access
HSDPA+	High Speed Downlink Packet Access Evolution
HSN	Hopping Sequence Number
HSPA	High Speed Packet Access
HSPA+	High Speed Packet Access Evolution
HS-PDSCH	High Speed Physical Downlink Shared Channel
HS-SCCH	High Speed Shared Control Channel
HSUPA	High Speed Uplink Packet Access
HTTP	Hyper Text Transfer Protocol

HEVC	High Efficiency Video Codec
HW	Hardware
IAF	Intra-frequency neighbor-related
IC	Interference Control
ICIC	Inter-Cell Interference Coordination
ID	Identity
IEC	International Electrotechnical Comission
IEEE	Institute of Electrical and Electronics Engineers
IEF	Inter-frequency neighbor-related
IGRP	Interior Gateway Routing Protocol
IM	Interference Matrix
IMT-Advanced	International Mobile Telecommunications - Advanced
IoT	Interfrence-over-Thermal
IP	Internet Protocol
iRAT	inter-Radio Access Technology
ISG	Inter-system (to GSM) neighbor-related
ISL	Inter-system (to LTE) neighbor-related
ISM	Industrial, Scientific and Medical
ISO	International Organization for standarization
IT	Information Technology
Itf-N	Northbound Interface
ITU	International Telecommunications Union
ITU-T	ITU-Standarization
JCT-VC	Joint Collaborative Team on Video Coding
KPI	Key Performance Indicator
LAC	Location Area Code
LBO	Load Balance Optimization
LIPA	Local IP Access
LLC	Link Layer Control
LMU	Location and Measurement Unit
LOS	Line-of-Sight
LTE	Long Term Evolution
MAHO	Mobile Assisted HandOver
MAIO	Mobile Allocation Index Offset
MAL	Mobile Allocation List
MC	Monte Carlo
MCS	Modulation and Coding Scheme
MDT	Minimization of Drive Tests
MGW	Media Gateway
MIB	Management Information Base
MIMO	Multiple Input Multiple Output
MLB	Mobility Load Balancing
mm	man month
MME	Mobility Management Entity
MML	Man-Machine Language
MMR	Mobile Measurement Recordings

MMS	Multi-media Messaging Service
MO	Managed Object
MP3	MPEG-1 Layer-3
MP4	MPEG-4 Part 14
MPEG-4	Motion Picture Expert Group Layer-4 Video
MR	Measurement Report
MRO	Mobility Robustness Optimization
MSC	Mobile Switching Centre
MSS	Mobile Satellite Service
MS-SPRing	Multiplex Section-Shared Protection Ring
NBP	National Broadband Plan
NE	Network Element
NEM	Network Element Manager
NGMN	Next Generation Mobile Networks
NII	National Information Infrastructure
NLOS	Non-Line-of-Sight
NM	Network Manager
NMS	Network Management System
NP-hard	Non-deterministic Polynomial-time hard
NPV	Net Present Value
NR	Neighbor Relation
NRT	Neighbor Relation Table
O&M	Operation And Maintenance
OFDM	Orthogonal Frequency Division Multiplexing
OFDMA	Orthogonal Frequency Division Multiple Access
OI	Overload Indicator
OIPF	Open IPTV Forum
OLPC	Outer-Loop Power Control
OPEX	Operational Expenditure
OS	Operations System
OSPF	Open Shortest Path First
OSS	Operations Support System
OTD	Observed Time Difference
OTDOA	Observed Time Difference Of Arrival
PCCPCH	Primary Common Control Physical Channel
PCG	Project Coordination Group
PCH	Paging Channel
PCI	Physical Cell Identifier
PCM	Pulse Code Modulation
PCU	Packet Control Unit
PD	Propagation Delay
PDF	Probability Density Function
PDH	Plesiochronous Digital Hierarchy
PDP	Packet Data Protocol
PLMN	Public Land Mobile Network
PM	Performance Management

PNNI	Private Network-to-Network Interaface
PRACH	Physical Random Access Channel
PRB	Physical Resource Block
PS	Packet Switched
PSC	Primary Scrambling Code
QAM	Quadrature Amplitude Modulation
QoS	Quality of Service
R6	Release 6
R99	Release 99
RAB	Radio Access Bearer
RAC	Routing Area Code
RACH	Random Access Channel
RAN	Radio Access Network
RAS	Remote Azimuth Steering
RAT	Radio Access Technology
RB	Resource Block
RET	Remote Electrical Tilt
RF	Radio Frequency
RLC	Radio Link Control
RLF	Radio Link Failure
RLS	Recursive Least Squares
RMU	RAN Measurement Unit
RNC	Radio Network Controller
RNTP	Relative Narrowband Transmit Power
ROI	Return On Investment
RRC	Radio Resource Control
RRM	Radio Resource Management
RS	Reference Signal
RSCP	Received Signal Code Power
RSL	Received Signal Level
RSRP	Reference Signal Received Power
RSRQ	Reference Signal Received Quality
RSSI	Received Signal Strength Indicator
RTMP	Real Time Messaging Protocol
RTP	Real-time Transport Protocol
RTSP	Real Time Streaming Protocol
RxLev	Received signal Level
RxQual	Received signal Quality
SA	Service and system Aspects
SA	Simulated Annealing
SACCH	Slow Associated Control Channel
SAE	System Architecture Evolution
SC	Scrambling Code
SC-FDMA	Single Carrier Frequency Division Multiple Access
SNCP	Subnetwork Connection Protection
SDCCH	Standalone Dedicated Control Channel

SDH	Synchronous Digital Hierarchy
SeNB	Source eNodeB
SFR	Soft Frequency Reuse
SGSN	Service GPRS Serving Node
S-GW	Serving Gateway
SHO	Soft HandOver
SIB	System Information Block
SINR	Signal to Interference-plus-Noise Ratio
SIR	Signal-to-Interference-Ratio
SMS	Short Message Service
SMSC	SMS Center
SNR	Signal to Noise Ratio
SON	Self-Organizing Networks
SONET	Synchronous Optical Network
SS7	Signalling System #7
SSC	Secondary Synchronization Codes
S-SCH	Secondary Synchronization Channel
STP	Signaling Transfer Point
SW	Software
TA	Timing Advance
TAC	Tracking Area Code
TCH	Traffic Channel
TCI	Target Cell Identifier
TDD	Time Division Duplex
TDMA	Time Division Multiple Access
TeNB	Target eNodeB
TM	TeleManagement
TNL	Transport Network Layer
TRX	Transceiver
TS	TroubleShooting
TSG	Technical Specification Group
TTI	Transmission Time Interval
TVWS	Television White Space
UARFCN	UTRA Absolute Radio Frequency Channel Number
UDP	User Datagram Protocol
UE	User Equipment
UICC	Universal Integrated Circuit Card
UL	Uplink
UMTS	Universal Mobile Telecommunications System
UTRA	Universal Terrestrial Radio Access
UTRAN	Universal Terrestrial Radio Access Network
VBR	Variable Bit Rate
VCEG	Video Coding Experts Group
VIP	Very Important Person
VLAN	Virtual Local Area Network
VoIP	Voice over IP

WAN	Wide Area Network
WCDMA	Wideband Code Division Multiple Access
WCS	Wireless Communications Service
WEP	Wired Equivalent Privacy
WG	Working Group
WI	Work Item
WiFi	Wireless Fidelity
WiMAX	Worldwide Interoperability for Microwave Access
WISPr	Wireless Internet Service Provider roaming
WLAN	Wireless Local Area Network
WMA	Windows Media Audio
WPA	Wi-Fi Protected Access
WPA2	Wi-Fi Protected Access 2
WVGA	Wide Video Graphics Array
XML	Extensible Markup Language

1

Operating Mobile Broadband Networks

Ken Stewart, Juan Ramiro and Khalid Hamied

1.1. The Challenge of Mobile Traffic Growth

The optimization of cellular network performance and the maximization of its efficiency has long been an objective of wireless network providers. Since the introduction of GSM in the late 1980s, the growth of traffic (and revenue per user) over wireless networks as the first 2G and 3G networks were deployed remained positive and relatively predictable. For those networks, voice and messaging services such as Short Message Service (SMS) and Multi-media Messaging Service (MMS) were dominating traffic. However, in the first decade of the twenty-first century, the deployment of high-performance wide-area wireless packet data networks, such as 3GPP HSPA and 3GPP2 HRPD, has combined with advances in Digital Signal Processing (DSP) capability, multi-media source coding, streaming protocols and low-power high-resolution displays to deliver the so-called smartphone. This device has fundamentally changed the trajectory of traffic growth over broadband wireless networks.

In June 2010, The Nielsen Company reported ([1], Figure 1.1) an annual increase between Q1-2009 and Q1-2010 of 230% in average smartphone data consumption. Nielsen further reported that some users were approaching 2 GB per month in total data usage, and that the top 6% of smartphone users were consuming nearly 50% of total data bandwidth. Therefore, as more users emulate the behavior of leading adopters, further growth in per-user data consumption is expected to follow. Most significant of all, from the perspective of future growth, Nielsen estimated that the penetration rate of smartphones into the US market was only 23%. Indeed, of those users, almost 1/4 generated zero data traffic, while 1/3 had simply not subscribed to a data plan at all. This suggests a latent demand for data connectivity and that networks are only beginning to see the onset of smartphone-induced load.

Self-Organizing Networks: Self-Planning, Self-Optimization and Self-Healing for GSM, UMTS and LTE,
First Edition. Edited by Juan Ramiro and Khalid Hamied.
© 2012 John Wiley & Sons, Ltd. Published 2012 by John Wiley & Sons, Ltd.

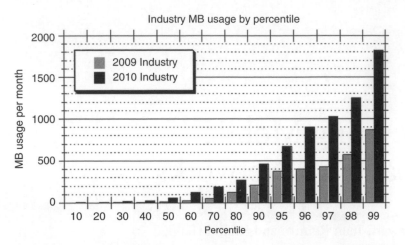

Figure 1.1   2009 and 2010 smartphone data usage distribution. Reproduced by permission of © 2010 The Nielsen Company.

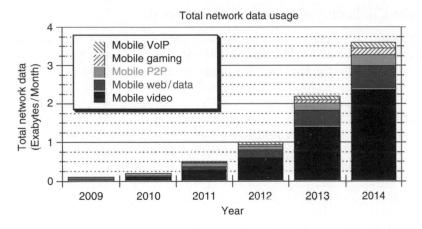

Figure 1.2   Total wireless mobile network traffic growth. Reproduced from © 2010 Cisco VNI Mobile.

The Nielsen Company's data is generally consistent with that reported by major network operators, particularly with respect to the wide distribution of user data consumption rates. For example, in June 2010, AT&T reported [2] that while the least active 65% of AT&T's smartphone subscribers used, on average, less than 200 MB of data per month, the top 2% of subscribers used more than 2 GB.

Although AT&T did not comment on future network traffic growth, others such as Cisco Systems have done so [3]. In Cisco's view, total wireless mobile network traffic growth (Figure 1.2) will exceed a Compounded Annual Growth Rate (CAGR) in excess of 100% per annum in the period 2010–2014, with video traffic providing as much as 2/3 of total traffic. In other words, on an annualized basis, total network traffic will double until at least the middle of the decade. This suggests that, as compared to 2009, if not prevented by

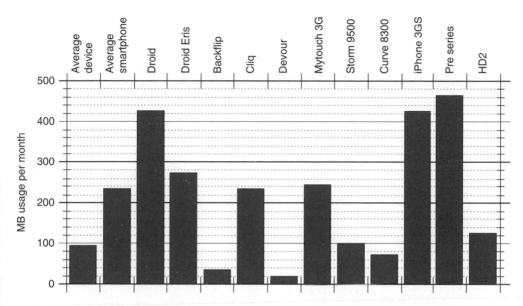

Figure 1.3 Traffic generation by smartphone type. Reproduced by permission of © 2010 The Nielsen Company.

factors such as limited-data subscription plans or insufficient spectrum, a 64-fold increase in total network traffic will result by 2015.

1.1.1. Differences between Smartphones

Even amongst those users who are equipped with smartphones, there is a wide disparity in data usage. There are a number of factors which influence the amount of data generated per device, including the user interface, applications available to the user (driven by operating system popularity), subscriber data plan, configuration of services using data link push and keep-alive techniques, etc. It is possible, however, to establish general trends by looking closely at measured data volumes on a per device basis. For example, in the 2010 Nielsen data depicted in Figure 1.3, along with the Palm Pre, the market-leading Motorola Droid and Apple iPhone 3GS devices both generated very significant data volumes, consistent with the rich set of experiences enabled by each platform. On average, both of these leading devices were generating around 400 MB of monthly data traffic per subscriber. This is well in excess of the average behavior of all devices (Average device) of around 90 MB per month, and even the average smartphone monthly consumption of 240 MB.

Figure 1.3 further suggests that the core capabilities of smartphone devices are also important in establishing data consumption. Table 1.1 lists selected capabilities of influential smartphone devices launched in 2009/10, including a subset of the most significant devices from Figure 1.3. A comparison of the Motorola Droid and Cliq devices shows a progressive increase in application processor, screen resolution and multi-media capabilities that would tend to drive the difference in user data consumption for each device observed in Figure 1.3.

Table 1.1 Contemporary smartphone capabilities

Device	Screen		Application Processor	Memory RAM, Flash, Max. μSD	Camera	Connectivity
	Size	Res.				
Motorola Cliq	3.1″	480×320	Qualcomm MSM7200 528 MHz	256 MB, 512 MB, 32 GB	5MP	UMTS-HSPA, EDGE, 802.11b/g
Motorola Droid	3.7″	480×854	TI OMAP 3430 600 MHz	256 MB, 512 MB, 32 GB	5MP	1xRTT, DO-A, 802.11b/g
Motorola Droid 2	3.7″	480×854	TI OMAP 3620 1 GHz	512 MB, 8 GB, 32 GB	5MP	CDMA 1x, DO-A, 802.11b/g/n
Motorola Droid X	4.3″	480×854	TI OMAP 3630 1 GHz	512 MB, 8 GB, 32 GB	8MP	CDMA 1x, DO-A, 802.11b/g/n
Apple iPhone 3GS	3.5″	480×320	Samsung ARM	256 MB, 16/32 GB, N/A	3MP	UMTS-HSPA, EDGE, 802.11b/g
Apple iPhone 4	3.5″	960×640	Samsung ARM	512 MB, 32 GB, N/A	5MP	UMTS-HSPA, EDGE, 802.11b/g/n
HTC Droid Eris	3.2″	480×340	Qualcomm MSM7600 528 MHz	288 MB, 512 MB, 32 GB	5MP	1xRTT, DO-A, 802.11b/g
HTC HD2	4.3″	480×800	Qualcomm Snapdragon 1 GHz	448 MB, 512 MB, 32 GB	5MP	UMTS-HSPA, EDGE, 802.11b/g

Source: Motorola 2010

The same general trend is observable in Table 1.1 for vendors other than Motorola such as Apple and HTC. It is worth noting that Table 1.1 spans a relatively short device launch period of only approximately two years.

1.1.2. Driving Data Traffic – Streaming Media and Other Services

The advent of streaming media services such as those offered by Pandora and YouTube has had a major impact on device data consumption.

Internet audio streaming (Internet radio) using, amongst others, streaming MPEG-1 Audio Layer-3 (MP3), Windows Media Audio (WMA), Flash Video (FLV) or Real Audio formats, and using protocols such as Real-time Transport Protocol (RTP), Real Time Streaming Protocol (RTSP), Real Time Messaging Protocol (RTMP), User Datagram Protocol (UDP) and HyperText Transfer Protocol (HTTP), has been deployed on the wired Internet since the late 1990s. Since 2005, however, despite the increasing enforcement of royalty-driven limitation on streaming, the advent of genre-based streaming services such as Pandora or even subscription services such as XM Radio Online has further increased the popularity of this type of service.

Depending on service type, server-client rate adaptation strategy and subscription policy, typical data rates for audio streaming services range from 56–192 kbps, yielding a per user consumption rate of ~25–85 MB/hr. This significant data consumption rate is most impactful when combined with the observed user behavior of invoking an audio streaming service and then permitting the stream to continue as a background audio service for an extended period (often several hours in duration) while executing other tasks.

Video streaming services represent another major source of network load. Services here are generally very well known, and include YouTube, Hulu, TV.com, etc. YouTube, which is a typical example of such a service, generally uses FLV or MP4 containers, plus MPEG-4 AVC (H.264) video encoding with stereo audio encoded using Advance Audio Coding (AAC). Typical served rates are 85–500 kbps (i.e. ~38–220 MB/hr), with a limit on total content duration (e.g. 10 min) and size (e.g. 2 GB) depending on the relationship between the entity uploading the source content and the streaming service provider.

Recently, the aforementioned services have become available for the wireless Internet due to the rich set of features implanted by smartphones. As a consequence, the large data volumes associated with these data services have to be carried by wireless radio networks, causing mounting pressure on the available wireless infrastructure.

1.2. Capacity and Coverage Crunch

Mobile data traffic is growing extensively and it is projected that a 64-fold increase in total network traffic will result by 2015 as discussed in Section 1.1. This explosive growth in mobile broadband places serious demands and requirements on wireless radio networks and the supporting transport infrastructure. The most obvious requirement is the massive capacity expansions and the necessary coverage extensions that need to be provided while meeting the required Quality of Service (QoS).

In general, traffic growth is healthy if network operators can charge for it proportionally and if they can provide sufficient network capacity to cope with that growth. It is worthwhile noting, however, that these capacity expansions are required at a time when operators' Capital

Expenditure (CAPEX) and Operational Expenditure (OPEX) budgets are limited and Average Revenue Per User (ARPU) growth is saturated.

The following sections provide an overview of techniques and solutions available to help wireless operators address the challenges associated with the explosive traffic growth.

1.3.   Meeting the Challenge – the Network Operator Toolkit

Fortunately, network operators have a wide variety of techniques available to deal with the challenge of mobile data growth. First, operators can employ economic incentives to modify user behavior by adjusting tariff structures. Another approach is to improve network capacity through the deployment of advanced Radio Access Technologies (RATs), such as 3GPP Long Term Evolution (LTE). This approach, and the significant CAPEX associated with it, is often combined with the acquisition of new spectrum. Interest in the use of WiFi companion networks and offloading techniques has recently grown, along with preliminary deployments of innovative network elements such as femto cells or home base stations. The optimization of protocol design and traffic shaping methods, together with the deployment of advanced source coding techniques, has recently become popular. Finally, and most significantly for the purpose of this book, there has been intensive interest in the optimization of existing radio network assets and there has been huge interest in expanding the scope of Self-Organizing Networks (SON) to cover 2G and 3G. This last approach has the added attraction of relatively low capital and operational investment.

1.3.1.   Tariff Structures

With the increasing adoption of smartphones, the era of unlimited data plans may be coming to an end. For example, in June 2010, AT&T publicly announced [2] two limited data plans: DataPlus and DataPro. Under the AT&T DataPlus plan, users were offered a total of 200 MB of data for US$15 per month, with an additional 200 MB of data available for use within the billing cycle for a further US$15 fee. Under the companion DataPro plan, 2 GB of data were included in the basic US$25 fee, with a further 1 GB available for use within the billing cycle at the cost of an additional US$10. New Apple iPad users were mapped to the AT&T DataPro plan, and the antecedent unlimited plan was phased out.

AT&T is, of course, not unique in taking this approach, and similar trends can be seen in other networks and geographic regions. For example, in June 2010, O_2 announced [4] that new and upgrading users would be mapped to a selection of data plans offering between 500 MB and 1 GB per month of data usage for £25–60, with additional data available for approximately £10 per GB, depending on the selected data bolt-on product. Notably, however, in Asia, after a period of expansion for data-limited plans, competitive pressure is re-establishing unlimited data offerings, at least for a period, by operators such as SK Telecom [5].

Limited data plans apply generically to all traffic transported by the network. However, new opportunities may also be emerging for network operators seeking to limit specific traffic flows from/to certain Internet Protocol (IP) addresses and port numbers, for technical or business reasons. These may include, for example, ports used by streaming media services or

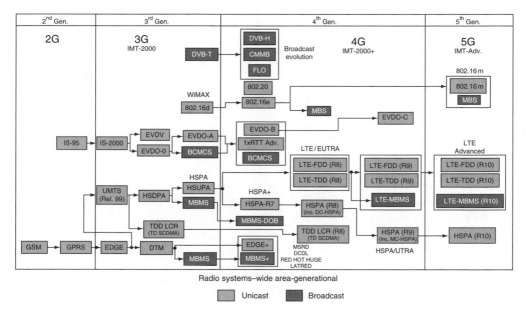

Figure 1.4 Evolution of WAN radio access. Reproduced by permission of © 2010 Motorola.

other data-intensive traffic sources. Alternatively, network operators may seek to limit traffic originating from particular applications, or indeed traffic exchanged with competing service providers could be limited or mapped to lower QoS classes. These approaches are the subject of intensive regulatory scrutiny. In the US, for example, the April 2010 Federal Court ruling on so-called net neutrality [6] may encourage more efforts by network operators to intervene in traffic flows, but further legislative or regulatory activity is very likely.

1.3.2. Advanced Radio Access Technologies

With the exception of green field deployments, the opportunity for network operators and device vendors to migrate towards network RATs with improved spectral efficiency is heavily dependent on existing commitments and compatible legacy technologies. This is illustrated in Figure 1.4, which shows the respective evolution of wide area RATs from the roots of GSM, CDMA and IEEE 802.16d into HSPA+, LTE, EV-DO and WiMAX. As the figure shows, the strategic landscape surrounding broadband wireless is in some ways becoming simpler with the deployment of 4G networks. For example, at present, the EV-DO family of technologies appears to have limited prospects for widespread deployment of the EV-DO Revision B (DO-B) and EV-DO Revision C (DO-C) variants, and consequently, although EV-DO technology will remain operational for many years, unless there is some shift in the strategic landscape, the evolutionary track for EV-DO is effectively terminating. Similarly, with the commencement of work in 3GPP [7] to support deployment of LTE in the U.S. 2.5 GHz band, further commitment to WiMAX 2.0 may be limited, although Q3-2010 commitments by Indian operators following the Indian 3G and Broadband Wireless Access (BWA) spectrum auctions to both WiMAX and

Table 1.2 HSPA+ and LTE evolution – device capability summary

Functional Support			3GPP HSPA			3GPP LTE		
Function	Units	Sub-Func.	Rel-7	Rel-8	Rel-9	Rel-8	Rel-9	Rel-10 (LTE-A)
Component Carrier Bandwidth	MHz	DL	5	5	5	{1.4, 3, 5, 15, 20}	{1.4, 3, 5, 10, 15, 20}	{1.4, 3, 5, 10, 15, 20}
		UL	5	5	5	{1.4, 3, 5, 15, 20}	{1.4, 3, 5, 10, 15, 20}	{1.4, 3, 5, 10, 15, 20}
Multicarrier	#Carriers	DL	1	2	2 (Non-Adj.)	1	1	5
		UL	1	1	2 (Adj.)	1	1	2
Link Bandwidth	MHz	DL	5	10	10	20	20	100
		UL	5	5	10	20	20	40
Max. Modulation	N/A	DL	64-QAM	64-QAM	64-QAM	64-QAM	64-QAM	64-QAM
		UL	16-QAM	16-QAM	16-QAM	16-QAM	16-QAM	64-QAM
MIMO	#Streams/Carrier	DL	2	2	2	4	4	8
		UL	1	1	1	1	1	4
Max. Data Rate (Terminal Capability)	Mbps	DL	28.0 (Cat. 16, 18)	42.2 (Cat. 20)	84.4 (Cat. 28)	10.3, 51, 102, 151, 302	10.3, 51, 102, 151, 302	10–1000
		UL	5.7 (Cat-6, 2ms)	11.5 (Cat-7, 2ms)	11.5 (Cat-7, 2ms)	5.2, 25.5, 51, 75	5.2, 25.5, 51, 75	5–200
Broadcast & Multicast	N/A	N/A	MBMS	MBSFN, DOB	MBSFN, DOB	N/A	EMBMS	EMBMS+

Source: Motorola 2010

Table 1.3 Comparison of HSPA+ and LTE spectral efficiency

Deployment Scenario	Downlink Spectral Efficiency (bps/Hz/cell)				
	HSDPA 1×2			LTE 1×2	LTE 2×2
Channel: TU 3 km/h	RAKE	MMSE	MMSE RxDiv	MMSE RxDiv	MMSE RxDiv
Urban Macrocell: 500 m SiteSite Dist.	0.30	0.59	1.09	1.23	1.83
Urban Macrocell: 1700 m SiteSite Dist.	0.29	0.58	1.05	1.3	1.75
Hotspot: 100 m SiteSite Dist.	0.30	2.42	NA	3.3	6.53

Source: Motorola 2010

to LTE Time Division Duplex (TDD) mode suggest that the long-term future of WiMAX may be undecided.

All of this appears to position 3GPP LTE, both Frequency Division Duplex (FDD) and TDD variants, and 3GPP HSPA+ as the critical Wide Area Network (WAN) RATs for the next decade, and even beyond, as enabled by the International Mobile Telecommunications – Advanced (IMT-Advanced) process, provided regional technology programs remain aligned to these central threads.

Focusing further on HSPA+ and LTE, it should first be noted that there is no formal definition for HSPA+. Since the High Speed Uplink Packet Access (HSUPA) companion specification to the 3GPP Release 5 High Speed Downlink Packet Access (HSDPA) was completed in 3GPP Release 6, it is reasonable to categorize networks or devices supporting one or more HSPA features from 3GPP Releases 7 through Release 9 to be HSPA+. Practically speaking, however, and leaving aside some relatively minor layer 2 efficiency improvements and the useful device power-consumption enhancements offered by the Continuous Packet Connectivity (CPC) feature, the major HSPA+ capacity-enhancing components offered by Release 7 are downlink dual-stream Multiple Input Multiple Output (MIMO) and 64-Quadrature Amplitude Modulation (QAM) capability (see Table 1.2), thus resulting in support for peak rates in excess of 10 Mbps. Notably, the deployment of MIMO-capable HSPA+ networks and devices appears increasingly unlikely due to infrastructure and legacy device equalizer limitations, leaving 64-QAM as the principal Release 7 HSPA+ enhancement.

This permits the comparison of spectral efficiency of a 10 MHz LTE Release 8 network and an HSPA+ network, which appears in Table 1.3. Here, Release 8 HSPA+ dual-carrier feature has been incorporated into HSPA+ results in order to permit direct comparison of the same 10 MHz FDD pair. It can be seen from the results that the performance of HSPA+ networks is comparable to that of LTE with the deployment of dual-port receivers denoted by 1×2 in Table 1.3 (one transmit antenna at the base station and two antennas at the handset providing receive diversity).

This defines the strategy of many 3G operators today – to execute selected upgrading of HSPA infrastructure, including the critical backhaul capacity elements, to support HSPA+ and hence to support device rates up to 21 Mbps (HSDPA Category 14) or more, while seeking to

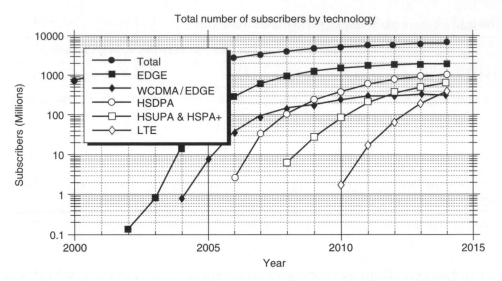

Figure 1.5   Total subscriber device capabilities by year. Reproduced by permission of © 2009
Strategy Analytics.

deploy LTE at the earliest date consistent with LTE device maturity and cost competitiveness.
Here, it is worth noting that (as indicated in Figure 1.5) the total installed base of devices
supporting at least one LTE band will not represent a truly significant fraction of total mobile
devices before 2015. Accordingly, it is reasonable to assume that the migration of data traffic
to higher performance wide area RATs will be gradual, and it may take until the end of decade
2010–2020 before the majority of worldwide operational terminals are LTE-capable.

1.3.3.   Femto Cells

A key component is the emergence of femto cells (or home base stations) deployed either in
enterprise or domestic environments. Although deployments of femto cells conformant to
conventional macro-cellular core specifications are completely feasible, several enhancements
to basic femto operation have been specified by 3GPP and other standards development
organizations (including WiMAX Forum and Femto Forum). In 3GPP, this has included the
definition of Closed Subscriber Groups (CSGs) who have been granted access to restricted femto
cells, methods for easily identifying femto cells during radio resource management procedures
(e.g. via CSG-specific synchronization sequences), and enhanced mobility procedures for
handing-off devices more reliably into a CSG femto cell. Upper-layer support for Local IP
Access (LIPA) to local network resources has been added to 3GPP Release 10. Nevertheless,
despite significant potential, the rollout of femto cells for domestic environments, such as
Vodafone's Sure Signal or AT&T's Femtocell brands, has had a limited impact from the
perspective of network load management and reduction. Rather, femto cell marketing to date has
emphasized enhancement of coverage-limited network access to voice services. Accordingly,
with limited adoption so far, and reuse of operators' core licensed spectrum required in any case,
femto cells are unlikely to make a significant contribution to network unloading before 2012–13.

1.3.4. Acquisition and Activation of New Spectrum

The identification, clearing and activation of new spectrum for mobile services, and the efficient refarming of existing spectrum, are problems receiving intense scrutiny by regulators in all three International Telecommunications Union (ITU) regions.

In ITU Region 2 (Americas), in March 2010, the United States announced, as part of the National Broadband Plan (NBP) [8], the intention to make available 500 MHz of additional spectrum for mobile broadband services by 2020, with 300 MHz to be made available by 2015. Assets in this case include a 20 MHz allocation in the 2.3 GHz Wireless Communications Service (WCS) band, disposition of the remaining 10 MHz (Block D) of spectrum from the 700 MHz auction of 2008, Federal Communications Commission (FCC) Auction 73, and a further 60 MHz of spectrum comprising mainly elements of the Advanced Wireless Services (AWS) band (generally, in the range of 1755–2200 MHz). In addition, a further 90 MHz of Mobile Satellite Service (MSS) spectrum from L- and S-bands would be made available under Ancillary Terrestrial Component (ATC) regulation (where devices supporting terrestrial broadband service must also support a satellite component). Perhaps most significant is the possibility of an additional 120 MHz of spectrum to be reallocated from broadcast use to mobile services.

In ITU Region 1 (Europe, Africa and Middle East), there is also considerable activity leading to new spectrum deployments. One example is the auction of 190 MHz of spectrum at 2.6 GHz (generally, in conformance to the band structure envisaged by the ECC/DEC(05)05 European directive) conducted in 2008–2010 by Norway, Sweden, Finland, Germany, Netherlands and Denmark, with other European countries expected to follow in 2010 or 2011. Perhaps most significant, however, is the German auction in May 2010 of 360 MHz of spectrum located mainly in the 800 MHz and 2.6 GHz bands yielding 2×10 MHz each at 800 MHz for Vodafone, T-Mobile and O_2 plus awards at 2.6 GHz to Vodafone (2×20 MHz FDD, 25 MHz TDD), T-Mobile (2×20 MHz FDD, 5 MHz TDD), O_2 (2×20 MHz FDD, 10 MHz TDD) and E-Plus (2×10 MHz FDD, 10 MHz TDD).

In ITU Region 3 (Asia), a similar narrative has evolved. In China, for example, band proposals for 700 MHz mobile operation (698–806 MHz) include options for allocation of the entire band for unpaired operation (i.e. TDD mode), or a split in allocation between paired (FDD, 698–746 MHz) and unpaired (TDD, 746–806 MHz) modes. It is unlikely, however, that this spectrum will be released before 2015. More immediate opportunities for new spectrum in China include 100 MHz available in the 2300–2400 MHz band plus up to 190 MHz of spectrum in the 2.6 GHz band (2500–2690 MHz). Of these, the 2300–2400 MHz spectrum was designated for unpaired operation as early as 2002, and has been used successfully for LTE-TDD trials at the Shanghai Expo of 2010. However, coexistence concerns with other services may limit future deployment in that band to indoor use. This has led to increased interest in the 2.6 GHz band, where, amongst other possibilities, alignment with the European or U.S. 2.6 GHz band plans has been considered, along with an alternative all-TDD designation favoring LTE-TDD mode. Similar commitment to release spectrum, albeit on a smaller scale, has emerged for the same bands in India, resulting most recently in the Indian 3G and BWA spectrum auction.

Clearly, the acquisition of new spectrum offers a major opportunity to enhance network capacity. Notably, however, the acquisition of spectrum can be highly capital-intensive. For example, the German auction of May 2010 yielded total bid amounts of €4300 million.

Further, the activation of new spectrum can involve costs to relocate users or services, and the provision of additional radio hardware and transmission backhaul at sites where the new spectrum is to be activated. All of this suggests that, while new licensed spectrum is a critical component to resolving network capacity shortages, it is generally a costly option, available only on a medium- to long-term basis.

1.3.5. Companion Networks, Offloading and Traffic Management

The cost of new licensed spectrum has led to renewed interest in the resources offered by unlicensed spectrum, such as the US 2.4 GHz Industrial, Scientific and Medical (ISM) band, the 5 GHz National Information Infrastructure (NII) band and the 700 MHz Television White Space (TVWS) band. Many major network operators now lease access to WiFi from WiFi network service aggregators (where one or more distinctive WiFi hotspot networks are gathered under a single brand and are accessible using a single set of access credentials), or operate public WiFi hotspot networks that are cobranded as companion networks to the operator's primary wide area broadband network. Authentication protocols such as Wireless Internet Service Provider roaming (WISPr) are usually applied, in combination with either Wired Equivalent Privacy (WEP) or, more frequently, WiFi Protected Access (WPA) and WiFi Protected Access 2 (WPA2) authentication methods using user- or operator-supplied credentials stored on the device or UICC-based[1] credentials. Significantly, access to such WiFi networks is increasingly offered without additional charge as an element of a broader wide area data plan.

Almost all contemporary smartphones support WiFi connectivity. This allows operators to offload a significant portion of growing data traffic from their primary WAN network onto their WiFi network. Leaving aside the consequent increased load on WiFi networks and the resulting interference, there are a number of obstacles here. First, enabling the device's WiFi interface on a continuous basis can lead to elevated device electric current drain and reduced battery life, with consequent user dissatisfaction. Second, the spatial density of the operator's hotspot network is often not sufficient to provide service coverage on a contiguous basis, despite hotspot collocation with transport or social centers (e.g. airports, cafes, etc.). For example, in one Chicago suburb, the spatial separation between WiFi hotspots associated with one 3G network was around 4 km, roughly twice that of the companion 3G network intersite separation, and therefore the opportunity to conveniently connect to a hotspot can be limited. Finally, some operator services, such as operator branded messaging or media services, referred to here as Carrier Branded Services (CBS), must have access on a trusted basis to the operator's core network in order to execute authentication functions.

The problem with low battery life can be overcome by using appropriate power management techniques in both the WiFi associated and non-associated states. The low spatial density of WiFi hotspots implies that many or even most data-offloading WiFi connections will be established in the user's home or enterprise. This raises questions about the capability[2] of those networks. Answers for these questions can be found through privacy-appropriate surveying techniques[3]. Summary results of a public WiFi Access Point (AP) beacon survey conducted by

---

[1] Universal Integrated Circuit Card

[2] For example, support for 802.11b or 802.11n, supported data rate, type of security support (if any), etc.

[3] The survey was conducted by examining only the system information in the 802.11 management frames transmitted by all nonhidden WiFi access points.

Table 1.4   Public WiFi AP survey summary – Q2-2010

AP Mode	%	Band	%	Bandwidth	%	802.11 Type	%	Security Mode	%
Ad Hoc	3	2.4GHz	98	20MHz	95	b	7	Open	20
Infrastructure	97	5GHz	2	40MHz	5	g	78	WEP	37
						n	15	WPA	24
								WPA2	19

Source: Motorola 2010

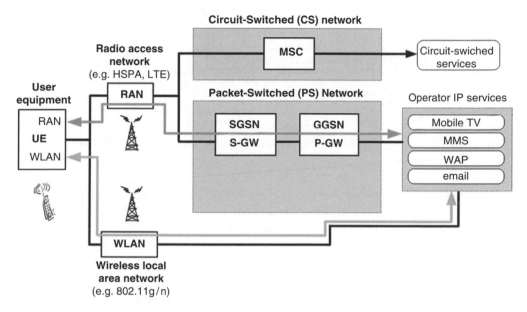

Figure 1.6   Offloading architecture – Type I. Reproduced by permission of © 2010 Motorola.

Motorola in Q2-2010 appear in Table 1.4. It can be seen that although more than 90% of public WiFi APs conform to the 802.11g amendment, few 802.11n APs were deployed at the time of the survey. Significantly, most APs were deployed in the 83 MHz of spectrum available in the 2.4 GHz ISM band, which, for practical purposes, can support a maximum of three 20 MHz 802.11 carriers. Almost no APs were operational in the 5 GHz band, which, at least in the US, offers a total of 550 MHz of spectrum, suggesting that, as smartphones increasingly support 5 GHz WiFi access, this approach to network offloading has real prospects for growth.

Finally, the difficulties related to accessing the operator's core network on a trusted basis can be resolved by using appropriate routing and tunneling techniques. There are several approaches to achieve this. In one of them, illustrated in Figure 1.6, the device maintains WAN (i.e. 3G/4G) and WiFi connections simultaneously. This permits the bulk of non-CBS data to transfer over the WiFi connection, while access to CBS data may occur over the secured WAN network. An evolution of this approach appears in Figure 1.7. In this architecture, the operator has invested in additional network-edge routers capable of terminating a secure

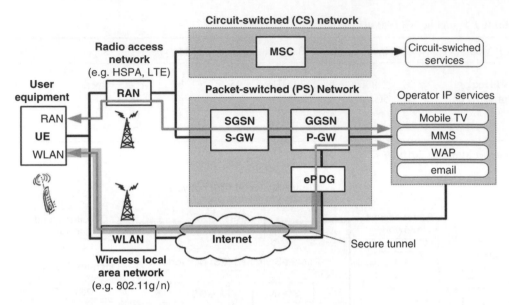

Figure 1.7   Offloading architecture – Type II. Reproduced by permission of © 2010 Motorola.

tunnel originating in the device over the unsecure WiFi network. As a result, additional CBS-specific traffic may enter the operator's core network over the unsecure WiFi connection, with the remaining traffic terminating directly in the Internet.

1.3.6.   Advanced Source Coding

New approaches to source coding also offer the prospect for improvements in network efficiency. Traditionally, cellular systems have looked into speech coding efficiency as a baseline measure of source coding performance. Here, notwithstanding the evolution of the CDMA Enhanced Variable Rate Codec (EVRC) family to the EVRC-C variant, the migration of CDMA operators to LTE appears likely to bring EVRC evolution to a close. At the same time, the need for improved voice quality means that a number of 3G operators, most notably T-Mobile International and France Telecom/Orange, are now migrating from the well-established 3GPP-specified Adaptive Multi-Rate NarrowBand (AMR-NB) codec (covering the audio range of 200 Hz to 3.5 kHz) to the evolved AMR WideBand (AMR-WB/G.722.2) codec (50 Hz to 7.0 kHz). This brings a corresponding increase in bit rate from 5.9–12.2 kbps (AMR-NB) to 12.65 kbps (AMR-WB) and above.

Fortunately, recent work in ITU-Telecommunication (ITU-T) SG-16 [9] on the G.718 codec, which can maintain bitstream compatibility with AMR-WB, indicates it is possible to achieve quality levels normally associated with AMR-WB at 12.65 kbps by using G.718 with 8 kbps. This is, in part, the motivation for the 3GPP Enhanced Voice Service (EVS) work [10]. Significantly, however, the 3GPP EVS specification is unlikely to be complete before 2011 and is unlikely to be operational before 2012. Accordingly, in the medium-term, speech source coding rates will not diminish in the period up to 2012, but may well increase as AMR-WB is deployed.

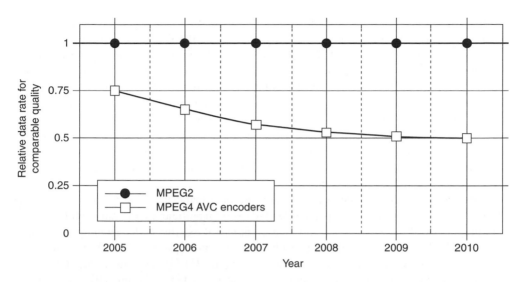

Figure 1.8 MPEG-4 and MPEG-2 quality versus time. Source: Motorola 2010.

When operating in the range of 32–64 kbps, the performance of the G.718 and EVS codecs begin to overlap with that of the AAC codec family specified by the International Organization for Standardization (ISO) and the International Electrotechnical Commission (IEC), most notably the AAC Low Delay (AAC-LD) and AAC Enhanced Low Delay (AAC-ELD) variants. Recent operator assessments (e.g. [11]) suggest, however, that neither of these codecs offer efficiency advantages over G.718 or the emerging EVS specification, with proprietary codecs such as SiLK (Skype) or Speex reported to operate at significantly poorer efficiencies. For medium- to high-rate audio coding applications (i.e. bit rates in the range of 32–256 kbps), smartphones such as the Motorola Droid X or the Apple iPhone make available MP3, AAC and AAC+ codecs. At these audio coding rates, there is little evidence at present that work in ISO/IEC or 3GPP will lead to significant improvements in efficiency in the near-term. Rather, the trend towards super-wideband (50 Hz to 14 kHz) and full-band (20 Hz to 20 kHz) codec operation, and potentially towards support for surround sound in WAN networks [12] suggests that in the next decade improved audio source coding will not lead to significant reduction in network load, but rather will emphasize improved quality and enhanced services.

Clearly, however, from the traffic growth information presented above, more efficient encoding of video traffic would have the greatest impact on total network load. Again, there appears to be limited opportunity for significant fundamental improvements in the near term. This is largely due to diminishing incremental improvements in the performance of the ITU/ISO/IEC MPEG 4 AVC/H.264 video codec (Figure 1.8).

Most significantly, while there is a clear recognition that further improvements in video coding efficiency are essential, it will clearly take time to achieve this. For example, ISO/IEC MPEG and ITU-T Video Coding Experts Group (VCEG) have established the Joint Collaborative Team on Video Coding (JCT-VC) to deliver a High Efficiency Video Coding (HEVC/H.265) specification [13], with the goal of achieving roughly a twofold improvement in encoding efficiency for the same or lower computational complexity. Subjective assessment, however, of initial proposals for HEVC commenced in March 2010 [14], with the date for completion of the specification targeted at Q3-2012. This suggests the earliest possible widespread deployment date of fully compliant HEVC codecs would be 2013.

At the same time, the advent of AVC/H.264-enabled 3D-video will tend to push streaming rates even higher. Moreover, and unrelated to 3D-video, smartphone and converged computing or tablet device (e.g. iPad) screen resolutions and rendering capabilities continue to increase. For example, the Motorola Droid X smartphone introduced in July 2010 offers a 4.3" (10.9 cm) Wide Video Graphics Array (WVGA) (480×854) display combined with a High-Definition Multi-media Interface (HDMI) port and the ability to render 720p AVC/H.264 content. The converged computing Apple iPad (launched in April 2010) correspondingly offers a 9.7" (24.6 cm) display supporting 1024×768 resolution, and AVC/H.264 video up to 720p format at 30 frames per second.

Accordingly, as such devices further penetrate broadband wireless markets, and in the obvious absence of a radical improvement in video coding efficiency, operators will continue to migrate network servers towards more efficient use of existing codec techniques (such as AVC/H.264) [15]. Opportunities for such advanced streaming procedures include proprietary methods such as Microsoft Smooth Streaming, Apple HTTP Live Steaming (HLS) and standardized approaches such as ongoing efforts in Open IPTV Forum (OIPF) and 3GPP [16]. Nevertheless, while such approaches will improve video rate adaptation (to better suit channel conditions or access technology) and will offer trick play features in an efficient way, they will not fundamentally reduce the growth of video traffic, although they may offer enhanced means of maintaining adequate video quality within specific data-rate constraints.

1.4.   Self-Organizing Networks (SON)

All capacity expansion techniques that have been discussed so far are valid paths, and operators' strategies need to rely on them to cope with growing data volumes and demanding customer expectations (in terms of QoS and service cost). Nonetheless, the techniques that are available today involve outstanding capital outlays, and therefore it is worth reflecting on whether the current infrastructure is being operated at its full performance potential before considering network expansions or evolutions.

Going back to basics, it is important to remember that, for example, the Universal Mobile Telecommunications System (UMTS) is a complex technology in which coverage, capacity and quality are deeply coupled to each other. There are many optimization levers that currently remain untouched or, at best case, fine-tuned at network level, i.e. with the same settings for all different cells. The bottom line is that, even though a UMTS network may be delivering acceptable Key Performance Indicators (KPIs), most likely there is still room for increasing its capacity, just by carefully tuning the different settings on a cell-by-cell basis.

The idea to carry out adaptive network optimization on a per sector (or even per adjacency) basis is part of the SON paradigm, which has been defined around a set of clear requirements formulated by the Next Generation Mobile Networks (NGMN) Alliance [17]. The objective of the SON proposal is to enable a set of functionalities for automated Self-Organization of LTE networks, so that human intervention is minimized in the planning, deployment, optimization and maintenance activities of these new networks. Subsequently, the support for this new network management paradigm is being translated into concrete functionalities, interfaces and procedures during the standardization of Evolved Universal Terrestrial Radio Access Network (E-UTRAN) in 3GPP.

The SON Use Cases can be structured in different ways. As will be discussed in Chapter 2, one of the possible high-level classifications is the following:

- Self-Planning: derivation of the settings for a new network node, including the selection of the site location and the specification of the hardware configuration, but excluding site acquisition and preparation.
- Self-Deployment: preparation, installation, authentication and delivery of a status report of a new network node. It includes all procedures to bring a new node into commercial operation, except for the ones included in the Self-Planning category, which generate inputs for the Self-Deployment phase.
- Self-Optimization: utilization of measurements and performance indicators collected by the User Equipments (UEs) and the base stations in order to auto-tune the network settings. This process is performed in the operational state, which is defined as the state where the Radio Frequency (RF) interface is commercially active (i.e. when the cell is not barred/reserved).
- Self-Healing: execution of the routine actions that keep the network operational and/or prevent disruptive problems from arising. This includes the necessary software and hardware upgrades or replacements.

Whereas current commercial and standardization efforts are mainly focused on the introduction and Self-Organization of LTE networks, there is significant value associated with the extension of the scope of Self-Planning, Self-Optimization and Self-Healing to cover GSM/GPRS/EDGE and UMTS/HSPA RATs. The implications of multi-technology SON are massive. On one hand, the adoption of a multi-technology approach allows operators to completely transform and streamline their operations, not only applying an innovative, automated approach to the new additional LTE network layer, but also extending the automation-related operational savings to all RATs, thereby harmonizing the whole network management approach and boosting operational efficiency. On the other hand, the availability of a multi-technology SON solution can lead to more comprehensive, holistic and powerful optimization strategies that deal with several RATs simultaneously.

Practical experience shows that the application of 3G SON technologies in current UMTS infrastructure can yield a capacity gain of 50% without carrying out any CAPEX expansion.

1.5. Summary and Book Contents

In summary, as smartphones continue to proliferate, there is a clear and present need to improve the efficiency and capacity of contemporary broadband networks. Fortunately, there is a wide variety of options available to network operators, ranging from evolution in network technology such as improved backhaul and the use of enhanced RATs such (e.g. HSPA+ and LTE), through acquisition of new spectrum, offloading to companion networks (e.g. WiFi) and the application of advanced source coding along with traffic shaping methods.

Equally clearly, however, no single approach will resolve the challenge caused by the exponential growth in network traffic. Critically for the present purpose, the approaches discussed in Section 1.3 are either capital intensive (e.g. new spectrum acquisition or network deployment) or are associated with extended time horizons (e.g. new source coding technology breakthroughs) or both. Therefore, SON techniques and functions have a unique role to play. They can be deployed today, at moderate to low cost, in contemporary 2G and 3G networks to increase operational efficiency with little or no delay. SON techniques will, of course, evolve to support LTE.

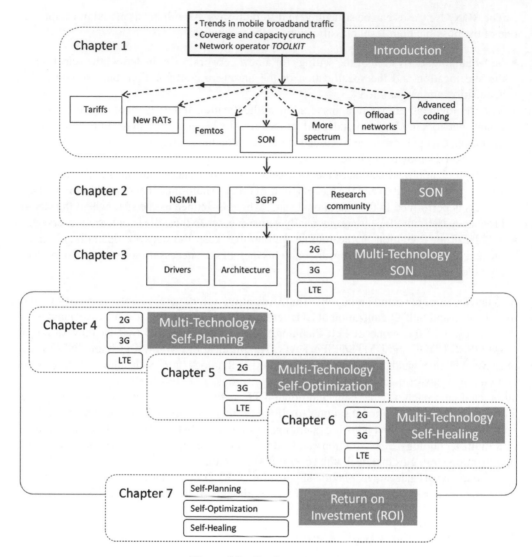

Figure 1.9   Book contents map.

The main purpose of this book is to describe multi-technology SON for 2G, 3G and LTE, and to cover the best practice deployment of Self-Organizing Networks that support multi-vendor and multi-technology wireless infrastructures. This will be done mainly from a technology point of view, but also covering some critical business aspects, such as the Return On Investment (ROI) of the proposed SON functionalities and Use Cases. Figure 1.9 provides a conceptual map summarizing the contents of the book. Chapter 2 provides an overview of the SON paradigm covering NGMN objectives and the activities in 3GPP and the research community. Chapter 3 covers the multi-technology aspects of SON, from main drivers to a layered architecture for multi-vendor support. Chapters 4, 5 and 6 cover the multi-vendor and multi-technology (2G, 3G and LTE) aspects of the Self-Planning, Self-Optimization and Self-Healing of wireless networks, respectively. Finally, critical business aspects, such as the ROI of the proposed SON functionalities and Use Cases are presented in Chapter 7.

1.6.   References

[1] The Nielsen Company (2010) *Quantifying the Mobile Data Tsunami and its Implications*, 30 June 2010, http:// blog.nielsen.com (accessed 3 June 2011).

[2] AT&T Press Release (2010) *AT&T Announces New Lower-Priced Wireless Data Plans to Make Mobile Internet More Affordable to More People*, 2 June 2010, http://www.att.com (accessed 3 June 2011).

[3] Cisco (2010) *Cisco Visual Networking Index: Global Mobile Data Traffic Forecast Update 2009–2014*, 9 February 2010, http://www.cisco.com (accessed 3 June 2011).

[4] O2 Press Release (2010) *O2 Introduces New Mobile Data Pricing Model*, 10 June 2010, http://mediacentre. o2.co.uk (accessed 3 June 2011).

[5] SK Telecom Press Release (2010) *SK Telecom Unveils Innovative Measures to Boost Customer Benefits*, 14 July 2010, http://www.sktelecom.com (accessed 3 June 2011).

[6] US Court of Appeals Ruling, District of Columbia (2010) *Comcast vs. US FCC*, 6 April 2010, http://pacer.cadc. uscourts.gov/common/opinions/201004/08-1291-1238302.pdf (accessed 3 June 2011).

[7] 3GPP, RAN Plenary Meeting #49, RP-100701 (2010) *Draft Report of 3GPP TSG RAN Meeting #48*, Section 10.4.9 LTE TDD in 2600MHz for US (Region 2), 1–4 June 2010, www.3gpp.org (accessed 3 June 2011).

[8] U.S. National Broadband Plan, http://www.broadband.gov/ (accessed 3 June 2011).

[9] ITU-T SG-16, TD 164 (PLEN/16) (2009) *Draft New Technical Paper GSTP-GVBR 'Performance of ITU-T G.718'*, 26 October - 6 November 2009, Geneva.

[10] 3GPP, Technical Report, Technical Specification Group Services and System Aspects (2010) *Study of Use Cases and Requirements for Enhanced Voice Codecs for the Evolved Packet System (EPS)*, 3GPP TR 22.813 Version 10.0.0, Release 10, 1 April 2010 http://www.3gpp.org/ftp/Specs/archive/22_series/22.813/22813-a00.zip (3 June 2011).

[11] 3GPP, SA WG4 Meeting #59, S4-100479 (2010) *Listening Tests Concerning Reference Codecs for EVS*, 21–24 June 2010, http://www.3gpp.org/ftp/tsg_sa/WG4_CODEC/TSGS4_59/Docs/S4-100479.zip (accessed 3 June 2011).

[12] 3GPP, Technical Report, Technical Specification Group Service and System Aspects (2010) *Study on Surround Sound Codec Extension for PSS and MBMS*, 3GPP TR 26.950 Version 1.3.2, Release 10, 27 June 2010, http:// www.3gpp.org/ftp/Specs/archive/26_series/26.950/26950-132.zip (accessed 3 June 2011).

[13] ISO/IEC JTC1/SC29/WG11, N11112 (2010) *Terms of Reference of the Joint Collaborative Team on Video Coding Standard Development*, 22 January 2010, Kyoto, Japan.

[14] ITU-T Q6/16, ISO/IEC JTC1/SC29/WG11, VCEG-AM91 (2010) *Joint Call for Proposals on Video Compression Technology*, 22 January 2010, Kyoto, Japan.

[15] AT&T, CTIA (2010) *AT&T's Rinne Campaigns for Spectrally Efficient Mobile Video*, 24 March 2010, http:// www.fiercewireless.com (accessed 3 June 2011).

[16] 3GPP, SA Plenary Meeting #47, SP-100032 (2010) *HTTP-Based Streaming and Download Services*, 22–25 March 2010, www.3gpp.org (accessed 3 June 2011).

[17] Next Generation Mobile Networks (NGMN) Alliance, White Paper (2006) *Next Generation Mobile Networks Beyond HSPA & EVDO*, Version 3.0, December 2006, www.ngmn.org (accessed 3 June 2011).

2

The Self-Organizing Networks (SON) Paradigm

Frank Lehser, Juan Ramiro, Miguel A. Regueira Caumel,
Khalid Hamied and Salvador Pedraza

2.1. Motivation and Targets from NGMN*

Reduction of cost and complexity is a key driver for Long Term Evolution (LTE), since with its deployment the new network layer needs to coexist with legacy systems without additional operating cost. Thus, it is of vital interest for operators to introduce automated engineering functions that minimize Operational Expenditure (OPEX) and, at the same time, increase network performance by dynamically adjusting the system configuration to the varying nature of wireless cellular networks. In this respect, technological advances allowing higher processing power with increased cost efficiency need to be leveraged in order to maintain or improve end-user performance while, at the same time, simplifying and improving network operability.

Deploying and operating cellular networks is a complex task that comprises many activities, such as planning, dimensioning, deployment, testing, prelaunch optimization, postlaunch optimization, comprehensive performance monitoring, failure mitigation, failure correction and general maintenance. Today, such critical activities are extremely labor-intensive and, hence, costly and prone to errors, which may result in customer dissatisfaction and increased churn.

Figure 2.1 shows the natural tendency towards a higher degree of automation and operational efficiency [1]. As can be seen, this trend is parallel to the evolution of wireless cellular systems. In the past, the first systems that were deployed required a large deal of manual operational effort, which was then gradually reduced as systems became more sophisticated and as more automated functionality was specified and/or made available.

* A significant share of the content of this section is based on the quoted documents by the NGMN Alliance. Permission for reproducing selected content has been granted by NGMN Ltd.

Self-Organizing Networks: Self-Planning, Self-Optimization and Self-Healing for GSM, UMTS and LTE,
First Edition. Edited by Juan Ramiro and Khalid Hamied.
© 2012 John Wiley & Sons, Ltd. Published 2012 by John Wiley & Sons, Ltd.

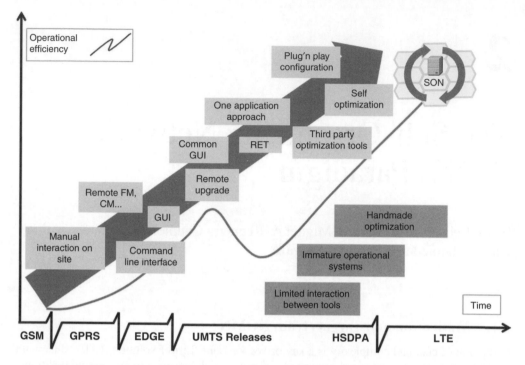

Figure 2.1 Operational efficiency versus time. Reproduced from © 2010 NGMN Ltd.

The Next Generation Mobile Networks (NGMN) Alliance[1] is an initiative by a number of leading mobile operators, whose mission is to complement and support the work within standardization bodies by providing a coherent view of what the operator community is going to require in the decade beyond 2010. Among other findings and conclusions, NGMN identified excessive reliance on manual operational effort as a significant problem and defined operational efficiency as a key target. In general, the requirements from the operator community are compiled in the NGMN White Paper entitled *Next Generation Mobile Networks Beyond HSPA & EVDO* [2].

Among the different projects and initiatives defined and started by NGMN, a project with its main focus on Self-Organizing Networks (SON) was initiated in 2006. Its main objective was to ensure that the operators' requirements are taken into account during the development of the Third Generation Partnership Project (3GPP) Operation And Maintenance (O&M) specifications, as well as in other specifications developed by similar groups of other standardization bodies, with the ultimate goal of automating operational tasks, such as planning, configuration, optimization and healing [2]. Along with the definition of SON Use Cases, additional effort has also been devoted to the definition of open O&M interfaces that facilitate a truly multi-vendor system.

After going through different phases, the outcome of the SON-related project is a set of public documents describing the envisioned SON Use Cases, which has provided the industry with inspiration to push for the development of SON solutions. These deliverables have had a strong influence on the work in 3GPP, for example, in the areas of Minimization of Drive

---

[1] In August 2010, the NGMN Alliance comprises 18 mobile network operators (Members), 37 vendors and manufacturers (Sponsors), and three universities or nonindustrial research institutes (Advisors). See www.ngmn.org.

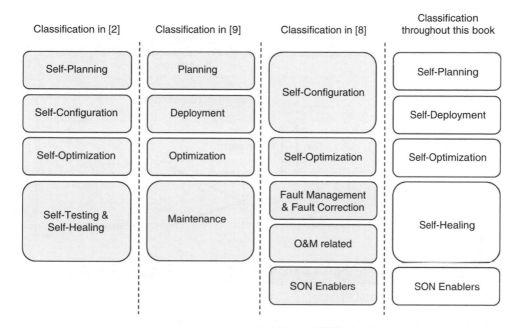

Figure 2.2   SON Use Cases grouped in different categories.

Tests (MDT) [3], Energy Saving (ES) [4], Handover Optimization [5], Automatic Neighbor Relation (ANR) management [6] and Load Balancing [5], as well as in other standardization bodies like the TeleManagement (TM) Forum [7].

2.2.   SON Use Cases*

2.2.1.   Use Case Categories

As summarized in Figure 2.2, different schemes have been proposed to categorize the SON Use Cases. Amid this lack of uniformity in the official way to name and define the different categories, the classification in the fourth column of Figure 2.2 has been adopted throughout the rest of this book. As can be seen, the Self-Configuration category, as defined in [8], has been divided into two groups, as suggested in [2] and [9], in order to clearly differentiate the derivation of the planning parameter values of new nodes (Self-Planning) from the actual plug and play character that is required for the deployment process (Self-Deployment). Moreover, the Self-Healing terminology has been adopted to cover the Fault Management, Fault Correction and O&M related categories in [8], i.e. the detection and correction/mitigation of problems, as well as the availability of functionalities that facilitate smooth system maintenance with minimal outage. This term, which has been adopted by 3GPP along the same lines [10], may look overly ambitious, since there are occasions in which the Self-Healing functionalities can only point out the existence (and possibly the root cause) of a problem, but the automatic generation of an immediate remedy that does not require user intervention is not feasible. In some cases, however, as will be discussed throughout the book, it is feasible to automatically derive and implement a

* A significant share of the content of this section is based on the quoted documents by the NGMN Alliance. Permission for reproducing selected content has been granted by NGMN Ltd.

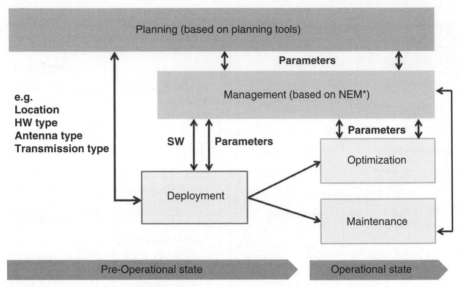

Figure 2.3 shows the diagram with labels:
- Planning (based on planning tools)
- Parameters
- e.g. Location HW type Antenna type Transmission type
- Management (based on NEM*)
- SW Parameters
- Parameters
- Deployment
- Optimization
- Maintenance
- Pre-Operational state
- Operational state

(*) NEM = Network Element Manager

Figure 2.3 Planning, deployment, optimization and maintenance processes. Reproduced by permission of © 2010 NGMN Ltd.

temporary solution that alleviates certain problems. This will be covered in Chapter 6. Furthermore, SON Enablers, which are support functionalities that facilitate the execution and availability of certain SON Use Cases, are considered separately, as already proposed in [8].

An exact definition of the Use Case categories that will be used throughout the rest of the book is presented in the following:

- Self-Planning. Derivation of the settings for every new network node, including the selection of the site location and the specification of the Hardware (HW) configuration, but excluding site acquisition and preparation.
- Self-Deployment. Preparation, installation, authentication and verification of every new network node. It includes all procedures to bring a new node into commercial operation, except for the ones already covered in the Self-Planning category, which generate inputs for the Self-Deployment phase.
- Self-Optimization. Utilization of measurements and performance indicators collected by the User Equipments (UEs, i.e. the handsets) and the base stations in order to auto-tune the network settings. This process is performed in the operational state, which is defined as the state where the Radio Frequency (RF) interface is commercially active (i.e. when the cell is neither barred nor reserved).
- Self-Healing. Execution of the routine actions that keep the network operational and/or prevent disruptive problems from arising. This includes the necessary Software (SW) and HW upgrades and/or replacements.

Although planning, deployment, optimization and maintenance do not necessarily need to be seen as sequential activities, Figure 2.3 puts them into perspective by means of a simplified workflow for the sake of clarity [9].

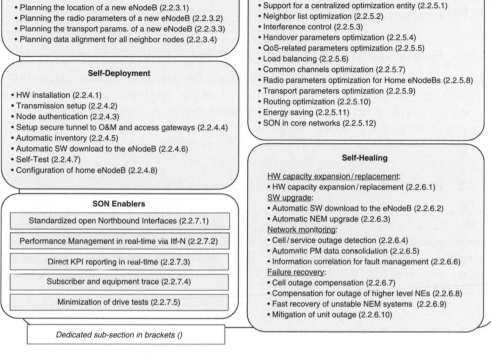

Self-Planning

- Planning the location of a new eNodeB (2.2.3.1)
- Planning the radio parameters of a new eNodeB (2.2.3.2)
- Planning the transport params. of a new eNodeB (2.2.3.3)
- Planning data alignment for all neighbor nodes (2.2.3.4)

Self-Deployment

- HW installation (2.2.4.1)
- Transmission setup (2.2.4.2)
- Node authentication (2.2.4.3)
- Setup secure tunnel to O&M and access gateways (2.2.4.4)
- Automatic inventory (2.2.4.5)
- Automatic SW download to the eNodeB (2.2.4.6)
- Self-Test (2.2.4.7)
- Configuration of home eNodeB (2.2.4.8)

SON Enablers

Standardized open Northbound Interfaces (2.2.7.1)

Performance Management in real-time via Itf-N (2.2.7.2)

Direct KPI reporting in real-time (2.2.7.3)

Subscriber and equipment trace (2.2.7.4)

Minimization of drive tests (2.2.7.5)

Dedicated sub-section in brackets ()

Self-Optimization

- Support for a centralized optimization entity (2.2.5.1)
- Neighbor list optimization (2.2.5.2)
- Interference control (2.2.5.3)
- Handover parameters optimization (2.2.5.4)
- QoS-related parameters optimization (2.2.5.5)
- Load balancing (2.2.5.6)
- Common channels optimization (2.2.5.7)
- Radio parameters optimization for Home eNodeBs (2.2.5.8)
- Transport parameters optimization (2.2.5.9)
- Routing optimization (2.2.5.10)
- Energy saving (2.2.5.11)
- SON in core networks (2.2.5.12)

Self-Healing

HW capacity expansion/replacement:
- HW capacity expansion/replacement (2.2.6.1)
SW upgrade:
- Automatic SW download to the eNodeB (2.2.6.2)
- Automatic NEM upgrade (2.2.6.3)
Network monitoring:
- Cell/service outage detection (2.2.6.4)
- Automatic PM data consolidation (2.2.6.5)
- Information correlation for fault management (2.2.6.6)
Failure recovery:
- Cell outage compensation (2.2.6.7)
- Compensation for outage of higher level NEs (2.2.6.8)
- Fast recovery of unstable NEM systems (2.2.6.9)
- Mitigation of unit outage (2.2.6.10)

Figure 2.4   Use Cases under each category.

In the rest of this Chapter, the Use Cases will be distributed among categories according to the scheme presented in Figure 2.4.

2.2.2.   *Automatic versus Autonomous Processes*

The functionalities that make it possible to execute the different Use Cases are structured around processes, which can be automatic or autonomous [9]:

- An autonomous activity is a process that is fully controlled by a machine without requiring any human interaction.
- An automatic activity is a process where a significant part of the action is handled by a machine but some human interaction is required (e.g. where an operator is required to activate machine specific macros, or required to verify results).

2.2.3.   *Self-Planning Use Cases*

As stated before, this corresponds to the Planning category presented in [9] (see Figure 2.2), and the included Use Cases constitute a subset of the ones presented in [8] under the Self-Configuration umbrella. The Self-Planning category groups those Use Cases that cover the derivation of the settings for a new network node, including, among others, the selection of the site location and the specification of the HW configuration, but excluding site acquisition and preparation. A brief description of these Use Cases is provided in the following subsections [8, 9].

2.2.3.1. Planning the Location of a New eNodeB and its Basic Characteristics

This Use Case involves planning the location of the new eNodeB, as well as the associated HW characteristics, transmission resources, number of sectors and basic parameters per cell (e.g. antenna and power settings). This activity is based on strategic guidelines, traffic forecasts, maps from planning tools, available measurements, coverage/capacity goals, list of available site locations and restrictions in terms of HW/transmission capabilities and budget.

2.2.3.2. Planning the Radio Parameters of a New eNodeB

This Use Case involves planning cell identifiers, power settings, neighbor lists, handover settings, Random Access Channel (RACH) parameters, Paging Channel (PCH) resource settings, (potentially) fractional resource allocation and any other relevant Radio Resource Management (RRM) parameters of the new eNodeB. In some cases, only initial planning is covered here, and there is room for later Self-Optimization when the system has real traffic.

2.2.3.3. Planning the Transport Parameters of a New eNodeB

This Use Case involves planning the transport parameters of the new eNodeB so that the required information is available before deploying the eNodeB and running the Transmission Setup Use Case. This means that the parameters that are needed to start a tunnel to other network entities and receive further network settings must be available. Some of these parameters are, for example, Internet Protocol (IP) addresses, Virtual Local Area Network (VLAN) identifier, transport Quality of Service (QoS) settings, etc.

2.2.3.4. Planning Data Alignment for all Neighbor Nodes

This Use Case manages the announcement of the new eNodeB to all neighboring nodes, as well as the corresponding alignment of their databases. White and black lists need to be considered and there is a wide range of possible operator policies: fully preplanned white and black lists, initially suboptimal neighbor lists as a starting point, and even initially empty neighbor lists, i.e. no neighbor relationships preconfiguration. Note that the last two possibilities are only advisable when the ANR functionality is in place.

2.2.4. Self-Deployment Use Cases

As stated before, the Use Cases presented in [8] under the Self-Configuration umbrella can also be split into two distinct categories: (Self-) Planning and (Self-) Deployment [9]. Following this division, the Use Cases in the Self-Deployment category cover preparation, installation, authentication and delivery of a status report for a new network node. In other words, this group includes all the procedures required to bring a new node into commercial operation, except for the ones included in the Self-Planning category.

A simplified example of the sequential way in which these Use Cases are meant to be combined is presented in the following. After the physical installation of its HW and cabling for antennas and networking, the new eNodeB needs to discover its Network Element Manager (NEM) and, in order to gain access to it, an authentication process is required. After the eNodeB has been successfully authenticated, it can carry out further actions, such as downloading relevant parameters and SW packages, as well as setting up transport (S1/X2) and establishing secure tunnels to O&M and access Gateways (aGWs). There may be some implications also in the neighboring eNodeBs as well as associated aGWs, e.g. neighbor cell information and tracking areas need to be configured. The eNodeB shall further run a Self-Test procedure with the configured parameters, followed by the corresponding status report to the network management entities. A brief description of these Use Cases is provided in the following subsections [8, 9].

2.2.4.1. Hardware Installation

This Use Case covers site preparation and HW installation, including error-free cabling and plug and play behavior for all components (antenna, boards, cages, etc.). The HW and the deployment process need to be designed so that installation time does not exceed 20 min, and the configuration of the new element is to be carried out with minimum human intervention.

2.2.4.2. Transmission Setup

The eNodeB establishes a secure tunnel to a configuration server that sends all the settings that are required to establish a full tunnel to the O&M system and the corresponding access nodes.

2.2.4.3. Node Authentication

The eNodeB receives the information that is needed to access a security gateway. After the process, the eNodeB is authorized to access the inner network and other network entities, such as the NEM.

2.2.4.4. Setup of a Secure Tunnel to O&M and Access Gateways

This Use Case covers generation of a secure tunnel to O&M and aGWs (optionally ciphered). This end-to-end connection needs to be bidirectional, stable and secure.

2.2.4.5. Automatic Inventory

This Use Case covers autonomous delivery of the configuration and status of all installed HW units to the Inventory System so that a comprehensive and up-to-date picture of the currently installed HW base is available.

2.2.4.6. Automatic Software Download to the eNodeB

This Use Case covers autonomous download of new SW packages. This can be done in a pull or push mode, and the process needs to include appropriate fallback mechanisms to take into account for potential problems during the update process. The process is not only suitable for eNodeB deployment, but also for SW updates of eNodeBs already in commercial operation. Similar processes need to be enabled for the eNodeB firmware as well.

2.2.4.7. Self-Test

This Use Case covers Self-Test of HW and SW, to verify whether the eNodeB is in the expected state and prepared for immediate commercial exploitation. If not, an illustrative report is generated. The Self-Test can be also commanded when the eNodeB is already on-air.

2.2.4.8. Configuration of Home eNodeB

Installation should follow a plug and play approach and human involvement should be minimized. This process covers all the required activities since the moment at which the Home eNodeB is switched on until all its radio parameters are properly set. Regarding transport parameters and configuration, the process depends on whether the router is packed together with the eNodeB.

2.2.5. Self-Optimization Use Cases

Mobile networks are dynamic structures, where continuously new sites are deployed, capacity extensions are made and parameter values are adapted to local traffic and environmental conditions. The optimization process to achieve superior performance is described in Figure 2.5 [9].

Network optimization is a continuous closed-loop process encompassing periodic performance evaluation, parameter optimization and redeployment of the optimized parameter values onto the network. The optimization decisions can be carried out by human subjects or computerized systems (referred to as controller in Figure 2.5). Input data for the optimization process shall be gathered by means of various techniques, involving several sources, such as O&M performance measurements, O&M alarms and traces from the main interfaces (e.g. Uu, Iub, Iu, etc.), as well as drive tests combining air interface measurements with location information. In Figure 2.5, all these sources of information are referred to as sensors, and the combination of these pieces of information, together with the current system configuration, is provided as an input to the optimization algorithm in charge of deriving the most appropriate parameter values are implemented in the network by means of an actuator (e.g. O&M or the Element Manager in Figure 2.5), thereby impacting the behavior of the network, which is referred to as control path in the model depicted in Figure 2.5. This optimization process will aim at tuning parameter values in order to achieve a well-defined goal, expressed in terms of coverage, capacity, quality or a controlled combination of them, taking into consideration that any optimization activity always involves implicit trade-offs between these key variables. The optimization algorithms may vary from one implementation to another, and

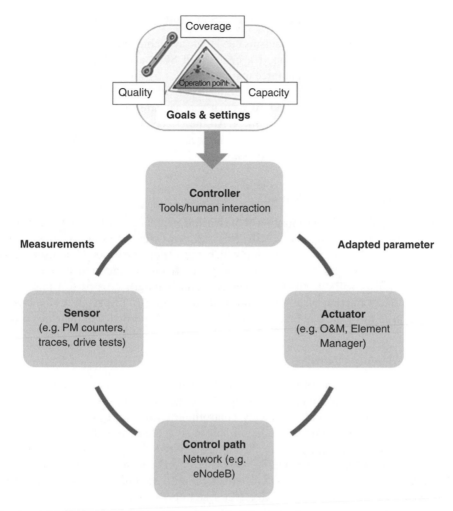

Figure 2.5   Closed-loop Self-Optimization process.

will be heavily impacted by the optimization policy envisioned by the operator. However, in order to facilitate advanced optimization processes, the necessary measurement functions and interfaces should be defined in the standards, especially to cope with multi-vendor scenarios.

Network equipment supplied by different vendors complicates the optimization process, since different vendors may support different performance counters and measurements in their O&M systems. As a consequence, network operators in a multi-vendor environment have traditionally had to develop adaptation layers to translate and harmonize the performance measurements from different vendors so that a common optimization platform can be used.

A brief description of the Self-Optimization Use Cases is provided in the following subsections [8, 9].

2.2.5.1. Support for a Centralized Optimization Entity

At network management level, mechanisms need to be available (e.g. through the Northbound Interface, denoted by Itf-N) to support a centralized SON architecture. In general, these mechanisms can be grouped around three main categories: (i) sufficiently comprehensive and flexible measurements and Key Performance Indicators (KPIs); (ii) sufficiently comprehensive and flexible events and alarms; and (iii) open mechanisms to read the current system configuration and modify it through the Itf-N. All these mechanisms need to be structured in such a way that third-party suppliers can provide centralized functionalities, either through standardized or open (proprietary) interfaces.

2.2.5.2. Neighbor List Optimization

This Use Case pertains to the optimization of the existing neighbor list of a cell plus the neighbor lists that are applied by the neighboring cells. This applies to cells that are already in a production environment, as well as to new cells, for which completely new neighbor lists are to be generated. The target scope covers intrafrequency, interfrequency and intersystem neighbors. White and black lists will eventually be generated and/or modified as part of this Use Case depending on the evaluation of different information sources. In this respect, the target is to minimize the need for white and black list configuration or, if needed, automate this process.

2.2.5.3. Interference Control

In order to improve signal quality and therefore throughput, LTE systems may need interference coordination, both in uplink (controlling the usage of the physical resource blocks in different sectors) and in downlink (mainly controlling the transmit power patterns). These procedures require the transmission of certain information elements through the X2 interface with the objective to facilitate packet scheduling decisions that take into consideration the interference situation in the surrounding cells [11]. Moreover, related RRM functionalities will be governed by parameters that indicate restrictions, preferences, thresholds and periods. Since the value of these parameters will affect system performance, they are also subject to Self-Optimization.

2.2.5.4. Handover Parameters Optimization

This Use Case covers the optimization of handover parameters with minimal operational effort in order to improve the quality of the handover process (i.e. to reduce handover failure rate due to too early or too late handovers, as well as handovers to a wrong cell), minimize undesired effects (such as ping-pong handovers) and cooperate with the actions to be taken as part of the Load Balancing Use Case.

2.2.5.5. QoS-Related Parameters Optimization

This Use Case covers the optimization of RRM parameters impacting QoS with minimal operational effort. This is to be done by means of a closed-loop scheme in which KPIs are measured in the real network and decisions are made in order to adapt the system configuration to the network environment, thereby achieving superior performance according to the operator's policies and guidelines. Different strategies can be used for different traffic classes.

2.2.5.6. Load Balancing

Tuning handover and/or cell reselection parameters to distribute traffic between sectors as a function of the load situation, with the ultimate goal to improve the trunking efficiency of the system and therefore enhance its capacity while maintaining or improving quality metrics.

2.2.5.7. Common Channels Optimization

The resource allocation for common channels, such as RACH, needs to be adjusted in order to find the right trade-off between accessibility and system capacity. The more resources allocated for RACH, the lower the number of collisions and therefore the better the call setup success rate. However, this will also mean lower uplink system capacity.

2.2.5.8. Radio Parameters Optimization in Scenarios with Home eNodeBs

Under the assumption that Home and Macro eNodeBs operate with the same spectrum, coordination activities and optimization are needed to minimize interference to and from the macro layer. Dynamic adaptation is especially important due to the fact that the conditions of Home eNodeBs can be changed discretionally.

2.2.5.9. Transport Parameters Optimization

This Use Case covers the optimization of transport-related parameters with minimal operational effort.

2.2.5.10. Routing Optimization

This Use Case covers the optimization of data routing in a meshed network.

2.2.5.11. Energy Saving

This Use Case covers minimization of the energy consumption by exploiting temporary redundancy in the base stations or in other network layers. Switching off unused parts of base stations or cells in low traffic periods (if these cells are backed up by other RAT/layer cells) reduces energy consumption.

2.2.5.12. SON in Core Networks

A lot of the aforementioned Use Cases in the Radio Access Network (RAN) domain can be mapped onto analogous Use Cases in the core network. For example, installation of new core network nodes and optimization of transport parameters.

2.2.6.   Self-Healing Use Cases

For further clarity, the Self-Healing Use Cases can be grouped into the following categories:

- HW capacity extension/replacement[2]. Within this category, any upgrade, extension or replacement of HW elements should require minimal operator attention.
- SW upgrade[3]. Similarly, SW updates shall need minimum operator attention and any functional outage (including loss of surveillance) must be minimized. The following Use Cases are part of this category:
    - o Automatic SW download to the eNodeB.
    - o Automatic NEM upgrade.
- Network monitoring. The system should provide sufficient measurements and analyses of the RAN performance in order to make qualified decisions about further improvements. In this context, multi-vendor scenarios should be supported without any extra complexity and call tracing shall be supported for troubleshooting purposes, as well as for special tasks (e.g. UE performance analysis). The following Use Cases are part of this category:
    - o Cell/service outage detection.
    - o Automatic Performance Management (PM) data consolidation.
    - o Information correlation for Fault Management.
- Failure recovery. Recovery of Network Element (NE) failures should not require complex manipulation by an expert. The following Use Cases are part of this category:
    - o Cell Outage Compensation (COC).
    - o Compensation for outage of higher level NEs.
    - o Fast recovery of unstable NEM systems.
    - o Mitigation of unit outage.

Having introduced the aforementioned subclassification, the detailed description of the Self-Healing Use Cases is provided in the following subsections.

2.2.6.1.   Hardware Capacity Expansion/Replacement

As previously stated, any upgrade, extension or replacement of HW elements should require minimal operator attention. Moreover, HW shall allow plug and play installation and technicians shall be able to handle HW from different suppliers without special training. The autonomous surveillance of the system utilization in order to suggest the convenience of HW expansions is also part of this Use Case.

2.2.6.2.   Automatic Software Download to the eNodeB

Similar to the Use Case with the same name defined in Section 2.2.4.6, extending its scope from network deployment and installation to system maintenance and SW updates.

---

[2] Some activities in this sub-group could also naturally fit within the Self-Deployment category, whereas others (e.g. the automatic suggestion of the need for an upgrade) could fit within the Self-Planning category.
[3] Both HW extension/replacement and SW upgrade were originally classified as Maintenance Use Cases by NGMN [9].

2.2.6.3.   Automatic NEM Upgrade

All the activities associated with the NEM upgrade must be carried out with minimum effort and disruption to daily operations. This procedure is structured in a safe way that allows fall-back to the previous configuration at any time during the upgrade process and the subsequent special surveillance period.

2.2.6.4.   Cell/Service Outage Detection

Mechanisms to detect outage in cells (e.g. sleeping cells) and/or services with poor quality should be in place and work reliably with minimal or no human intervention. For that, the system must combine pieces of information from all available sources and generate a report that serves to consolidate information and also trigger potential automatic/autonomous reactions to solve or, at least, temporarily alleviate the problem.

2.2.6.5.   Automatic PM Data Consolidation

This Use Case covers automated analysis of PM counters and consolidation of these into higher level KPIs at eNodeB level. These KPIs will be used for reporting purposes, for surveillance and as input to automatic/autonomous routines that take care of solving/alleviating network incidences.

2.2.6.6.   Information Correlation for Fault Management

In order to simplify and (partly) automate Fault Management, the generation of unambiguous alarms, containing the root cause and the location is critical. For this, as many sources of information as possible need to be combined and correlated at the lowest possible level. Furthermore, there should be mechanisms that allow the operator to customize the information correlation process to generate alarms, as well as the logic that defines the subsequent actions to be taken, which can be either autonomous or require human intervention and/or approval.

2.2.6.7.   Cell Outage Compensation

After detecting cell outage, execute actions to solve or, at least, alleviate the problem. This process is meant to restore the system back to normal operation as soon as possible and, therefore, needs to be remarkably fast, i.e. near real-time. When the root cause of the problem cannot be autonomously found in a short time frame and an immediate full solution cannot be applied, a trouble-ticket needs to be issued after implementing temporary solutions that at least alleviate the problem. In addition, for any implemented solution, automated verification needs to be carried out and its outcome needs to be reported. Among the possible solutions, the following can be highlighted: creation, banning or favoring of certain neighbors, reduction of the transmission power in the affected cell in order to shrink its footprint and decrease its load, RF/power retuning of the surrounding cells and triggering of Self-Optimization activities that allow full exploitation of the new environment until a final solution for the problem is implemented. When this happens, automated/autonomous processes need to be in place to fall back to the configuration before the outage incident.

2.2.6.8.   Compensation for Outage of Higher Level NEs

Outage of higher level NEs (e.g. Mobility Management Entity, denoted by MME) shall be compensated by other elements at the same hierarchical level. This can be purely triggered by an outage event or made available through pooled resources exploited by means of dynamic load sharing.

2.2.6.9.   Fast Recovery of Unstable NEM Systems

Enable fast recovery of unstable NEM systems by means of updated backup elements, autonomous takeover mechanisms, inherent load sharing functionality and/or high availability architectures.

2.2.6.10.   Mitigation of Unit Outage

When a replaceable part of a network node (i.e. a unit) enters an outage state, this situation has to be detected and the corresponding alarms and warnings need to be generated. Furthermore, autonomous mechanisms must be in place in order to temporarily reconfigure the equipment in such a way that performance degradation is minimized. The availability of backup units in hot standby will help, as well as the ability to reallocate the load to different network resources, both in the NEs suffering from outage, as well as in the neighboring ones.

2.2.7.   SON Enablers

SON Enablers are support functionalities that facilitate the execution and availability of certain SON Use Cases. However, they are not SON Use Cases by themselves. A list of relevant SON Enablers is presented in the following subsections [8].

2.2.7.1.   Standardized Open Northbound Interfaces

Interfaces between NEM and tools operating at network management level have to be open for simpler and less costly integration with third-party tools, as well as lower SW production costs.

2.2.7.2.   Performance Management in Real-Time via Itf-N

It is useful to support network monitoring and SON. Apart from being able to monitor general statistics within a time interval of 15–30 min, it will be possible to monitor, with a more restricted scope, certain performance alarms in near real-time (i.e. with a delay of around 1 min). Moreover, it will be possible to carry out data gathering within configurable intervals that range from 10 s to 5 min.

2.2.7.3. Direct KPI Reporting in Real-Time

For a set of KPIs, which are defined based on PM counters, it will be possible to activate alarms that will be raised when the values of these KPIs, which are monitored continuously at the eNodeBs, fulfil certain conditions (e.g. when a KPI at a certain eNodeB exceeds a threshold that is configurable by the operator).

2.2.7.4. Subscriber and Equipment Trace

Implement inherent trace functionality (e.g. for specific subscribers or cells) that cover the following interfaces: S1-MME, S1-U, X2 and Uu. It must be possible to store this information locally at the eNodeB, transfer it somewhere else, filter it according to the operator's requirements and make it available to third-party solutions.

2.2.7.5. Minimization of Drive Tests

Use inherent trace functionality in terminals, eNodeBs and other network nodes, and enrich these with location information in order to minimize the need for dedicated drive tests. For this Use Case, different subcategories can be defined with different targets, such as coverage, capacity, QoS and mobility optimization (including the parameterization of common channels), as well as automatic troubleshooting to identify and heal specific network issues. The key functionality is to bind location information, such as cell identifiers, radio fingerprints (based on timing advance and signal strength values) and eventually Global Positioning System (GPS) coordinates, with trace and performance measurement data in a standardized way.

2.3. SON versus Radio Resource Management

When discussing about SON functionalities, especially within the scope of real-time Self-Optimization features, conceptual questions may be asked about the exact difference between these algorithms and the ones embedded in traditional RRM, since the border may become unclear for real-time functionalities.

RRM algorithms are responsible for the dynamic allocation of spectrum, power and transmission intervals within the radio domain in a multi-user and multi-cell environment, with the objective to maximize network performance in varying propagation and traffic conditions. In this respect, network performance comprises capacity, coverage, quality and mobility. For the Universal Mobile Telecommunications System (UMTS), examples of RRM algorithms are fast power control, outer-loop power control, handover control, admission/ load/congestion control and packet scheduling.

On the other hand, as previously stated, Self-Optimization is defined in this context as the utilization of measurements and performance indicators collected by the UEs and the base stations in order to auto-tune the settings of a network in operational state.

Thus, one first approximation is to claim that RRM algorithms are governed by parameters and Self-Optimization functionalities take care of dynamically adjusting the value of these parameters in order to adapt the system behavior to the current environment. Let us focus on a UMTS example with two classical RRM algorithms: Fast Closed-Loop Power Control (CLPC) and Outer-Loop Power Control (OLPC), which can be defined as follows [12]:

- CLPC is responsible for controlling the transmit power at each link via a closed-loop feed-back scheme, in order to fulfill the Signal to Interference-plus-Noise Ratio (SINR) target in all the links while minimizing the total amount of transmitted power. In Uplink (UL), such algorithm is essential in order to overcome the so-called near-far effect [13], and its use in Downlink (DL) increases the capacity [14].
- OLPC adjusts the SINR target for CLPC in order to maintain the quality of the communication in terms of Frame Erasure Rate (FER) at the desired level.

Both algorithms are tightly coupled to each other, since the output for OLPC is the value of a key parameter (the SINR target) affecting the behavior of CLPC (see Figure 2.6).

In this respect, OLPC could be deemed as a real-time Self-Optimization algorithm focused on adjusting the parameters governing CLPC with the objective to ensure the desired quality with the minimum resource utilization. Based on this example, two key characteristics can be listed in order to differentiate RRM algorithms from real-time Self-Optimization functionalities:

- The output of Self-Optimization algorithms affects the value of the parameters impacting key system processes (e.g. the power that is allocated to the primary Common Pilot Channel (CPICH) in UMTS, the RACH settings in LTE, the neighbor definitions, etc.) or governing

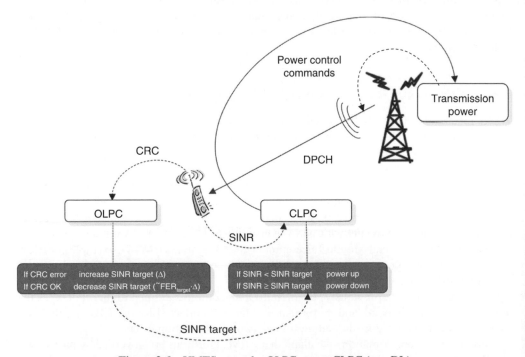

Figure 2.6 UMTS example: OLPC versus CLPC (e.g. DL).

RRM functionalities, which take care of the actual allocation of system resources (e.g. the load thresholds for admission control).

- The parameters governing RRM algorithms are typically low level, engineering-oriented magnitudes (e.g. power thresholds, time periods, etc.), whereas the inputs for Self-Optimization functionalities are typically higher level, service-oriented parameters (e.g. desired FER in the case of OLPC, or targeted trade-off between capacity, coverage and quality when applying Self-Optimization of UMTS RRM parameters, as described in Chapter 5).

2.4. SON in 3GPP

2.4.1. 3GPP Organization

3GPP is an alliance and a standards body that works within the scope of the International Telecommunication Union (ITU) to develop 3rd Generation (3G) and 4th Generation (4G) specifications based on evolved Global System for Mobile communications (GSM) standards. The project was created in December 1998 by signing The 3rd Generation Partnership Project Agreement [15].

Originally, the scope of 3GPP was to produce the Technical Specifications and Technical Reports for a global 3G cellular system. Later, the responsibility of 3GPP was extended to include the maintenance and development of GSM Technical Specifications and Technical Reports [16]. Currently, 3GPP is supporting the evolution of 3G by developing the Technical Specifications and Technical Reports for LTE and LTE-Advanced.

The structure of 3GPP consists of a Project Coordination Group (PCG) and some Technical Specification Groups (TSGs) [17]. The PCG manages the overall work progress of 3GPP and there are four 3GPP TSGs, each one of them being divided into several Working Groups (WGs):

- TSG GERAN: evolution and interoperability of GSM/GPRS/EDGE radio access technologies (RATs). Note that GERAN stands for GSM EDGE Radio Access Network, GPRS stands for General Packet Radio Service and EDGE stands for Enhanced Data rates for GSM Evolution. This group is divided into three WGs: WG1 (Radio Aspects), WG2 (Protocol Aspects) and WG3 (Terminal Testing).
- TSG RAN: radio aspects. It is divided into five WGs: WG1 (Radio Layer 1), WG2 (Radio Layer 2 and 3), WG3 (UTRAN O&M Requirements), WG4 (Radio Performance) and WG5 (Mobile Terminal Conformance Testing). Note that UTRAN stands for Universal Terrestrial Radio Access Network and E-UTRAN stands for Evolved UTRAN.
- TSG SA: overall architecture and service capabilities. This TSG is divided into five WGs: WG1 (Services), WG2 (Architecture), WG3 (Security), WG4 (Codec) and WG5 (Telecom Management). Note that, in this context, SA stands for Service and system Aspects.
- TSG CT: core network and terminal specifications. Note that, in this context, CT stands for Core network and Terminals.

The groups responsible for the evolution of SON specifications are mainly RAN WG3 and SA WG5, sometimes with collaborative work from other WGs, such as RAN WG2, GERAN WG2 and SA WG2.

2.4.2.   SON Status in 3GPP (up to Release 9)

Since Release 8, features started to be added to the 3GPP specifications in order to progressively support the SON Use Cases envisioned by NGMN. A summary of the most relevant standardized functionalities is provided in this section, taking the final version of the Release 9 specifications as a reference point [18, 19]. In general, the SON concepts and requirements are captured in [20], which also includes the different architectures (centralized, distributed and hybrid) and the business level requirements.

2.4.2.1.   Self-Configuration in 3GPP (up to Release 9)

The specifications for Self-Configuration of new eNodeBs are provided in [21, 22, 23], which describe a framework that covers all the necessary steps to put an eNodeB in operational state with minimal human intervention. For this process, which is focused on LTE networks, the precondition is that the eNodeB has been physically installed and connected to the operator's IP network. According to [21], the whole sequence can be initiated either manually or automatically after the initial Self-Test of the installed equipment. This process includes the transfer of configuration data to the eNodeB, but does not include, in general, the derivation and preparation of such information. Among others, the key steps of this process are: allocation of an IP address to the new eNodeB, provision of basic information about the transport network, announcement (by the eNodeB) of its key characteristics to the O&M system, connection of the eNodeB to the O&M subsystems in charge of Self-Configuration, connection of the eNodeB to the O&M systems in charge of normal O&M functionality, automatic SW download, automatic configuration of initial transport and radio parameters, establishment of S1 links, establishment of planned X2 links, announcement of the new eNodeB to the inventory system, execution of Self-Test functions after selected steps, generation of status reports for the operator, announcement of the new eNodeB to entities above the Itf-N and SW installation and activation.

In Chapter 4, which is focused on Self-Planning of multi-technology networks, emphasis is put on the derivation of the initial network configuration, which is outside the scope of the 3GPP specifications.

Even though a framework has been provided to support Self-Configuration procedures up to Release 9, many aspects in this area remain open and need to be solved by means of proprietary procedures. In this respect, it has been deemed a challenge to find standardized scenarios and models to really achieve the driving vision behind standardization, which aspires to a situation in which integration work and operator-specific solutions are not needed any longer in the area of Self-Configuration.

2.4.2.2.   Self-Optimization in 3GPP (up to Release 9)

Up to Release 9 of 3GPP, specifications have been provided for the Self-Optimization functionalities listed below. This does not necessarily mean that the SON functions will be implemented by all vendors in the same way, since in most cases the specifications only provide standard mechanisms and procedures that enable the construction of the SON functionalities under discussion with a certain degree of freedom. Such mechanisms primarily ensure the availability of certain UE measurements and the possibility to exchange useful information between UEs and eNodeBs, as well as between eNodeBs. The standardized SON functionalities are the following:

- ANR, which covers the NGMN Use Case on Neighbor List Optimization, is described in [6] and [24], encompassing intra- and interfrequency LTE neighbors, 2G neighbors and 3G neighbors of an LTE cell. Management functionalities have also been defined to control basic settings, such as white and black lists [25].
- Mobility Load Balancing (MLB), which covers the NGMN Use Case on Load Balancing, is introduced in [11], [24], [26], [27] and [28] by making the required set of standard procedures available.
- Mobility Robustness Optimization (MRO), which covers the NGMN Use Case on Handover Parameters Optimization, is described in [11], [24] and [26].
- RACH optimization, which covers the NGMN Use Case on Common Channels Optimization, is described in [11], [24], [26] and [28].
- Interference Control is supported through the standardization of procedures to exchange certain information elements through the X2 interface, such as High Interference Indicator (HII) and Overload Indicator (OII) in the uplink, and Relative Narrowband Transmit Power (RNTP) in the downlink, with the objective to facilitate packet scheduling decisions that take into consideration the interference situation in the surrounding cells [11].

For all of them, details on the standardized mechanisms are provided in Chapter 5 (Self-Optimization).

Apart from the specification of the aforementioned Use Cases, initial discussions have been held and requirements have been formulated for Capacity and Coverage Optimization (CCO). The provision of optimal coverage and capacity has been identified as a key objective in [26], some generic situations to be solved have been described in [5] (E-UTRAN coverage holes with 2G/3G coverage, E-UTRAN coverage holes without any other radio coverage and E-UTRAN coverage holes with isolated island cell coverage) and some configuration parameters to materialize this Use Case have been standardized in [25].

Additionally, management capabilities have been specified for MLB and MRO, allowing the O&M system to activate/deactivate the functionality, set targets with different priorities and collect related performance measurements [29, 30].

2.4.2.3. Self-Healing and Others in 3GPP (up to Release 9)

On a different note, a technical study about Self-Healing has been produced in Release 9 [31], in which a general workflow for Self-Healing is provided and different recovery actions are identified for different types of system problems. Additionally, three concrete Self-Healing Use Cases are defined: Self-Recovery of NE SW, Self-Healing of board faults and Self-Healing of cell outage.

Another important technical study in Release 9 is related to MDT [3]. This function is envisioned to be a key enabler, utilizing commercial UEs to log and report measurements, together with the location (if available) at which each measurement was taken. These traces are considered to be paramount for network monitoring and other SON Use Cases, such as MRO, CCO and parameterization of common channels.

2.4.3.   SON Objectives for 3GPP Release 10

For Release 10, the most important SON-related objectives for the different WGs are presented in the following subsections [32].

2.4.3.1.   Objectives for TSG SA WG5

The most relevant SON-related objectives for TSG SA WG5 are the following:

* Work on the management aspects of Interference Control (IC), CCO and RACH optimization.
* Continue working on the coordination between different SON functions.
* Based on the completion of the Self-Healing study [31]: (i) specification of the requirements for Self-Healing; (ii) definition of inputs and outputs for Self-Healing functions; and (iii) specification of the O&M support for such functions. This work is to be captured in [10].
* Based on the completion of a study on ES [4], support ES management functions through the definition of the associated O&M requirements, the coordination with other functions and the definition of measurements that can be used to assess ES actions. The associated work will be captured in 3GPP TS 32.551 (*Telecommunication management; Energy Saving Management (ESM); Concepts and requirements*).
* Complete a study on integration of device management information with Itf-N [33], which focuses on how to manage measurement data collection at the UEs and how to make that information available for MDT and SON.

2.4.3.2.   Objectives for TSG RAN WG1

The most relevant SON-related objective for TSG RAN WG1 is to carry out a study on ES, focused on the identification of solutions to enable ES within UMTS NodeBs and conduct preliminary evaluations [34].

2.4.3.3.   Objectives for TSG RAN WG2

A paramount SON-related objective for TSG RAN WG2 is to specify a solution for MDT with coverage optimization as a priority Use Case. For that purpose, new UE measurements and logs, as well as new MDT functionality related to configuration and reporting, need to be defined. That work will be captured in 3GPP TS 37.320 (*Radio Measurement Collection for Minimization of Drive Tests (MDT); Overall description; Stage 2*).

2.4.3.4.   Objectives for TSG RAN WG3

The main SON-related objectives for TSG RAN WG3 are the following:

* Enhance MLB and MRO, which need to support multi-RAT environments.
* Carry out further work on CCO.

- Specify ANR for UTRAN, supporting intra-UTRAN and iRAT handover scenarios, and thereby enabling UTRAN NodeBs to manage autonomously their neighbor relationships with 2G/3G/LTE.
- Carry out a study entitled *Network Energy Saving for E-UTRAN* [35], which is a continuation of the original work on the ES Use Case. The objective of the study is to identify and evaluate potential solutions for ES in E-UTRAN. Specifically, the study will consider intra-eNodeB, inter-eNodeB and iRAT ES Use Cases. The first Use Case (intra-eNodeB ES), however, will be covered by RAN WG2 while the rest will be handled by RAN WG3.

2.5. SON in the Research Community

Since infrastructure vendors are developing SON solutions, whitepapers from those vendors depicting their vision on SON and the architecture of their proposed solutions are widely available. In addition to vendors, the following consortiums and projects are actively researching many aspects of SON.

2.5.1. SOCRATES: Self-Optimization and Self-ConfiguRATion in wirelEss networkS

The SOCRATES project was started in January 2008 and was completed in December 2010. The project is funded by the European Union under the 7th Framework Program [36].

The project is focused on the development of Self-Organization methods to enhance the operation of wireless access networks. They have chosen the 3GPP LTE radio interface as the central radio technology for the studies. The project's objectives are to develop novel concepts, methods and algorithms for SON, and to define new measurement elements that would be needed by those new methods. The project's objectives are also to validate the developed concepts and methods through extensive simulations, and to evaluate the implementation and operational impact of the proposed schemes.

The expected impact of the SOCRATES project is to influence 3GPP standardization and NGMN work, and to reinforce the European leadership in the development of standards. The findings of the project will create new business opportunities within the management and control area.

The SOCRATES consortium consists of equipment vendors (Ericsson and Nokia Siemens Networks), a mobile operator (Vodafone), a spin-off developing tools for planning, configuring and optimizing networks (atesio) and research organizations (IBBT, TNO Information and Communication Technology and TU Braunschweig).

The Use Cases considered by the project are listed in [37], and [38] describes the selected criteria and methodologies to assess the solutions for Self-Organization. The framework and guidelines for the development of Self-Organization functionalities within the project are described in [39]. Updates of the Use Cases and framework are available in [40, 41].

The SOCRATES project is in its final stage, and therefore it has started to publish results of its investigation. For example, an algorithm for MRO that optimizes handover failure, call drop and ping-pong by tuning hysteresis and time-to-trigger parameters has been presented in a workshop in Athens, in February 2010 [42]. In the same workshop, a similar study for Load

Balance Optimization (LBO) was presented. The proposed algorithm distributes the load of a cell among the neighbors by tuning the handover offsets [43].

Other similar documents have been presented for other Use Cases: COC [44], QoS optimization [45], Self-Optimization of call admission control for LTE downlink [46] and packet scheduling parameters optimization [47]. Additional documents are being made available at [48] as they are presented in conferences.

2.5.2.   Celtic Gandalf: Monitoring and Self-Tuning of RRM Parameters in a Multi-System Network

The Celtic Gandalf project is part of Celtic[4], a publicly funded research and development program that was created as a cluster of Eureka[5] (Σ!) to foster European leadership in telecommunications [50]. The project started in April 2005 and ended in April 2007, and it was primarily focused on automating RRM tasks in a multi-system environment including GSM, GPRS, UMTS and Wireless Local Area Network (WLAN). More specifically, the project covered: (i) automated diagnosis for troubleshooting; (ii) auto-tuning of network parameters; and (iii) advanced and joint RRM [51, 52].

In 2008, the project received the Celtic Excellence Award, which distinguished six projects for their outstanding achievements.

The official partners of the project consortium were two universities (University of Limerick, Ireland, and University of Málaga, Spain) and four companies (Moltsen Intelligent Software, France Telecom R&D, Ericsson and Telefónica I+D) [51].

2.5.3.   Celtic OPERA-Net: Optimizing Power Efficiency in mobile RAdio Networks

The OPERA-Net project is also part of Celtic [53]. It was created in June 2008 and it is expected to be concluded in May 2011. The project aims to constitute a task force that considers mobile radio networks in a holistic manner, adopting an end-to-end approach involving all relevant NEs and their mutual dependencies at different levels. One of the main objectives is to achieve energy efficiency at system, infrastructure and terminal levels, in order to allow the European industry to take a leadership role in green mobile networks. According to [53], the project partners are the following: Alcatel-Lucent (Ireland), Cardiff University (UK), France Telecom R&D (France), Freescale (France), IMEC (Belgium), Nokia Siemens Networks (Finland), Thomson Grass Valley (France), VTT (Finland), MITRA I (Belgium) and I2R (Singapore). Moreover, Efore (Finland) is reported to be joining the project. Some of

---

[4]Celtic is a research and development European project aiming at strengthening Europe's competitiveness in telecommunications. Celtic was launched in November 2003 and will be completed in 2011. It brings together interested partners from industry, telecom operators, small/medium sized enterprises, universities and research institutes to set up and carry out joint international, collaborative R&D projects [49]. The costs for Celtic projects are shared between national funding agencies and private investment.

[5]EUREKA is an intergovernmental network launched in 1985, composed of 39 members, including the European Community.

the results of the project can be found in [54]. Related to the SON paradigm, it has been reported, for example, that 33% ES is achievable by selectively putting individual sectors into sleep mode.

2.5.4. E3: End-to-End Efficiency

E3 is an integrated project of the 7th Framework Program funded by the European Commission [55]. It was started in January 2008 and was concluded in December 2009. It was an ambitious project with many partners (public institutions and private companies) that aimed to identify and define new designs to evolve current heterogeneous wireless systems into a scalable and efficient beyond 3G (B3G) cognitive radio system.

The key objective of the project was to design, develop, prototype and showcase solutions to guarantee interoperability, flexibility and scalability between existing legacy and future wireless systems, manage the overall system complexity and ensure convergence across access technologies, business domains, regulatory domains and geographical regions. The idea is to distribute the management functions among different NEs and levels to optimize the use of the radio resources, following the concept of cognitive radio networks.

Among many documents on requirements, architectures, scenarios and reports for cognitive radio systems [56], it is worth pointing out a document that presents some algorithms and simulation results for some SON Use Cases [57]. In this document, a study has been presented comparing a rule-based algorithm and a genetic algorithm for COC and handover parameter optimization. Moreover, a rule-based algorithm for Load Balancing has been evaluated. Finally, a simulation framework for ES and Inter-Cell Interference Coordination (ICIC) has been presented.

2.6. References

[1] Next Generation Mobile Networks (NGMN) Alliance, White Paper (2010) *NGMN Technical Achievements 2007–2010*, Version 2.0, 16 December 2010, http://www.ngmn.org/uploads/media/NGMN_Technical_ Achievements_2007-2010. pdf (accessed 3 June 2011).

[2] Next Generation Mobile Networks (NGMN) Alliance, White Paper (2006) *Next Generation Mobile Networks Beyond HSPA & EVDO*, Version 3.0, 5 December 2006, http://www.ngmn.org/uploads/media/Next_Generation_ Mobile_Networks_Beyond_HSPA_EVDO_web.pdf (accessed 3 June 2011).

[3] 3GPP, Technical Report, Technical Specification Group Radio Access Network (2010) *Study on Minimization of Drive-Tests in Next Generation Networks*, 3GPP TR 36.805 Version 9.0.0, Release 9, 5 January 2010, http:// www.3gpp.org/ftp/Specs/archive/36_series/36.805/36805-900.zip (accessed 3 June 2011).

[4] 3GPP, Technical Report, Technical Specification Group Service and System Aspects (2010) *Study on Energy Savings Management (ESM)*, 3GPP TR 32.826 Version 10.0.0, Release 10, 6 April 2010, http://www.3gpp.org/ ftp/Specs/archive/32_series/32.826/32826-a00.zip (accessed 3 June 2011).

[5] 3GPP, Technical Specification, Technical Specification Group Service and System Aspects (2010) *Self-Organizing Networks (SON) Policy Network Resource Model (NRM) Integration Reference Point (IRP): Requirements*, 3GPP TS 32.521 Version 9.0.0, Release 9, 6 April 2010, http://www.3gpp.org/ftp/Specs/ archive/32_series/32.521/32521-900.zip (accessed 3 June 2011).

[6] 3GPP, Technical Specification, Technical Specification Group Service and System Aspects (2009) *Automatic Neighbour Relation (ANR) Management; Concepts and Requirements*, 3GPP TS 32.511 Version 9.0.0, Release 9, 31 December 2009, http://www.3gpp.org/ftp/Specs/archive/32_series/32.511/32511-900.zip (accessed 3 June 2011).

[7] TM Forum (2011) www.tmforum.org (accessed 3 June 2011).

[8] Next Generation Mobile Networks (NGMN) Alliance, Requirement Specification (2008) *NGMN Recommendation on SON and O&M Requirements*, Version 1.23, December 2008, http://www.ngmn.org/uploads/media/NGMN_Recommendation_on_SON_and_O_M_Requirements.pdf (accessed 3 June 2011).

[9] Next Generation Mobile Networks (NGMN) Alliance, Deliverable (2008) *NGMN Use Cases Related to Self Organising Network, Overall Description*, Version 2.02, December 2008, http://www.ngmn.org/uploads/media/NGMN_Use_Cases_related_to_Self_Organising_Network_Overall_Description.pdf (accessed 3 June 2011).

[10] 3GPP, Technical Specification, Technical Specification Group Services and System Aspects (2010) *Self-Healing Concepts and Requirements*, 3GPP TS 32.541 Version 1.4.0, Release 10, 6 August 2010, http://www.3gpp.org/ftp/Specs/archive/32_series/32.541/32541-140.zip (accessed 3 June 2011).

[11] 3GPP, Technical Specification, Technical Specification Group Service and System Aspects (2010) *X2 Application Protocol (X2AP)*, 3GPP TS 36.423 Version 9.3.0, Release 9, 15 June 2010, http://www.3gpp.org/ftp/Specs/archive/36_series/36.423/36423-930.zip (accessed 3 June 2011).

[12] Ramiro, J. (2003) *System Level Performance Analysis of Advanced Antenna Concepts in WCDMA*, Ph.D. Thesis, Aalborg University.

[13] Lee, C. and Steele, R. (2006) Closed-loop power control in CDMA systems, *IEE Proceedings on Communications*, **143**(4), pp. 231–239.

[14] Jalali, A. and Mermelstein, P. (1993) Power control and diversity for the downlink of CDMA systems, *Conference Records on the 2nd International Conference on Universal Personal Communications*, **2**, pp. 980–984, October 1993.

[15] 3GPP (1998) *The 3rd Generation Partnership Project Agreement"*, 4 December 1998, http://www.3gpp.org/ftp/Inbox/2008_web_files/3gppagre.pdf (accessed 3 June 2011).

[16] 3GPP (2011) *About 3GPP*, http://www.3gpp.org/About-3GPP (accessed 3 June 2011).

[17] 3GPP (2010) *Specification Groups*, http://www.3gpp.org/Specification-Groups (accessed 3 June 2011)

[18] 3GPP (2010) *Overview of 3GPP Release 8*, Version 0.1.0, April 2010, http://www.3gpp.org/ftp/Information/WORK_PLAN/Description_Releases/Previous_versions/Rel-08_description_20100421.zip (accessed 3 June 2011).

[19] 3GPP (2010) *Overview of 3GPP Release 9*, Version 0.1.0, June 2010, http://www.3gpp.org/ftp/Information/WORK_PLAN/Description_Releases/Previous_versions/Rel-09_description_20100621.zip (accessed 3 June 2011).

[20] 3GPP, Technical Specification, Technical Specification Group Service and System Aspects (2009) *Self-Organizing Networks (SON); Concepts and Requirements*, 3GPP TS 32.500 Version 9.0.0, Release 9, 31 December 2009, http://www.3gpp.org/ftp/Specs/archive/32_series/32.500/32500-900.zip (accessed 3 June 2011).

[21] 3GPP, Technical Specification, Technical Specification Group Service and System Aspects (2010) *Self-Configuration of Network Elements; Concepts and Requirements*, 3GPP TS 32.501 Version 9.1.0, Release 9, 6 April 2010, http://www.3gpp.org/ftp/Specs/archive/32_series/32.501/32501-910.zip (accessed 3 June 2011).

[22] 3GPP, Technical Specification, Technical Specification Group Service and System Aspects (2010) *Self-Configuration of Network Elements Integration Reference Point (IRP); Information Service (IS)*, 3GPP TS 32.502 Version 9.2.0, Release 9, 18 June 2010, http://www.3gpp.org/ftp/Specs/archive/32_series/32.502/32502-920.zip (accessed 3 June 2011).

[23] 3GPP, Technical Specification, Technical Specification Group Service and System Aspects (2010) *Self-Configuration of Network Elements Integration Reference Point (IRP); Common Object Request Broker Architecture (CORBA) Solution Set (SS)*, 3GPP TS 32.503 Version 9.1.0, Release 9, 18 June 2010, http://www.3gpp.org/ftp/Specs/archive/32_series/32.503/32503-910.zip (accessed 3 June 2011).

[24] 3GPP, Technical Specification, Technical Specification Group Radio Access Network (2010) *Overall Description; Stage 2*, 3GPP TS 36.300 Version 9.4.0, Release 9, 18 June 2010, http://www.3gpp.org/ftp/Specs/archive/36_series/36.300/36300-940.zip (accessed 3 June 2011).

[25] 3GPP, Technical Specification, Technical Specification Group Service and System Aspects (2010) *Evolved Universal Terrestrial Radio Access Network (E-UTRAN) Network Resource Model (NRM) Integration Reference Point (IRP); Information Service (IS)*, 3GPP TS 32.762 Version 9.4.0, Release 9, 8 October 2010, http://www.3gpp.org/ftp/Specs/archive/32_series/32.762/32762-940.zip (accessed 3 June 2011).

[26] 3GPP, Technical Specification, Technical Specification Group Radio Access Network (2010) *Self-Configuring and Self-Optimizing Network (SON) Use Cases and Solutions*, 3GPP TS 36.902 Version 9.2.0, Release 9, 15 June 2010, http://www.3gpp.org/ftp/Specs/archive/36_series/36.902/36902-920.zip (accessed 3 June 2011).

[27] 3GPP, Technical Specification, Technical Specification Group Service and System Aspects (2010) *S1 Application Protocol (S1AP)*, 3GPP TS 36.413 Version 9.3.0, Release 9, 15 June 2010, http://www.3gpp.org/ftp/Specs/archive/36_series/36.413/36413-930.zip (accessed 3 June 2011).

[28] 3GPP, Technical Specification, Technical Specification Group Radio Access Network (2010) *Radio Resource Control (RRC); Protocol Specification*, 3GPP TS 36.331 Version 9.3.0, Release 9, 18 June 2010, Release 9, http://www.3gpp.org/ftp/Specs/archive/36_series/36.331/36331-930.zip (accessed 3 June 2011).

[29] 3GPP, Technical Specification, Technical Specification Group Service and System Aspects (2010) *Radio Self-Organizing Networks (SON) Policy Network Resource Model (NRM) Integration Reference Point (IRP): Information Service (IS)*, 3GPP TS 32.522 Version 9.0.0, Release 9, 6 April 2010, http://www.3gpp.org/ftp/Specs/archive/32_series/32.522/32522-900.zip (accessed 3 June 2011).

[30] 3GPP, Technical Specification, Technical Specification Group Service and System Aspects (2010) *Self-Organizing Networks (SON); Policy Network Resource Model (NRM) Integration Reference Point (IRP); Common Object Request Broker Architecture (CORBA) Solution Set (SS)*, 3GPP TS 32.523 Version 9.0.0, Release 9, 18 June 2010, http://www.3gpp.org/ftp/Specs/archive/32_series/32.523/32523-900.zip (accessed 3 June 2011).

[31] 3GPP, Technical Report, Technical Specification Group Service and System Aspects (2009) *Study on Self-Healing*, 3GPP TR 32.823 Version 9.0.0, Release 9, 1 October 2009, http://www.3gpp.org/ftp/Specs/archive/32_series/32.823/32823-900.zip (accessed 3 June 2011).

[32] 3GPP (2010) *Overview of 3GPP Release 10*, Version 0.0.7, June 2010, http://www.3gpp.org/ftp/Information/WORK_PLAN/Description_Releases/Previous_versions/Rel-10_description_20100621.zip (accessed 3 June 2011).

[33] 3GPP (2010) Technical Report, Technical Specification Group Service and System Aspects, *Integration of Device Management Information with Itf-N*, 3GPP TR 32.827 Version 10.1.0, Release 10, 22 June 2010, http://www.3gpp.org/ftp/Specs/archive/32_series/32.827/32827-a10.zip (accessed 3 June 2011).

[34] 3GPP (2009) Document presented at Radio Access Network Group Plenary Meeting, Meeting #46, *SID on Study on Solutions for Energy Saving within UTRA Node B*, RP-091439, Release 9, 4 December 2009, http://www.3gpp.org/ftp/tsg_ran/TSG_RAN/TSGR_46/Docs/RP-091439.zip (accessed 3 June 2011).

[35] 3GPP (2010) RAN Plenary Meeting #48, RP-100674, *SID on Study on Network Energy Saving for E-UTRAN*, 3 June 2010,http://www.3gpp.org/ftp/tsg_ran/TSG_RAN/TSGR_48/Docs/RP-100674.zip (accessed 3 June 2011).

[36] SOCRATES Project (2008) http://www.fp7-socrates.eu (accessed 3 June 2011).

[37] SOCRATES Project (2008) Deliverable, *Use Cases for Self-Organising Networks*, March 2008, http://www.fp7-socrates.eu/files/Deliverables/SOCRATES_D2.1%20Use%20cases%20for%20self-organising%20networks.pdf (accessed 3 June 2011).

[38] SOCRATES Project (2008) Deliverable, *Assessment Criteria for Self-Organising Networks*, June 2008, http://www.fp7-socrates.eu/files/Deliverables/SOCRATES_D2.3%20Assessment%20criteria%20for%20self-organising%20networks.pdf (accessed 3 June 2011).

[39] SOCRATES Project (2008) Deliverable, *Framework for the Development of Self-Organising Networks*, July 2008, http://www.fp7-socrates.eu/files/Deliverables/SOCRATES_D2.4%20Framework%20for%20self-organising%20networks.pdf (accessed 3 June 2011).

[40] SOCRATES Project (2009) Deliverable, *Review of Use Cases and Framework*, March 2009, http://www.fp7-socrates.eu/files/Deliverables/SOCRATES_D2.5%20Review%20of%20use%20cases%20and%20framework%20(Public%20version).pdf (accessed 3 June 2011).

[41] SOCRATES Project (2009) Deliverable, *Review of Use Cases and Framework II*, December 2009, http://www.fp7-socrates.eu/files/Deliverables/SOCRATES_D2.6%20Review%20of%20use%20cases%20and%20framework%20II.pdf (accessed 3 June 2011).

[42] SOCRATES Project (2010) Presentation, *Handover Parameter Optimization in LTE Self-Organizing Networks*, February 2010, http://www.fp7-socrates.eu/files/Publications/SOCRATES_2010_COST%202100%20TD(10)10068.pdf (accessed 3 June 2011).

[43] SOCRATES Project (2010) Presentation, *Load Balancing in Downlink LTE Self-Organizing Networks*, February 2010, http://www.fp7-socrates.eu/files/Publications/SOCRATES_2010_COST%202100%20TD(10)10071.pdf (accessed 3 June 2011).

[44] SOCRATES Project (2010) Presentation, *Automatic Cell Outage Compensation*, March 2010, http://www.fp7-socrates.eu/files/Presentations/SOCRATES_2010_NGMN%20call%20-%20cell%20outage%20compensation.pdf (accessed 3 June 2011).

[45] SOCRATES Project (2010) Presentation, *SOCRATES QoS Optimisation: an Introduction*, February 2010, http://www.fp7-socrates.eu/files/Presentations/SOCRATES_2010_NGMN%20call%20-%20QoS%20optimisation%20intro.pdf (accessed 3 June 2011).

[46] SOCRATES Project (2010) Presentation, *Self-Optimising Call Admission Control for LTE Downlink*, February 2010, http://www.fp7-socrates.eu/files/Presentations/SOCRATES_2010_NGMN%20call%20-%20admission%20control%20optimisation.pdf (accessed 3 June 2011).

[47] SOCRATES Project (2010) Presentation, *D3.1B Packet Scheduling Parameter Optimization Use Case*, February 2010, http://www.fp7-socrates.eu/files/Presentations/SOCRATES_2010_NGMN%20call%20-%20packet%20 scheduling%20optimisation.pdf (accessed 3 June 2011).

[48] SOCRATES Project (2011) *List of publications,* http://www.fp7-socrates.eu/?q=node/10 (accessed 3 June 2011).

[49] Celtic-Plus Initiative (2010) http://www.celticplus.eu/ (accessed 3 June 2011).

[50] GANDALF Project (2010) http://www.celtic-initiative.org/Projects/Celtic-projects/Call2/GANDALF/gandalf-default.asp (accessed 3 June 2011).

[51] GANDALF Project (2010) http://www.celtic-initiative.org/Projects/Celtic-projects/Call2/GANDALF/gandalf-default.asp (accessed 3 June 2011).

[52] Stuckmann, P., Altman, Z., Dubreil, H., Ortega, A., Barco, R., Toril, M., Fernandez, M., Barry, M., McGrath, S., Blyth, G., Saidha, P. and Nielsen, L.M. (2005) The EUREKA Gandalf project: monitoring and self-tuning techniques for heterogeneous radio access networks, *IEEE Proc. 61st Vehicular Technology Conference*, **4**, pp. 2570–2574, May 2005.

[53] OPERA-Net Project (2008) http://opera-net.org/ (accessed 3 June 2011).

[54] OPERA-Net Project (2010) Presentation at the Celtic Event 2010 in Valencia, *OPERA-Net Project Stand 21*, April 2010, http://opera-net.org/Documents/5024v0_Opera-Net_Celtic%20Event%20Valencia%202010_ Demos%20Presentation.pdf (accessed 3 June 2011).

[55] E3 Project (2009) https://ict-e3.eu (accessed 3 June 2011).

[56] E3 Project (2009) *List of deliverables*, https://ict-e3.eu/project/deliverables/deliverables.html (accessed 3 June 2011).

[57] E3 Project (2009) Deliverable, *Simulation based recommendations for DSA and self-management*, https://ict-e3. eu/project/deliverables/full_deliverables/E3_WP3_D3.3_090631l.pdf (accessed 3 June 2011).

3

Multi-Technology SON

Rubén Cruz, Khalid Hamied, Juan Ramiro, Lars Christoph Schmelz,
Mehdi Amirijoo, Andreas Eisenblätter, Remco Litjens,
Michaela Neuland and John Turk

3.1. Drivers for Multi-Technology SON

Whereas current commercial and standardization efforts are mainly focused on the introduction and Self-Organization of Long Term Evolution (LTE) networks, there is significant potential in the extension of the scope of the Self-Organizing Networks (SON) paradigm to also cover GSM, GPRS, EDGE, UMTS and HSPA radio access technologies, as well as of course LTE. Throughout the rest of the book, such extended concept is referred to as multi-technology SON.

The implications of multi-technology SON are manyfold. On one hand, the adoption of a multi-technology approach allows operators to completely transform and streamline their operations, not only applying an innovative, automated approach to the new additional LTE network layer, but also extending the key benefits from SON to all radio access technologies, thereby harmonizing the whole network management approach and boosting operational efficiency. The aforementioned key benefits from SON are the following:

- Reduced Operational Expenditure (OPEX) through: (i) automation; (ii) energy saving; and (iii) lower need for leased transmission lines due to optimized resource utilization.
- Reduced Capital Expenditure (CAPEX) through: (i) more technically sound capacity planning processes; and (ii) higher effective system capacity due to the application of intelligent optimization techniques.
- Better Quality of Service (QoS) through faster reaction to disruptive network incidences and the massive application of state-of-the-art optimization techniques.

Self-Organizing Networks: Self-Planning, Self-Optimization and Self-Healing for GSM, UMTS and LTE,
First Edition. Edited by Juan Ramiro and Khalid Hamied.
© 2012 John Wiley & Sons, Ltd. Published 2012 by John Wiley & Sons, Ltd.

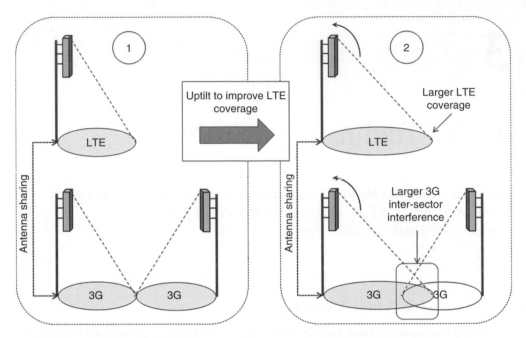

Figure 3.1   Multi-technology implications in RF optimization (one simple example).

Additionally, the availability of a multi-technology SON solution enables more comprehensive optimization strategies that deal with several radio access technologies simultaneously. Such multi-technology awareness is crucial in those cases in which there are cross-layer restrictions in the involved optimization processes, e.g. when the same antenna is used for different radio access technologies. In that case, if the parameters to be planned or optimized cannot be tuned separately for different radio access technologies (e.g. if the antenna model does not allow the configuration of different electrical tilt values for UMTS and LTE), it is not appropriate to carry out automated optimization that only considers one technology, since the LTE performance might be improved at the expense of degrading UMTS. One clear example, illustrated in Figure 3.1, is a mature UMTS network with contiguous coverage that coexists with an LTE network deployed on selected locations. As depicted in Figure 3.1, the LTE traffic level and quality indicators may suggest an antenna uptilt in order to improve LTE coverage, especially if the system is still far from its capacity limit and user throughput is satisfactory. However, if that action also implies an uptilt in the UMTS layer due to the fact that antennas are shared, the overall convenience of this adjustment may become questionable due to increased intercell interference problems in UMTS. In this case, it is necessary to look at the network in a combined multi-technology fashion when executing these optimization actions, which become joint multi-technology processes. More details are provided in Chapter 4.

Another important area in this respect is the orchestration of smart load balancing strategies between different technologies in order to boost the trunking efficiency of the composite multi-layer system and therefore increase the overall network capacity, which results in lower congestion, i.e. better Grade of Service (GoS) and reduced and/or delayed needs for additional capital outlays.

However, some key considerations need to be made about the application of multi-technology SON:

- SON is not a native characteristic of 2nd Generation (2G) and 3rd Generation (3G) networks.
- The integration of an external centralized entity that provides SON functionalities is likely to be required. Such entity can be provided by the infrastructure vendor or a third-party supplier. See Section 3.2 for more details.
- When translating the LTE SON Use Cases to 2G and 3G networks, the characteristics, Operation And Maintenance (O&M) functionality and limitations of the different vendors need to be closely evaluated and taken into consideration due to the lack of standardized support.
- A clear business case must be built that justifies the addition of SON to existing legacy networks.

3.2. Architectures for Multi-Technology SON

The architecture of a system is the model that defines its structure and basic components, and describes the relationship among those components. In this section, the high level system architecture for multi-technology SON is presented.

3.2.1. Deployment Architectures for Self-Organizing Networks

The following SON architectures have been defined in the Third Generation Partnership Project (3GPP) for LTE networks [1]. In this section, these architectures will be referred to as deployment architectures.

- Centralized: all SON algorithms and functions are located in the O&M system. In this architecture SON functions reside in a small number of locations at a high hierarchical level.
- Distributed: all SON algorithms and functions are located at the Network Element (NE) level, e.g. Enhanced NodeB (eNodeB). In this architecture SON functions reside in many locations at a relatively low hierarchical level.
- Hybrid: some SON algorithms and functions are located in the O&M system, while others are located at the NE level.

The three SON architectures for LTE are shown in Figure 3.2. In the following sections, the deployment architectures for LTE are extended to cover GSM and UMTS. It is noteworthy to mention, however, that since GSM and UMTS are legacy radio access technologies, a distributed SON architecture is only feasible through 3GPP standardization or using vendor specific proprietary solutions. Although 3GPP standardization started to specify some SON Use Cases for UTRAN, e.g. Automatic Neighbor Relation (ANR), it is believed that distributed 2G and 3G SON support will be limited to a few Use Cases.

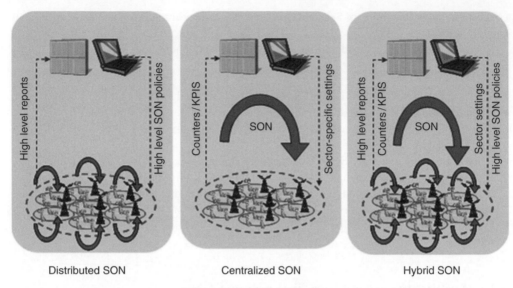

Distributed SON   Centralized SON   Hybrid SON

Figure 3.2   SON architectures.

3.2.2.   Comparison of SON Architectures

3.2.2.1.   Pros and Cons of a Centralized SON Architecture

In a centralized SON architecture, the decision making entity is located in the O&M system, which enables access to Configuration Management (CM) data, Performance Management (PM) data and other enterprise data sources (e.g. planning databases) since those are centralized by nature. The following advantages are mainly attributed to the privileged position of the SON function which allows access to all kinds of performance indicators from different cells. In the following, the pros and cons will be given in the context of Self-Optimization although the same arguments, to some extent, apply to Self-Planning and Self-Healing. A centralized SON architecture has the following pros:

- Capability and flexibility to support sophisticated optimization strategies that involve cross-correlation of parameters and performance indicators from different cells. Therefore, a centralized architecture enables global optimization in which the decision to change a parameter of a given target cell will be taken considering global metrics and parameters of other cells that are not necessarily located in close proximity of the target cell.
- Capability and flexibility to support powerful optimization strategies that involve cross-correlation of parameters and performance indicators from cells with different radio access technologies. A centralized architecture enables multi-technology joint optimization, which means the simultaneous optimization of two or more technologies taking into consideration the impact of changing parameters on the performance of the different technologies. This is usually needed when resources are shared (e.g. antenna sharing as shown in Figure 3.1). Moreover, multi-technology joint optimization is also needed even if resources are not shared (e.g. for traffic balance Use Case).

- Flexibility in modifying the optimization strategy and algorithms. Since the decision making entity has access to a large base of information, modifying the optimization strategy or algorithms (which may require using additional parameters and performance indicators) is relatively simple and unconstrained by the need for additional information exchange, since that information is readily available.
- Easy to deploy since the SON functions are located in small number of locations. It will also be easy to upgrade existing SON functions for the same reason.
- Ability to align optimization strategies across multi-vendor networks. Another key advantage of the centralized SON architecture is the handling of multi-vendor networks. By means of a centralized SON solution, it is easier to make sure that the optimization strategies across the entire network are fully aligned.
- Allows decoupling of SON functionality from network infrastructure and, therefore, enables competitive third party SON solutions to be plugged into the network. This will increase innovation and reduce time to market.

A centralized SON architecture has the following cons:

- Requires large bandwidth and large data exchange in order to have CM and PM data available to the SON functions. However, this data is available to the Network Management System (NMS) anyway (see Section 5.13) since it is required for daily network operation.
- Centralized SON functions are not able to react to network incidences nor are they able to adapt to traffic changes in real-time. This is due to the fact that the responsiveness of centralized SON functions is constrained by the delays associated with the availability of CM and PM data, and the delays associated with writing (modifying) CM parameters (i.e. the difference between the time when the parameter change was requested and the time when the actual change of that parameter took place). It is important, however, to distinguish between a Self-Organizing function and a Radio Resource Management (RRM) functionality, which is a real-time control routine that involves certain strategies and algorithms for controlling system behavior such as handover, admission control, etc. The distinction between RRM and SON can be unclear sometimes but they should not be confused (see Section 2.3 for more details). Centralized Self-Organization is suitable for slow processes that monitor network performance and modify network configuration to optimize performance and to adapt to slow changes such as variations in the radio propagation environment or traffic patterns. In this context, a centralized SON function can be used to optimize the performance of an RRM algorithm, such as admission control, by dynamically optimizing the parameters and thresholds used in that algorithm.
- Single point of failure. This is in general a characteristic of centralized systems and, in this context, it means that the failure of the SON system will impact all network elements or nodes under its control. Two types of failures will be distinguished: (i) an outage or loss of availability; and (ii) a malfunction, i.e. the SON function made a network change that resulted in significant network degradation. However, if centralized SON is controlling a slow process that optimizes network performance and adapts to slow network changes, then it is not critical for network availability and SON failure will not cause network outage and downtime. The only consequence of a centralized SON outage is that system adaptability will be halted and the network will be operating in a slightly suboptimum manner. An

exception to the previous fact is the Energy Saving (ES) Use Case where an outage in that SON function could result in keeping some sites turned off and that would result in significant performance degradation later when the reason for turning those sites off (e.g. very small traffic) ceased to exist. The malfunction of centralized SON can also cause significant network degradation.

In general, the 'single point of failure', disadvantage can be addressed and mitigated by the following:

- Use of redundancy and high availability infrastructures. This will minimize the downtime of the centralized SON system.
- Mechanisms to turn on all sites upon detecting an outage in the ES SON function. This can be provided by a separate, high availability, entity that monitors the status of the ES SON function and takes control when an outage is detected.
- Mechanisms to fallback to the last known good configuration. Fallback functionality should be provided by a separate entity that takes control when a malfunction of the regular SON system is detected. A malfunction can be identified when degradation in the network performance exceeds a predefined threshold and when that degradation is correlated with network changes.

3.2.2.2. Pros and Cons of a Distributed SON Architecture

In a distributed SON architecture, the decision making entity is located at the NE, which enables access to real-time information from the NE hosting the SON function as well as information from direct and immediate neighboring network elements through the X2 and S1 interfaces. A distributed SON architecture has the following advantages:

- Allows real-time execution. SON functions are able to react to local network incidences and adapt to sudden changes in traffic patterns in real-time.
- Does not require large bandwidth and large exchange of data since is collected locally and exchanged with direct neighbors.
- Does not have a single point of failure. Therefore, at system level, distributed SON has the characteristic of failure resilience. If a SON function breaks or malfunctions, it will impact one node only (and perhaps a few neighboring nodes), not a large number of network elements.

A distributed SON architecture has the following cons:

- Optimization in a distributed architecture is typically local and usually involves few NEs. However, support of global optimization strategies can be extended to distributed architectures if exchange of additional information and coordination between a larger number of cells are facilitated by standardized or vendor specific functionality.
- Does not support optimization strategies that involve cross-correlation of parameters and performance indicators from cells with different radio access technologies since that requires information exchange between them. However, that support can be extended if

exchange of information and coordination between cells with different technologies are facilitated by standardized or vendor specific functionality.

- Difficult to modify the optimization strategy and algorithms since this may require the standardization of additional information exchange or the implementation of vendor specific mechanisms at NE level.
- Lack of ability to align optimization strategies across multi-vendor networks except for those cases where the optimization algorithm is fully standardized.
- Difficult to coordinate between SON functions that reside on different nodes since exchange of information is limited to neighboring nodes. However, this disadvantage can be circumvented if implicit coordination is used. Exploiting implicit coordination is an important design paradigm for Self-Organization in communication networks [2].

3.2.2.3.   Pros and Cons of a Hybrid SON Architecture

A hybrid architecture is, by definition, a mixture of centralized and distributed. Therefore, it has the best of the two architectures. For example, it allows sophisticated optimization strategies using a centralized decision making entity, and at the same time supports cases that require real-time optimization. The hybrid architecture, however, has the disadvantage of complex coordination (or perhaps lack of coordination) between the centralized and distributed SON functions. That can be mitigated by splitting the responsibilities between the two sets of SON functions (i.e. the centralized and the distributed).

3.2.3.   Coordination of SON Functions[1]

A SON system will offer multiple SON functions to perform several tasks related to optimization, planning and healing. In a given SON deployment, it is highly likely that some of the SON functions are applied to the same geographical area of the network at the same time. This implies a management complexity for operators, who have to manage the configuration of SON functions, as well as the potential interwork aspects between SON functions overlapping in area and time. Many factors need to be considered to address this management complexity. Those factors, as well as other conceptual solutions, are discussed in this section for any architectural choice (centralized, distributed or hybrid). The SON entity that implements the management of SON functions, as described in this section, is the SON coordinator (or the SON manager). The SON coordinator provides a framework that assists the various SON functions to work in parallel (see also [4]).

3.2.3.1.   Benefits of the SON Coordinator

The SON coordinator acts as an interface between the SON functions, the operator and the network as shown in Figure 3.3.

---

[1] The work presented in this section was carried out within the FP7 SOCRATES project [3], which is partially funded by the Commission of the European Union.

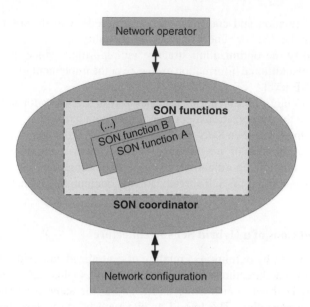

Figure 3.3 SON coordinator as an interface between SON functions, the operator and the network.

The operator no longer has to configure each SON function individually, while making sure that there are no conflicts between the objectives given to the various SON functions. As such, this avoids giving detailed objectives to different SON functions that are mistakenly in conflict. This task is performed by the coordinator based on a single set of high-level objectives from the operator. This is further described in Section 3.2.3.2.1. The coordinator also analyzes the outputs of all the SON functions for conflict over parameter changes (or the effect of parameter changes) and takes actions when conflict is detected. This is further described in Section 3.2.3.2.2. When different SON functions make decisions on the same processed input data, the coordinator provides information to the SON functions and other parts of the coordinator, as described in Section 3.2.3.2.3. The coordinator is able to detect if SON actions behave inappropriately and is further able to take urgent rectifying actions as described in Section 3.2.3.2.4.

3.2.3.2.   Components of the SON Coordinator

The conceptual structure of the SON coordinator is depicted in Figure 3.4. Components depicted in the figure are described in the following subsections.

3.2.3.2.1.   *Policy Function*
Configuring SON functions may be cumbersome to the operator and risk-prone. The policy function provides an interface to the network operator. Through it, the operator furnishes high-level performance targets expressing user satisfaction and/or business-level objectives, for example, minimum requirements on cell edge performance, average user throughput and

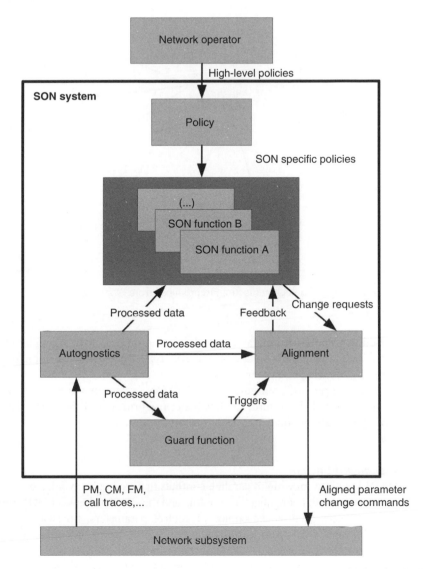

Figure 3.4   SON coordinator conceptual structure (simplified). Reproduced from © 2011 IEEE.

coverage. The policy function translates these high-level targets to SON function-specific performance targets as shown in Figure 3.5. This provides several benefits (which may be considered requirements for the policy function), namely:

- It ensures that the objectives given to each SON function are coherently aiming towards a common objective.
- Inter-SON function conflicts can be addressed due to the intervention of the policy function by intelligent application of SON-specific policies.

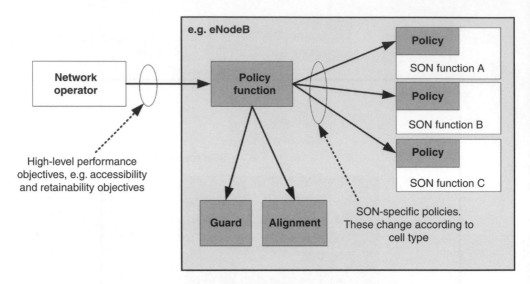

Figure 3.5   The policy function. Reproduced from © 2011 IEEE.

- Each SON function is configured (parameterized) in an optimal manner, taking into account the interaction with other SON functions, the cell type to which it is applied (urban, suburban, rural, indoor, etc.) and other factors.
- The operator needs to provide configuration to the SON system that is both minimal and meaningful for the user. This, in turn, implies better performance of the SON system and lower probability of malfunction due to incorrect configuration.

3.2.3.2.2. Alignment Function
Since several SON functions may run in parallel, conflicts that cannot be avoided by the policy function may arise. It is worth noting that the coupling and the conflicts between SON functions depend on various factors such as the choice of control parameters, timescale, target and objective of the SON functions, input measurements, etc. These factors will have an impact on the coupling and dependency between the SON functions. In this respect, the SON functions can either be independently designed and subsequently coordinated (if there is a need to do so), or they can be designed from the beginning such that the coupling between the SON functions is minimized. Therefore, a careful design of the SON functions may render no coupling or dependencies at all. From hereafter, we assume that there are dependencies and conflicts between at least two SON functions. Such conflicts are resolved by the alignment function. Conflicts can be caused by different SON function proposing changes in the same parameter of a given managed object, or they may pertain to the effect of parameter changes, for example, pilot power and antenna tilt affecting coverage. When conflicting change requests are detected, alignment can either accept one and reject the other, or create a compromise solution. The conflicting requests do not necessarily have to be simultaneous, but can occur within a defined period. Alignment must also enforce rules, such as permitted parameter ranges or value combinations.

3.2.3.2.3.   *Autognostics Function*

The autognostics function supplies information to SON functions and other parts of the coordinator. It processes network performance, network configuration and network fault data on behalf of all SON functions that request this service. Having a common, independent unit for processing this data has three advantages. First, it is more efficient to perform the calculations once in the autognostics function rather than having each SON function repeat such calculations. Second, this ensures that each SON function bases its decisions on a common data set, and third, it is possible to detect and filter out unreliable data, and to determine the statistical accuracy of the estimates.

3.2.3.2.4.   *Guard Function*

The guard function detects and takes actions when extreme, potentially catastrophic SON behavior occurs. It receives input from autognostics and tests for poor performance such as large oscillations in Key Performance Indicators (KPIs), extreme absolute values in network KPIs and unusual combinations of KPIs, e.g. high random access channel load and low carried traffic. The output of the guard function is the triggering of the necessary actions to tackle the problem, such as to mandate the alignment function to slow down, stop or revert recent parameter changes. The capability to carry out fallback operation is described in 3GPP specifications as some means for the SON system to monitor the network after a new configuration is implemented, and to set the network back to a previous, safe configuration if the network does not perform as expected [5]. The guard function plays no part in assisting SON features to achieve performance targets; neither should its actions be triggered by fluctuations in subscriber traffic.

3.2.3.3.   Grouping

Managing SON functions as described earlier in this section may be easier to achieve if SON functions are grouped according to their location (i.e. managed object or domain) and interdependencies (i.e. the potential conflicts between SON functions). The means implemented to coordinate SON functions would exist per group. Thus, SON functions in different locations, i.e. centralized SON functions residing at O&M level and distributed functions for eNodeBs may be separately coordinated. The same applies to SON functions within a location but with no interdependency. For example, handover optimization and Physical Cell Identifier (PCI) allocation SON functions at the eNodeBs have no shared parameters and no common influence on performance, so it is not necessary for them to share the same policy and alignment functions. The concept of grouping is illustrated in Figure 3.6. The following describes the interaction between the different SON coordinators.

- It is prohibitively complex to coordinate between centralized SON functions and distributed ones. It is therefore recommended to split the responsibilities between centralized and distributed functions.
- As stated in Section 3.2.2.2, it is difficult to coordinate between SON functions that reside in different locations. It is therefore recommended to have minimum coordination between the SON functions that reside on different nodes.

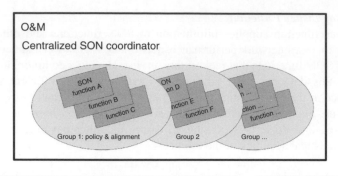

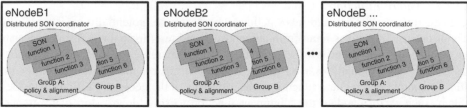

Figure 3.6 Grouping of SON functions for modular coordination.

3.2.3.4. Implementation Considerations

Management of SON functions, as described above, must implement policy, alignment, autognostics and guard functions facilitated by grouping. There are two implementation alternatives regarding how to achieve effective SON management:

1. In the first one, all coordination intelligence is placed in modules outside the SON functions, which would remain self-contained modules, unaware of each other. Figure 3.4, depicted to illustrate the components and functions to be considered in a SON coordinator at a conceptual level, also depicts this implementation option. This alternative implies naturally that policy, alignment, autognostics and guard modules include much of the intelligence already contained in the SON functions. Let us illustrate this with an example of the alignment module. If the alignment module is to resolve a conflict caused by two different SON functions trying to change the same parameter, it would be appropriate that the decision takes into account the effect of the proposed parameter changes in the domain of each conflicting SON function. For instance, if the optimization carried out by the SON functions was based on cost (penalty) functions, where different cost functions are used for different SON functions, then the relative change in the cost functions caused by the parameter changes may be substantially different. Therefore, the decision from the alignment module would favor the parameter change of the SON function whose relative cost varies more. This would imply that the alignment function is aware of the cost function or, in general, the optimization, planning and healing logic of each SON function. This leads to the conclusion that the approach that utilizes independent coordinating modules must be carefully devised so as to avoid duplication of logic and to have clear hierarchical separation of concerns between the SON functions and the coordinating agents, without sacrificing performance.

2. The second approach is to implement the coordination logic together with the SON functions (i.e. simultaneously). Resulting SON functions would be composite. They would include the logic of unitary SON functions with joint impact on the target parameter. Thus, collaboration between unitary SON functions would be facilitated via predesigned flows, and the conflicts would naturally take into account the internal logic of each concurrent SON function (e.g. regular ANR and Mobility Optimization).

3.2.4. Layered Architecture for Centralized Multi-Technology SON

In this section, a reference architecture for multi-technology SON is presented. A layered approach to the architecture is used to provide clear separation of responsibilities, that is to present each layer with a clear and definite responsibility. Moreover, the architecture follows a centralized paradigm for the obvious multi-vendor and multi-technology advantages associated with the centralized SON architecture[2]. Actual realizations of SON solutions will architecturally differ from what is presented here, but the concepts and functions described herein aim at providing a benchmark upon which comparative and taxonomic considerations can be built. The architecture is depicted in Figure 3.7.

3.2.4.1. Network Layer and Support Systems

The lowest layers in the model are the network layer and the support systems layer. These exist independently of a centralized SON system. They may provide SON-focused information (e.g. special performance counters defined in the context of SON standardization), but in technologies other than LTE this is not the case. In general, for a centralized SON system, the network layer and the support systems layer are not considered as part of the SON system itself although they are included in the architecture to provide clarity to the context of the SON system. The network layer includes NEs, which represent the telecommunications network itself and can be regarded as the object of the SON functions. The network layer also includes the essential operations systems required to manage the NEs, that is the Element Managers (EMs). The description of these elements can be found in Section 5.13. The network layer includes equipment from various vendors and different technologies. The interaction of the SON system with the network layer occurs in four distinct domains, following the nomenclature in the lower part of Figure 3.7:

- CM-write and manage: CM data (network parameters) must be managed and able to be modified. Different SON functions may require altering the configuration of the network (change parameters, instantiate managed objects such as creating a new neighbor relation, etc.), or carrying out complex CM functions (e.g. a fallback to a predefined safe configuration).

---

[2] In this context, there is no specific consideration about the possibility to run centralized SON algorithms in the BSC and RNC for 2G and 3G networks, respectively. However, in practice, those scenarios are similar to the centralized deployments under analysis, with the following differences: (i) when running closer to the base stations, algorithms can react faster to network changes; (ii) when running at lower level in the network architecture, visibility is limited to a lower number of network nodes, which is likely to lead to alignment problems in borders between adjacent BSCs/RNCs; and (iii) the implementation of SON algorithms in the BSC/RNC makes it significantly harder to support powerful optimization strategies that involve cross-correlation of parameters and performance indicators from cells with different radio access technology. It will also be harder to deploy third-party SON solutions.

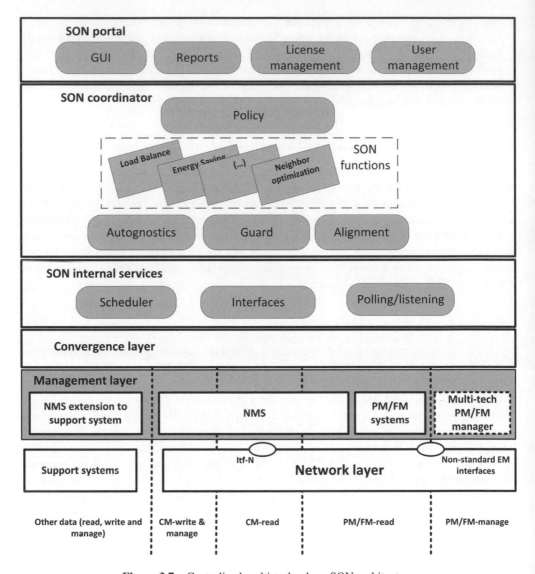

Figure 3.7   Centralized multi-technology SON architecture.

- CM-read: live network topological and configuration data are obvious inputs to SON functions.
- PM/FM-read: live network performance information such as PM data (e.g. counters and call traces[3]) and Fault Management (FM) data such as system alarms are also obvious inputs to SON functions[4].

---

[3] Call Traces data is classified as PM data.

[4] System alarms are included as part of the PM data in 3GPP TS 32.401 [6]. However, a distinction was made in the same document between Performance Management and Fault Management. System alarms will be classified as FM data throughout the book.

- PM/FM-manage: performance measurements and alarms must be managed (initiated, configured, paused/stopped, etc). The EM provides the necessary functions to achieve this [7].

As stated in Section 5.13, EM services are provided through the standardized Northbound Interface (Itf-N), although other nonstandard interfaces may be offered by the EM and used by the management layers (see Section 3.2.4.2).

The support systems layer comprises all systems not included in the network layer, with which the SON system may be required to interwork (i.e. to read, write and manage), such as planning databases (holding locations of sites information, antenna types and orientation, etc.), propagation path-loss predictions, ticketing systems, Charging Data Records[5] (CDRs), etc. The data domain in the figure is labeled 'other data'.

3.2.4.2. Management Layer

The management layer groups systems providing various management services to the network layer. Such systems may belong to the SON system deployment, or may exist previously. Depending on this, the offered management services may or may not be homogeneous across vendors and technologies in the network layer. Furthermore, the interfaces provided by those management services may be used by non-SON applications, and are therefore not specialized for SON functions. Per data domain, the systems involved in the management layer are:

- NMS is meant to provide functionality to carry out CM changes, read CM and topological information and perform other CM management tasks, such as consistency enforcement, etc. The reader may refer to Section 5.13 for details of NMS functions and interwork with SON functions. It is potentially possible, however, that EM interfaces are directly used by upper layers (bypassing the NMS); this approach is not recommended because of potential inconsistencies in configuration data that SON functions may cause in the NMS (this is further developed in Section 5.13). Obviously, if no NMS is available, or if it offers unsuitable services to upper layers, the EM interfaces will have to be tapped into, but this is not depicted in Figure 3.7 for the sake of simplicity. Although options exist for deploying a centralized SON system without an independent NMS (e.g. in deployments involving limited Use Cases), an NMS layer has been included in Figure 3.7 due to the added value provided by an NMS system and in particular due to the architectural capability to provide multi-vendor and multi-technology CM functions.
- NMS can offer interfaces to manage the support systems through extensions. This is due to the varied nature of support systems.
- PM/FM functionality, such as storage, recovery, peak hour calculation (only for PM), object aggregation, etc. can be realized in NMS systems, or in dedicated PM/FM systems (developed by infrastructure vendors, specialized vendors or inhouse). The assumption is that PM/FM data is made available in databases.

---

[5]CDR is classified in this book as charging management data or billing management data according to 3GPP TS 32.298 [8].

- PM/FM management (e.g. activation/deactivation of measurements counters and call traces) resides, as mentioned above, in EMs. It is not mandated that EMs offer interfaces for external applications to manage PM measurements and alarms. Therefore, building interfaces to handle PM/FM management in the SON system may not be possible. The related functional entity in Figure 3.7, labeled 'Multi-tech PM/FM manager', is dashed to indicate this.

3.2.4.3. Convergence Layer

The purpose of the convergence layer is to ensure that the services of the management layer are provisioned in each domain in a manner that is appropriate for the SON functions. 'Appropriate' means that the operational aspects of the interfaces are harmonized for all vendors and technologies in the network layer (for example, same database engine and schema for data retrieval, same interface for management of PM, etc.), and that the services satisfy the needs of the SON functions in the following domains:

- CM-write and manage (see Section 5.13 for more details):
  - o Schedule parameter change requests (i.e. CM modification).
  - o Ensure consistency of Management Information Base (MIB) between EMs and NMS after CM modification.
  - o Enable fallback functionality, that is, capability to revert to a predefined safe configuration.
  - o Obtain information about implementation changes with parameter and managed object granularity (e.g. to ensure successful implementation of the parameter change requests).
- CM-read:
  - o Retrieve CM information with desired granularity. Sufficient historical information must be available.
  - o Force update of the NMS MIB to ensure subsequent CM retrieval operations refer to the live network.
- PM/FM-read:
  - o Retrieve PM and FM (counters, alarms and call traces) for selected objects and time period. Sufficient historical information must be available.
  - o Enable surveillance of near real-time performance.
- PM/FM-manage: management of performance measurements and alarms on demand and on a common interface if possible.
- Other data: retrieval of information from support systems, such as propagation path-loss predictions, drive measurements, clutter and terrain information, antenna radiation patterns, etc.

As mentioned before, it is to be noted that the presence of an independent convergence layer is tied to the existence of a standalone NMS. Although this may not be required in certain circumstances, it is the most general architectural approach for multi-vendor and multi-technology centralized SON, and therefore it is discussed in this section.

The convergence layer must support administration functions allowing management of the databases, monitoring of collection functions, etc.

The detailed functionality of the convergence layer depends on the services provided by the management layer. Also, the detailed requirements of the SON functions for the convergence layer may depend on the SON algorithms themselves. Finally, as explained in Section 5.14, it is possible to carry out a further normalization across vendors (e.g. a semantic normalization that would calculate vendor-agnostic KPIs, parameters, etc.). For vendor-specific information that is difficult to normalize, the SON functions need to include vendor-specific subfunctions.

3.2.4.4. SON Internal Services Layer

This layer provides services potentially useful for SON functions, such as an Application Programming Interface (API) to access the data wrapped by the convergence layer. Also, this layer includes scheduling capabilities allowing recurrent execution of SON functions and polling/listening mechanisms to trigger SON processes when certain events take place.

3.2.4.5. SON Coordinator

A SON 'function manager', referred to in this chapter as the SON coordinator (see Section 3.2.3), supports the management, configuration, scheduling and execution of SON functions. The actual SON algorithms reside in SON functions. Each SON function is responsible for a specific SON Use Case or responsible for an atomic, self-contained part of a Use Case (Load Balancing, Energy Saving, etc.). In a centralized SON system, the SON functions may be assembled into composite flows. A flow is a process involving several SON functions[6]. The flow describes an ordered sequence of SON function executions, with the proper control flow mechanisms (loops, conditional execution, etc.).

For instance, integration of a new UMTS NodeB is desirably followed by a specific optimization task (e.g. neighbor list optimization). Therefore, in a modular design, a separate SON function may be defined for new NodeB integration, another one for neighbor list optimization and the two can be combined in a specific flow. In this context, the new NodeB integration, the neighbor list optimization and the combined flow can be referred to as a SON function, and the SON coordinator can configure, manage, schedule and execute those SON functions. For example, a periodic (e.g. daily) process may be configured to optimize the neighbor lists. While sharing the same internal logic, the neighbor list optimization in the new NodeB integration may be based on propagation predictions (i.e. to define an initial neighbor list before turning the NodeB on), whereas the daily function may utilize performance counters and/or call traces.

3.2.4.6. SON Portal

The functionality provided by the SON coordinator and the SON system to the end user is offered by the SON portal. The SON portal implements a Graphical User Interface (GUI) as well as support functions, such as user administration, license management and reporting. The SON portal enables users to configure and administer the centralized SON system for the whole network. The network can be broken into several regions or geographical areas; each

---

[6]The SON functions themselves can be made available, for example, through a library of functions.

area can be administered independently. The following are examples of the services provided by the SON portal:

- Allows remote and secure access to users (administrators for example), preferably through a web interface.
- Enables the user to configure all interfaces between the SON system and the network to enable data collection from all available sources (CM, PM and FM data), and to enable the execution of parameter change commands.
- Enables the user to activate/deactivate SON functions and Use Cases, and enables the user to control when SON functions and Use Cases are activated (always, periodic, event-based, etc.) and where (i.e. ability to exclude certain network elements or areas).
- Enables the user to provide high-level policies, strategies and parameter settings for all SON functions.
- Provides real-time status of all active SON functions (e.g. alive or down).
- Provides high-level reporting on the status of previous executions of SON functions and whether those executions where successful or not. This includes allowing access to log files and showing KPIs impacted by the corresponding SON function.
- Provides real-time or near-real-time surveillance of KPIs impacted by the active SON functions.

3.3. References

[1] 3GPP, Technical Specification, Technical Specification Group Services and System Aspects, (2010) *Self-Organizing Networks (SON); Concepts and Requirements*, 3GPP TS 32.500 Version 10.0.0, Release 10, 18 June 2010, http://www.3gpp.org/ftp/Specs/archive/32_series/32.500/32500-a00.zip (accessed 3 June 2011).
[2] Prehofer, C. and Bettstetter, C. (2005) Self-organization in communication networks: Principles and design paradigms, *IEEE Comm. Mag.*, **43**, pp. 78–85.
[3] SOCRATES Project (2010) www.fp7-socrates.eu (accessed 3 June 2011).
[4] Schmelz, L.C., Amirijoo, M. et al. (2011) A Coordination Framework for Self-Organisation in LTE Networks, *IEEE IM 2011 Conference*, pp. 193–200, May 2011.
[5] 3GPP, Technical Specification, Technical Specification Group Services and System Aspects (2010) *Self-Organizing Networks (SON) Policy Network Resource Model (NRM) Integration Reference Point (IRP); Requirements*, 3GPP TS 32.521 Version 9.0.0, Release 9, 6 April 2010, http://www.3gpp.org/ftp/Specs/archive/32_series/32.521/32521-900.zip (accessed 3 June 2011).
[6] 3GPP, Technical Specification, Technical Specification Group Services and System Aspects (2010) *Performance Management (PM); Concept and Requirements*, 3GPP TS 32.401 Version 10.0.0, Release 10, 8 October 2010, http://www.3gpp.org/ftp/Specs/archive/32_series/32.401/32401-a00.zip (accessed accessed 3 June 2011).
[7] 3GPP, Technical Specification, Technical Specification Group Services and System Aspects (2010) *Principles and High Level Requirements*, 3GPP TS 32.101 Version 10.0.0, Release 10, 8 October 2010, http://www.3gpp.org/ftp/Specs/archive/32_series/32.101/32101-a00.zip (accessed 3 June 2011).
[8] 3GPP, Technical Specification, Technical Specification Group Services and System Aspects (2010) *Charging Data Record (CDR) Parameter Description*, 3GPP TS 32.298 Version 10.2.0, Release 10, 8 October 2010, http://www.3gpp.org/ftp/Specs/archive/32_series/32.298/32298-a20.zip (accessed accessed 3 June 2011).

4

Multi-Technology Self-Planning

Josko Zec, Octavian Stan, Rafael Ángel García, Nizar Faour,
Christos Neophytou, Khalid Hamied, Juan Ramiro, Gabriel Ramos,
Patricia Delgado Quijada, Javier Romero and Philippe Renaut

4.1. Self-Planning Requirements for 2G, 3G and LTE

Network planning is defined as the process of identifying the parameter settings of new network elements, including site locations and hardware configuration [1]. Self-Planning is defined as the automation of the network planning process to reduce the associated cost and operational effort. This can be achieved by utilizing functions and components that are highly automated. The degree of automation depends on the functions and components used in the planning process. Examples will be provided throughout the chapter. The following is a list of requirements for Self-Planning:

- Accuracy of the models and input data: planning processes usually utilize propagation path-loss predictions, traffic volume forecasts, call models, traffic spatial distributions, etc. The accuracy of those models can be improved by using model tuning and calibration based on drive data, Operations Support System (OSS) performance counters and mobile measurements. See Section 4.5 for more details.
- Optimality of the derived parameter settings: mathematical optimization techniques should be used in order to provide optimum parameter settings, i.e. the ones that provide best results. See Section 4.4 for more details.
- Automation: ability to activate the new network elements and provide autonomous derivation and implementation of the parameter settings.
- Multi-vendor: ability to derive vendor-specific parameter settings for the new network elements and ability to generate vendor-specific commands to activate the new network elements and implement the derived parameter values.

Self-Organizing Networks: Self-Planning, Self-Optimization and Self-Healing for GSM, UMTS and LTE,
First Edition. Edited by Juan Ramiro and Khalid Hamied.
© 2012 John Wiley & Sons, Ltd. Published 2012 by John Wiley & Sons, Ltd.

• Multi-technology: ability to handle all relevant technologies separately and jointly if necessary. The choice of certain parameter settings for a given technology may impact the performance of another technology. Cross-technology coverage, capacity and quality must be taken into account during the planning process.

4.2.   Cross-Technology Constraints for Self-Planning

Multi-technology networks need to be planned carefully when they coexist in the same geographical area since they are usually designed to provide combined coverage, quality and capacity for the entire offered traffic and service profiles. In general, layers with different radio access technologies, such as GSM, UMTS and LTE, usually share the same site locations and sometimes share transmit and receive antennas. The footprint and coverage of a given technology layer, as well as its capacity, affect the performance of other technology layers. The following are examples of cross-technology constraints that condition the planning process:

• Antenna sharing: the planning process needs to derive the antenna settings of a new network element (e.g. azimuth, mechanical tilt and electrical tilt). If antennas are shared among different radio access technologies, the planning process has to take into consideration the coverage, capacity and quality of those technologies that share antennas. The suggested antenna settings for the new network element should not, for example, cause degradation to the performance of other technology layers beyond an allowed or acceptable threshold.
• Traffic share: when new network elements are planned, assumptions are made on the traffic split among the different radio technologies. The planned coverage and capacity of new network layers should be consistent with those of legacy layers, and take into account the forecasted traffic migration between layers (subject, for example, to availability of capable terminals) in order to avoid overinvestment in new infrastructure.

4.3.   Self-Planning as an Integrated Process

The Self-Planning process is illustrated in Figure 4.1. The first step in the planning stage is to detect automatically that new sites are required. This task can be carried out following two different approaches:

• A proactive approach that relies on a roll-out plan generated beforehand, based on a traffic forecast and detailed system models.
• A reactive approach based on the surveillance of system coverage, quality and capacity and the trigger of network expansions based on predefined rules. This implies that this reactive process must be always on.

The multi-technology Self-Planning process needs to provide all necessary functionality to follow any of the two aforementioned approaches: traffic forecasting, system surveillance and processes to configure rules that are triggered when the deployment of additional infrastructure is deemed necessary. The detailed definition of the Key Performance Indicators (KPIs) to be monitored and the rules to be applied differ for each radio access technology.

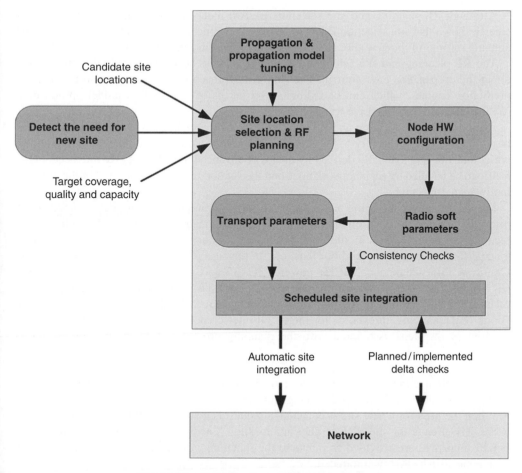

Figure 4.1   Self-Planning as an integrated process.

After identifying the need for a new site, the derivation of its Radio Frequency (RF) parameter settings and location are carried out using the following processes:

- RF planning: this process refers to the derivation of the antenna settings (antenna type, azimuth, mechanical tilt, electrical tilt and height) and the transmit powers of new sectors. RF planning needs to consider a weighted combination of high level KPI objectives in terms of target coverage, quality and capacity. In case of cross-technology constraints (e.g. if antennas are shared between different radio access technologies), the planning process needs to consider all affected layers simultaneously.
- Site selection: this process refers to the derivation of the site location, which is usually done by selecting the optimum location from a list of candidate locations. Site selection also needs to consider a weighted combination of high level KPI objectives.

Although RF planning and site selection are two independent functions, experience has clearly shown that site selection needs to be executed concurrently with the derivation of the antenna settings of the new site, as well as those of the surrounding ones.

The RF planning and site selection calculation process needs to be fed with information about the propagation environment. The simplest approach is to use traditional propagation prediction engines, which can be calibrated by means of propagation model optimization routines that are fed with drive test data.

After deciding the location and RF settings of the new site, the next step is to determine the base-band hardware capacity to be deployed (in case of modular equipment configuration), which must be automatically derived by means of advanced models that take into account the traffic to be handled by each network node and the vendor-specific constraints for the modular hardware architecture.

Moreover, before bringing new nodes on-air, additional soft parameters need to be properly planned in an automated manner. These parameters depend on the radio access technology. For 2G, they include power settings, frequency plans, Base Station Identity Code (BSIC) allocations, neighbor lists, etc. In the case of 3G, such parameters set covers Common Pilot Channel (CPICH) powers, 3G-3G and inter-Radio Access Technology (iRAT) neighbor lists, scrambling codes, etc. For LTE, the planning process needs to consider intra-LTE and iRAT neighbor lists, Physical Cell Identifiers (PCIs), Resource Block (RB) planning, thresholds for Inter-Cell Interference Coordination (ICIC), etc.

Most of the steps associated with the planning process can be highly automated as follows:

- The detection of the need for a new site can rely on automated functions that trigger network expansions due to capacity and coverage reasons.
- When the need for a new site is detected, a request is made to the function responsible for the derivation of the site's RF settings and location. Candidate site locations and high level KPI objectives are provided as inputs to the functions in charge of the calculations. This step can also be highly automated.
- Additional soft parameters: some parameters, such as admission control thresholds, can initially be assigned default values. Assigning default values can be performed automatically by means of user-defined templates, where multiple templates are used to distinguish between cell types or categories. For other parameters, however, default values cannot be assumed and different settings must be applied for different sectors. For example, scrambling codes for 3G and PCI for LTE need to be adapted explicitly per sector. Such parameter values can be derived automatically using specialized optimization and planning functions. This step can be highly automated.
- Next, the site can be scheduled for integration. This step requires human intervention to select the date to install the hardware and bring the site on-air.
- Consistency checks are performed before the site is integrated. Some examples are:
    - Missing children: identify objects having no child objects (e.g. sites without cells).
    - Orphaned elements: identify elements with no parent elements (e.g. cells without sites).
    - Missing parameters: identify missing parameter values.
    - Ensure internal consistency and compliance with operator's configuration rules.

If issues are identified that cannot be corrected automatically, human intervention is needed in this step.

- An automated process or function needs to be in place in order to integrate the new sites into the network landscape. This function, referred to as site integration, should automatically generate and execute all scripts and commands that are needed in order to create all logical entities and managed objects associated with the activation of the site into the live network.
- Finally, discrepancies between planned parameter values and what has been implemented in the network need to be identified and corrected.

4.4. Planning versus Optimization

Planning and optimization are two key areas in wireless mobile communication engineering. In a typical mobile wireless operator, these two aspects are usually separated into two distinct departments. Despite organizational separation, principles and required expertise applied in both planning and optimization processes, within a given technology, are the same. However, there are certain differences in scope. The crucial difference is that the domain of the planning process is restricted to new cells to be built, whereas optimization focuses on already operational cells and clusters. The objective of this section is to remove any possible confusion in material presented elsewhere in the chapter when discussing the terms of planning and optimization.

Radio network planning can be defined as the process of proposing locations, configurations and settings of the new network nodes to be rolled out in a wireless network. The general objectives of radio network planning are:

- Obtaining sufficient coverage over target area, ensuring satisfactory service quality and low bit-error rates.
- Providing the demanded network capacity with low service blocking, satisfactory user throughputs and low dropped call rates.
- Preserving profitability by implementing an economically efficient network infrastructure (i.e. minimum number of sites and transceiver units required to satisfy coverage, quality and capacity requirements).

These objectives are fulfilled by proper choice of site locations and derivation of cell settings and parameters, including antenna models, antenna radiation center heights, pointing directions (azimuth and tilt angles), etc. The next planning step is to specify technology-dependent parameters such as:

- GSM: Broadcast Control Channel (BCCH) and Traffic Channel (TCH) frequencies, Base Station Color Codes (BCCs), Hopping Sequence Numbers (HSNs), Mobile Allocation Index Offsets (MAIOs), etc.
- UMTS: CPICH powers, Primary Scrambling Codes (PSCs), etc.
- LTE: PCIs, Downlink (DL) and Uplink (UL) assigned RBs, etc.

Deriving neighbor lists for new sites is also part of the planning process, especially for 2G and 3G. All other numerous standardized and vendor-specific parameters are typically set at nominal values at rollout and continuously tuned when cells are operational and KPIs become available. Good planning is reflected in satisfied subscribers, profitable network (low infrastructure cost) and lower optimization efforts following the network rollout.

Radio network optimization can be defined as the set of steps and measures that are required in an already active wireless communications network in order to improve or maintain performance. Improvements can be defined in terms of any combination of coverage, capacity and service quality. The ultimate goal of the optimization process is the same as in planning: to satisfy subscribers through preventing and resolving complaints while avoiding unnecessary investments in costly infrastructure expansions and new site additions. The general way to achieve this goal is the same as in planning: proper choice of parameter settings.

Besides significant overlap in scope and target, an additional source of confusion emerges between Self-Planning and Self-Optimization, since both processes rely on applied and customized generic mathematical optimization techniques. These techniques (encompassing definitions of configuration spaces, cost functions and search strategies) apply equally to network planning and network optimization. The crucial clarification comes from understanding the difference between underlying optimization techniques and optimization processes. Optimization techniques are commonly deployed in a Self-Organizing Network to automatically derive cell parameter values. However, if used in preoperational stage, they are referred to as a part of planning while applying the same methods to already operational cells and clusters is typically called optimization.

Thus, discussing optimization techniques can be largely interleaved between planning and optimization stages with an obvious difference in the data available to feed the automated solutions when planning new sites and when optimizing existing sites. While the process to plan new cells relies mostly on propagation predictions and network simulations, operational networks can be optimized with much more accurate input information coming from OSS counters and call traces. Various optimization techniques applied to Self-Planning are discussed in this chapter, while Chapter 5 focuses on optimization techniques applied to Self-Optimization.

4.5.   Information Sources for Self-Planning

4.5.1.   Propagation Path-Loss Predictions

In a mobile radio environment, the propagation path-loss depends not only on the distance, but also on the natural terrain, man-made structures, antenna heights and others. Scatterers and reflectors are randomly distributed even in the same environment. Therefore, the value of the path-loss at a given distance is a random variable with a mean and variance that depend heavily on the local terrain characteristics. A propagation path-loss estimate from the transmitter to the receiver is needed to compute received signal strength, interference and Signal to Noise Ratio (SNR). Propagation path-loss information is critical to derive most of the RF parameters of new sites. The calculation of propagation path-loss estimates is referred to as path-loss predictions.

Several propagation path-loss prediction models exist such as Hata [2], COST231 [3], Lee [4], etc. These models are based on empirical methods and, therefore, use different parameters to model path-loss in different environments and for different radio frequencies. These mathematical models are usually inaccurate and require calibration using drive test measurements.

4.5.2. Drive Test Measurements

Drive testing is the process of measuring and recording sets of received signal strengths from several transmitters at different geographical locations. This is usually done using a vehicle with measurement equipment that can scan several transmitters, measure and record the received signal strength from the active transmitters. The drive test equipment also provides Global Positioning System (GPS) information for location logging.

Drive test measurements are very accurate compared to propagation path-loss predictions, and can be used in the planning process in different ways:

- Drive test measurements alone: if the new node is not turned on yet, a Continuous Wave (CW) transmitter is typically used to radiate an unmodulated carrier from the node location and center of radiation (height). This method is both tedious and costly. If, however, the new node is going to colocate with another node using another radio access technology (a legacy technology for example), the drive test measurements on the legacy network can be used along with antenna scaling to derive the measured propagation path-loss for the new node. Antenna scaling is the process of using gain computation to strip out the antenna gain associated with the corresponding transmit antenna in the legacy network from the measured signal strength, thus normalizing the received signal strength.
- Drive test measurements can be combined with propagation path-loss predictions and used to calibrate the predictions data. A combination of smoothing and two-dimensional, intelligent interpolation between measured data points can be used to calibrate predictions data. The combined data set (i.e. the drive test measurements at those locations where drive measurements exist, and the calibrated path-loss data elsewhere) provides remarkable accuracy.
- Propagation model tuning: the parameters of the path-loss prediction model are tuned to minimize the error between predictions and measurement data [5]. Those parameters are tuned for different environments and morphologies, thereby improving the accuracy of propagation predictions.

4.6. Automated Capacity Planning

Capacity planning is defined as the science and art of calculating the physical, hardware and transmission resources that will be required in order to serve the forecasted traffic over a certain period of time with the desired Quality of Service (QoS). It is a critical activity, since it determines the extent and scope of the operators' investments, clearly stating where and when new resources need to be deployed. Under-dimensioned network resources will impact the delivered QoS negatively, whereas unnecessarily large or premature investments will jeopardize the operator's financial performance.

In practice, these calculations are not only triggered to plan the network deployment in a new service area, but also to reinforce existing infrastructure in order to ensure the desired

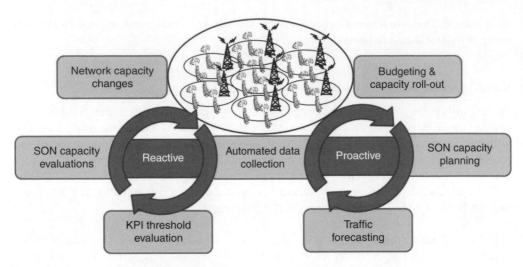

Figure 4.2   Reactive versus proactive capacity planning.

QoS across the network, anticipating the impact of the operator's marketing initiatives (e.g. the launch of an especially appealing handset) or changes in service profiles (e.g. increasing data volumes due to lower tariffs).

In today's networks, and especially with the explosive growth of mobile broadband traffic [6], it is paramount to have an end-to-end approach for capacity planning activities. Indeed, it is key to apply dimensioning methodologies and models that are detailed enough to properly characterize the need for resources in the Radio Access Network (RAN), transmission and Core Network (CN) domains. Similarly, it is important that the applied processes take into account all key properties of the services that are being delivered over these networks, as well as their associated QoS requirements.

Traditional capacity planning techniques used to rely on basic, widely accepted analytical models, such as Erlang-B or Kaufman-Roberts [7], which can be easily applied through computer spreadsheets with a limited set of inputs and a large number of assumptions and simplistic generalizations. However, increasing system complexity and the changing nature of the services landscape make it advisable to apply more advanced and complex methodologies, which match perfectly with some of the requirements of the Self-Planning category of the Self-Organizing Networks (SON) paradigm. Apart from the obvious Operational Expenditure (OPEX) savings derived from the automation of capacity planning activities, the application of these techniques will bring the following key benefits:

• Streamlined capital outlays, through a more technically sound decision process.
• Harmonization of the investment process across the entire network.
• Availability of an instrument (the process itself) to unambiguously justify the need for capacity expansions.

In general, there are two approaches for capacity planning (see Figure 4.2):

- Reactive approach: monitoring KPIs on a continuous basis and triggering selective capacity expansion processes when some conditions are met (e.g. when the utilization of a certain Iub link or the radio load factor in a given sector exceeds a certain value).
- Proactive approach: analyzing current network performance, forecasting traffic growth and predicting exactly where and when capacity expansions are required. By nature, this approach is intimately related to the budget estimation process.

In this scenario, the fact that current networks are typically multi-vendor and multi-technology makes them more challenging and cumbersome to effectively maintain, configure and expand. In order to facilitate automated capacity planning in such environment, vendor-specific data collection processes are essential in order to provide the actual analysis and dimensioning engines with harmonized inputs. Along the process, some capacity planning functions will be treated in a unified, vendor-agnostic way, although the generation of specific, local recommendations will require some vendor-specific particularization. Moreover, when dealing with a multi-technology environment (e.g. with GSM, UMTS and LTE), there are critical aspects that need to be considered, such as the inter-system traffic sharing and inter-technology mobility patterns. This section addresses the critical aspects of automated capacity planning in vendor- and technology-agnostic manner unless explicitly stated.

4.6.1.   Main Inputs for Automated Capacity Planning

Data collection needs to be fully automated, and information must be collected directly from the operator's main configuration and performance database(s), guaranteeing reliable, fast, periodical and frequent updates. Since this information is typically spread over different systems and databases, the data collection process needs to consolidate all the involved sources and guarantee their consistency, thereby building a faithful, end-to-end inventory of the mobile network.

The most significant inputs for automated capacity planning are:

- Network physical topology: for existing networks, the network physical topology information can be obtained automatically from the OSS. This information needs to be complemented with additional input from the operator's planning database in order to capture nonexisting, planned network elements. The network physical information is especially relevant for the definition of the logical interfaces and network hierarchy, which are key inputs to the RAN and CN capacity analyses. Therefore, they should be updated periodically to follow all network element additions or rehosting operations.
- Network configuration: the configuration parameters that have an impact on the capacity at different levels, such as number of Transceivers (TRXs), channel configuration for 2G and number of Channel Elements (CEs) for 3G. This information can also be obtained from the OSS. Each vendor defines its own set of configuration parameters. Some parameters are standard, whereas others correspond to vendor-specific features.
- Current network load: this is the most relevant information for capacity analysis. The term network load includes all those metrics that can be implicitly or explicitly benchmarked against the network equipment capacity limits. Typically, all these metrics are defined through traffic

figures (Erlangs or kbps) or number of active subscribers. In this respect, the operator's strategy and potential service level agreements will determine the aggregation level of the different analyses, i.e. how detailed they are. Depending on this, traffic figures may or may not be broken out into different categories (voice, web browsing, video streaming, etc.) or service classes (real-time, non-real-time, background, etc.). Similarly, active subscribers can be broken out into categories such as Very Important Person (VIP) users, roamers, prepaid users, etc.

- Network equipment capacity limits: normally provided by the operator based on equipment vendor specifications or catalogue. The specifics of these capacity limits may vary for each part of the network (RF, RAN and CN) and one possible general categorization is provided below:

  o Hard capacity limits: depends directly on the hardware space (network element cards or racks installed), interface capacity or floor space for the network equipment. Examples include:

    ▪ RF: maximum number of TRXs per GSM Cabinet, maximum number of CEs per NodeB.
    ▪ RAN: maximum number of Abis E1/T1 links, maximum number of Packet Control Units (PCUs) per Base Station Controller (BSC).
    ▪ CN: maximum number of packet processing units per Service GPRS Serving Node (SGSN), maximum number of transcoder cards per Media Gateway (MGW).

  o Soft capacity limits: they are specified for some reference service mixes or specific user profiles, typically the ones existing in the network at the time of the analysis. Examples of soft capacity limits are the maximum number of busy hour call attempts per BSC or Radio Network Controller (RNC), the maximum Signaling System #7 (SS7) signaling load, or the maximum percentage of Central Processing Unit (CPU) utilization. While hard capacity limits can be easily found in the equipment vendor documentation, soft capacity limits are, whenever provided by the infrastructure vendor, always associated with certain test environment conditions. In some other cases, the network operator is computing those real soft capacity limits based on historical statistics.

- Vendor-specific network features: these features are introduced by different infrastructure vendors in a proprietary manner and have a direct impact on network capacity limits. For automated capacity planning, it is extremely important to identify such active features in each network area, automatically collect their associated settings and apply a technically sound model to factor in the capacity increase that they provide. Examples of standard features are the introduction of Half Rate (HR) or Adaptive Multi-Rate (AMR) in GSM, Enhanced Data rates for GSM Evolution (EDGE) modulation and coding schemes in EGPRS or High Speed Packet Data (HSPA) support in UMTS, where each of the individual vendors provide their own proprietary settings and implementation.

4.6.2. Traffic and Network Load Forecast

The first step to execute a proper capacity planning process is to identify the target network load. In order to do so, a forecast process needs to be in place. In order to achieve acceptable prediction accuracy, the following areas need to be covered:

- Individual forecast per network element (in this respect, sector level is the lowest recommended level), based on historical load samples. For this, different forecast methods can

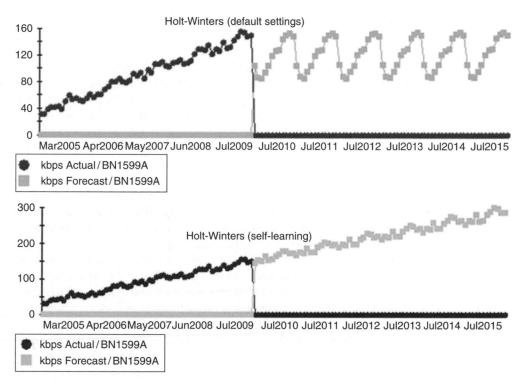

Figure 4.3 Comparison of self-learning and original forecast for the Holt-Winters algorithm.

be applied [8, 9] and it is recommended that the selected technique is based on rolling windows with seasonal-pattern considerations (e.g. by using the Holt-Winters algorithm [9]). Applying smoothing filtering for spurious peaks of load is also advisable. Moreover, self-learning, i.e. the automatic adjustment of the algorithm settings on a per element basis, will dramatically improve the accuracy of the forecasted load. See Figure 4.3.

- Additional information from the operator's marketing department should be provided in order to fine-tune the final forecasted figure; taking into consideration expected events (such as new tariffs, the launch of a new service or the exclusive release of an especially appealing handset) that may have a drastic impact on the previous per element fore-casted figures.
- Finally, it is essential to consider the spatial overlap between different radio access technologies. In the frequent case of significant overlap, the operator may choose to pri-oritize the utilization of one technology over the other, on a per service type basis, thereby resulting in a clear impact on traffic growth for each technology.

4.6.3. Automated Capacity Planning Process

To describe the automated capacity planning process, the first step is to define the triggers and the design criteria driving the automated capacity planning actions, which assist operators in identifying when, where and what should be deployed in the network to address capacity issues. Four main scenarios are identified:

- Coverage-driven capacity planning: the main objective is either expansion of the current coverage footprint or improvement of the RF coverage in current service areas, even if capacity is not exhausted (e.g. due to the existence of different stages in the business plan of the operator, due to commitments with the regulator, etc.). In this scenario, the driving metrics will be pure RF coverage distributions such as Received Signal Level (RxLev) in GSM, Received Signal Code Power (RSCP) and *Ec/Io* in UMTS, Reference Signal Received Power (RSRP) and Reference Signal Received Quality (RSRQ) in LTE, or in some cases statistics of the achievable peak throughput under some load conditions. Moreover, it is worth noting that, as a result of an expansion in the effective coverage footprint, more traffic will be attracted to the system and, as a consequence, some capacity-driven expansions may be required, not only on the RF side, but also in the RAN, CN and Transmission domains.
- Capacity-driven capacity planning: in this case, the main trigger for the capacity expansion is current or forecasted lack of capacity, i.e. lack of sufficient resources in the network to handle the current or forecasted network load with the desired QoS. The capacity analysis for the current or forecasted traffic needs to be performed at all network levels (RF, RAN and CN) and for all network technologies. Depending on the technology under consideration, this lack of capacity may lead to different specific symptoms such as:
    - o In GSM and General Packet Radio Service (GPRS), lack of capacity causes voice calls to drop either during call setup phase or handover procedures, while at the same time the data territory will be reduced to the minimum, causing degradation in the data throughput per user or even causing blocking in the Packet Switched (PS) domain. Excessive signaling load may also be observed.
    - o When UMTS is approaching its capacity limit, excessive interference will lower Channel Quality Indicator (CQI) values and therefore HSDPA throughput will degrade due to low SNR and excessive multiplexing. CE capacity may also be exhausted. In the Circuit Switched (CS) domain, blocking will increase and the data rates of Release 99 (R99) packet traffic (if any) will suffer.
    - o On the RAN side, when the system is excessively loaded, interfaces (particularly the last mile Abis/Iub) are likely to become congested, causing degradation of EGPRS throughput due to lack of pooled resources [10] and service degradation (including high latency) in UMTS, especially on the background (nonreal-time) services. In certain cases, even the signaling interfaces and elements will become congested, thereby increasing the amount of dropped calls and sessions. In general, the practical appearance of congestion in these interfaces will ultimately depend on whether they are the bottleneck of the system.
    - o On the CN side, typical congestion symptoms are caused when the maximum number of attached subscribers (both for CS and PS) is reached. More specifically, on the PS Core, these limitations are typically related to the amount of active Packet Data Protocol (PDP) contexts and the associated data throughput, consuming the capacity of packet processing units (with regards to CPU load) in both the SGSN and the GGSN. In the same way, on the CS core, the limitations come from the number of busy hour call attempts and processed Erlangs, associated with the different codec utilizations.
- Technology-driven capacity planning: in this case, after making decisions related to the rollout of a new technology, the rearrangement of the spectrum resources or the traffic share

between technologies, the network operators need to evaluate capacity implications of those decisions and revisit the resource allocation accordingly. Some examples:

- o When introducing new spectrum and a new radio access technology with significant spatial overlap with an existing technology, (e.g. when introducing UMTS or LTE in an area previously covered by GSM), the sites supporting more modern radio access technologies will absorb a significant share of the data traffic on the new technology layer, which means that their capacity in the legacy layer needs to be redimensioned to take into account the fact that they will most likely carry lower data volume (primarily voice).
- o After carving spectrum to accommodate new technologies (e.g. to rollout UMTS on spectrum previously used for GSM, or deploy LTE on spectrum previously used for UMTS), the capacity of the GSM layer will be reduced and some hardware and transmission elements will be decommissioned. In this case, it is important to understand exactly how the traffic is going to be partially offloaded from the legacy technology to the new one, and specify the decommissioning plan accordingly.
- o When changing certain network configuration parameters (e.g. iRAT handover settings) during optimization processes (including Self-Optimization), the effective footprint of one technology layer (3G, for example) will change, and so will the amount of corresponding served traffic. Most likely, this may have a direct impact on the required capacity at all levels (RF, RAN, CN and transmission).

- Marketing-driven capacity planning: expansions associated with specific tariff campaigns, the launching of an especially appealing mobile device or the introduction of a new service in the network (a killer service or application). In these cases the definition of the expected timeframe for the traffic growth, as well as the consideration of different what-if scenarios that reflect reasonable uncertainties in the impact of these marketing actions on the actual traffic, is paramount to anticipate and size the network expansion properly.

The following subsections describe important settings that are associated with the automated capacity planning process and need to be properly configured.

4.6.3.1. Timeframe and Busy Period Definition

One relevant criterion in this methodology is related to the timeframe for the analysis. The reactive strategy will be focused on the current or short-term network load situation and the current network resources. It is important for this operational case at least to define the process with a minimum forecasting period so that there is enough time for requesting and installing the equipment once the need for expansion has been identified. The proactive approach will look at longer timeframes, and will be oriented to budgeting and mid- or long-term planning.

Another important aspect is the definition of the busy period, in which the network is offered the traffic that determines its design in terms of capacity. The proper selection of the busy period is an important engineering decision. Typically, the individual sector busy hour will be selected on the RF side, whereas for the RAN analysis (i.e. BSC/RNC level busy hour) and CN analysis (i.e. MSC/MGW/SGSN busy hour), the system busy hour will be selected instead. Another potential approach for medium- and long-term planning is to identify weekly or monthly aggregated figures, i.e. to perform capacity planning without using daily traffic forecast, but instead to use one traffic sample per week or per month. In such cases, engineers need to decide whether to use the busy hour of the whole week/month (a very conservative approach) or the average of the daily busy hours for that week/month.

4.6.3.2. Financial Constraints

Financial limitations typically impose a constraint into the process through the available budget. In that case, it is important that the automated capacity planning process can prioritize its recommendations within the available budget in order to make the most out of the available resources. It is up to the operator to decide on the prioritization strategy, which will be likely based on a mixture of performance factors (e.g. favor network spots that deviate more from the performance targets), discretional criteria (e.g. favoring certain subscriber profiles), Return On Investment (ROI) considerations, commercial/organizational visibility (e.g. impact on service level agreements for high-priority contracts), etc.

The SON automated capacity planning function should also consider prioritization of the proposed actions that can be implemented automatically and do not require any human intervention (such as parameters or configuration settings modification) over those actions that need site visits or real hardware expansions. In this way, both OPEX and Capital Expenditure (CAPEX) figures will be minimized.

4.6.4.   Outputs of the Process and Implementation of Capacity Upgrades in the Network

As a result of the automated capacity planning process, different recommendations will be obtained and the ultimate target is to implement them in the network. Whenever possible (certainly not in every case), the availability of routines that automate the implementation of certain recommendations will be extremely helpful and contribute to keep OPEX down throughout the process. Other actions involve the installation or decommissioning of hardware elements, which cannot be done remotely.

4.6.4.1. RF Upgrades

In GSM, there are recommendations that can be implemented remotely and in an automated manner (e.g. the reconfiguration of the resources that are devoted to the PS territory and the expansion of half-rate capable timeslots). Other actions, such as the addition of TRXs or the upgrade of a certain cabinet model, may require site visits and human intervention.

In UMTS, capacity-related actions include reevaluation of the service QoS requirements and settings (e.g. Eb/No targets and guaranteed bit rates) and the reconfiguration of power allocations (e.g. pilot/signaling and HSDPA+ power), which can be carried out remotely by means of scripts and specialized routines. However, other actions, such as the addition of extra channel elements, require sites visits and human intervention.

4.6.4.2. RAN Upgrades

In GSM, Abis capacity needs to be sized properly, based on real LapD signaling and EGPRS resources consumption [10]. For the Iub interface in UMTS, it is critical to dimension the transmission capacity accurately, as well as to properly configure the selected underlying technology, such as Asynchronous Transfer Mode (ATM) virtual circuits/virtual paths, Internet Protocol (IP) tunnels, etc.

Concerning BSCs and RNCs, load balancing actions may be recommended to increase the overall capacity of the system, mainly through the execution of rehome plans that redefine the network topology in terms of relationships between sites and BSC/PCU/RNCs. However, in many situations, BSC/PCU/RNC capacity upgrades may be required.

4.6.4.3. Network Upgrades

When planning the CS Core Network (MSC, MGW), one of the most critical capacity limits is the CPU load factor due to the number of call attempts or ongoing calls. Thus, safety margins must be planned in each platform in order to secure adequate QoS during the expected busy period, although eventually the process will end up recommending selected capacity upgrades in the elements facing excessive load. Moreover, CPU load is also heavily affected by mobility patterns and the associated procedures (e.g. handovers, location updates, etc.). Therefore, proper planning of neighbor lists and location area borders is also paramount when defining the scope of an automated planning solution. Finally, recommendations about dimensioning the SS7 signaling network is also a typical result of the automated capacity planning process, including the sizing of the Signaling Transfer Point (STP) platform and the corresponding A/E and B/D interfaces.

For the PS Core Network, a typical recommendation may be to upgrade SGSN and GGSN capacity, mainly on the signaling side (PDP context handling) or on the packet processing side (depending on the detected type of overload pattern). In other platforms, Short Message Service (SMS) traffic may require special attention, eventually triggering corrective actions in the SMS Center (SMSC) or even at MSC level.

4.7. Automated Transmission Planning

The transmission area comprises different technologies and mediums that are used to connect the radio and core networks. The technologies covered in this section are the most commonly used in the cellular environment: IP, Synchronous Digital Hierarchy (SDH), ATM and Plesiochronous Digital Hierarchy (PDH). A summary of those technologies is found in Table 4.1.

Regarding SON, it is interesting to note that this concept has been in place for transmission networks long before being proposed and developed for the radio access. One of the reasons is that, in transmission networks, it is rather common to have big, meshed and continuously changing topologies, with thousands of links that have to be routed through. Additionally, as hundreds or thousands of links may be concentrated in a few trunks (e.g. a connection between MSCs), any mistake in the configuration will have a large impact on the network. Finally, the complexity of creating static routes depending on variable factors (like capacity or performance) forced the transmission technologies to evolve towards a Self-Organizing philosophy several years ago, as it will be shown in Section 4.7.1. Protocols and features depicted in Table 4.1 provide a wide range of automation and Self-Organizing features.

To illustrate this idea, two examples in the case of IP networks can be highlighted. Firstly, when inserting a new node in an IP network, a unique IP address has to be assigned to this new node, together with some additional parameters like network submask, default gateway, name

Table 4.1 Transmission technology summary

Technology	Description	Main use	Protocols/Features
IP	Internet Protocol	Internet and general use networks	Address Resolution Protocol (ARP), Dynamic Host Configuration Protocol (DHCP), Open Shortest Path First (OSPF), Enhanced Interior Gateway Routing Protocol (EIGRP)...
PDH	Plesiochronous Digital Hierarchy	Level 2 protocol, up to some Mbps	N/A
SDH	Synchronous Digital Hierarchy	Level 2 protocol, starting from 155 Mbps	Protection Rings, Subnetwork Connection Protection (SNCP)
ATM	Asynchronous Transfer Mode	Basic transmission layer for UMTS networks. Typically deployed over PDH or SDH	ATM Inter-Network Interface (AINI), Private Network-to-Network Interface (PNNI)

server address, physical address of all neighbor nodes, etc. Considering the ubiquity of IP nodes connected to the Internet, it is not practical that every end-user configures all those parameters manually. A mistake in some of those parameters may cause a failure in a section of the network. To solve those problems, protocols like ARP [11] or DHCP [12] have been developed; they provide the entire configuration automatically to the node, which simplifies the configuration of the home equipment that is connected to the Internet.

The other example of Self-Organization consists of protocols like OSPF [13] or EIGRP [14], which are able to define the routing tables autonomously, based on network topology, link status, failures, etc. The main advantage of those protocols is their scalability, since they divide the network into smaller sections that communicate among themselves, keeping the routing tables and the signaling traffic in a manageable state. Other technologies, like SDH/ Synchronous Optical Network (SONET) or ATM have similar autoconfiguration characteristics that will be covered later in this section.

4.7.1. Self-Organizing Protocols

With the expansion of the transmission networks, several algorithms and protocols have been developed for most transmission technologies. Their main target is to automate the configuration of the nodes connected to the network, especially the definition of the routing tables. In the following, the main protocols that fit the SON concept are described. The technology each protocol is applicable to is shown in parentheses.

4.7.1.1. DHCP (IP)

The algorithm used in DHCP [12] is designed to provide the newly connected nodes with the configuration parameters required to start functioning. This includes the IP address, subnet mask, Domain Name System (DNS) server, etc. After obtaining those parameters, the new node should be able to communicate with any other host on the Internet without manual intervention.

For this algorithm, the DHCP server employs a pool of IP addresses which are leased (or temporarily assigned) to the nodes as they are connected. Clients use the broadcast address to request configuration parameters.

4.7.1.2. OSPF (IP)

OSPF [13] is a link-state routing protocol commonly used in IP networks. OSPF is an open protocol (its specification is publicly available), and the OSPF algorithm is installed in every IP router. The algorithm follows a distributed self-organized pattern, works autonomously without human intervention and it is able to adapt quickly to any change in the network.

The OSPF routing policies for constructing a routing table are governed by link cost or penalty factors (external metrics) associated with each routing interface. Cost factors may be the distance to a router (round-trip time), the throughput of a network link, or the link availability and reliability, expressed as simple dimensionless numbers.

Every router constructs a database of different candidate routes along with their corresponding costs based on quality measurements on the adjacent links. Those measurements are typically communicated to all other nodes in the network, allowing these nodes to calculate the best route to any destination by using Dijkstra's algorithm [15].

In order to improve scalability, the network is divided into smaller areas so that nodes only have to contain routing information for other nodes in the same area. Interarea routing calculation will be done by the area border routers which conform the transmission.

The main advantage of OSPF protocol is that it adapts quickly to changes in the network.

4.7.1.3. (E)IGRP (IP)

The Interior Gateway Routing Protocol (IGRP) [14] is another widely used protocol for automatic route definition, proprietary of Cisco Systems, Inc. As opposed to OSPF, it is a distance-vector protocol, which means that nodes exchange all or part of their routing databases with their neighbors only. The information used for constructing the routing database, includes multiple metrics, like bandwidth, delay, load, reliability, etc. The main drawback of this algorithm is that it is less responsive to network changes as compared to OSPF.

This protocol was improved by the EIGRP, which adds some concepts of link-state algorithms that optimize the time response.

4.7.1.4. PNNI (ATM)

In ATM networks, the route specification for every virtual path can be done statically by using the ATM AINI protocol [16]. However, the PNNI protocol [17] allows self-configuration of routes between every pair of nodes in the ATM networks.

PNNI works in a similar way as OSPF: the nodes hold a database that contains the cost of choosing certain routes. It also has a hierarchical structure, in which one router from every group is designated as representative for taking care of the intergroup routing updates.

4.7.1.5. Protection Rings (SDH)

SDH technology [18] is another clear example of SON, since its design is based on the automatic protection and restoration of the communications in case of failure. Furthermore, the basic element in SDH networks is the ring: a group of nodes connected in a closed loop, which allows keeping the connection alive even in the case of link failure. Thus, a typical SDH network will be composed of several interconnected SDH rings.

There are two types of protection rings. The first type, the Multiplex Section-Shared Protection Ring (MS-SPRing), is configured automatically to send the information through the shortest path. In case of link failure, the data is sent in the opposite direction. The second type of protection rings is the SNCP, which sends the information in both directions of the ring simultaneously. This is less efficient in terms of required capacity, but it is more robust and less vulnerable to node failure.

4.7.2. Additional Requirements for Automated Transmission Planning

Although the protocols presented before provide Self-Organizing features, there is still room for adding more functionality:

* Some technologies do not have the previously mentioned Self-Organizing features, so an external solution is required to get the benefits of automation. For example, 2G access transmission is typically a PDH network, which does not embed any automatic feature, as it was shown in Table 4.1.
* In multi-technology[1]/multi-vendor environments, it may not be possible to communicate between different networks using built-in automatic configuration protocols. Currently, it is required to integrate dedicated gateways that have to be configured manually.
* All algorithms described earlier work in a reactive way, which means they will act only after a network change (e.g. link failure or overload, new node/link addition, etc). Even though it is hard to predict network problems, many issues can be anticipated and fixed in a proactive and efficient manner by using simple forecasting algorithms.
* Forecasting can also help to detect, in advance, when and where new hardware equipment is required, allowing operators to acquire and install the necessary equipment exactly when it is going to be needed, and with the exact capacity that is going to be required.
* An external process can be used to evaluate different what-if scenarios before implementing a network change. This allows, for example, considering different options before adding or removing a node from the network.

Considering that operators' transmission networks typically contain several transmission technologies supplied by several vendors, together with the time required to install new hardware and the ever growing traffic volumes, the addition of an automatic transmission planning process with embedded forecasting algorithms will positively impact the overall efficiency of network management.

---

[1] Transmission technology in this specific context.

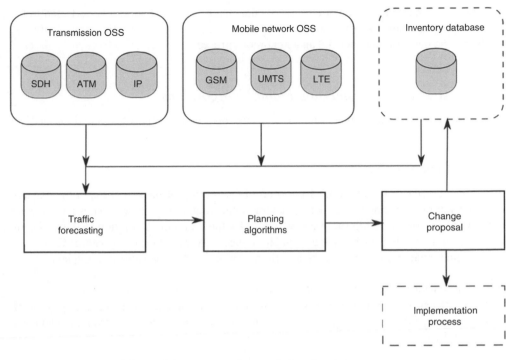

Figure 4.4   Automated transmission planning process.

4.7.3.   Automatic Transmission Planning Process

Figure 4.4 shows the automatic transmission planning process. The process extracts information from OSS databases, which include GSM/UMTS/LTE OSS systems, transmission specific database and inventory database. Obviously, keeping the transmission databases fully updated with the actual equipment in the network is essential for obtaining good results with any automatic transmission planning process.

Once the data is collected, the forecast algorithm estimates the traffic growth so that the planning algorithms are able to propose meaningful network changes. As stated before, it should not only propose what to change, but also when to implement those changes.

After the automatic transmission planning functionality recommends the changes to be carried out, this information should be made available to the entities in charge of the implementation process. In such a heterogeneous area as transmission planning, the implementation process will likely encompass different aspects. For example, if the change consists of updating a routing table, it can be performed autonomously just by accessing the designated router, which would not require any human intervention. On the other hand, if the solution is to create a new physical link between two nodes, it implies that some hardware has to be ordered and installed; in this case, the implementation process may consists of sending a notification to the appropriate department with a description of the items to be purchased and deployed.

Together with the implementation process, the change proposal should be reflected in the transmission database as well, so that it is fully synchronized with the actual network. Updating the database may also be an autonomous process.

4.7.4. Automatic Transmission Planning Algorithms

In the following sections some automatic transmission planning algorithms are described.

4.7.4.1. Link Detection

Protocols described in Section 4.7.1 are used in the routing of packets through the existing network, but they are not able to detect lack of connectivity between two nodes and they are not able to propose the addition of a new link to fix the problem. A solution to this problem, however, can be achieved by using a graph connectivity calculation algorithm. Given two nodes which require connectivity, the algorithm can proceed in the following sequence:

1. Check if a path between the two nodes exists.
2. If no path exists, identify new possible connections that would fix the connectivity problem.
3. Based on a predefined criteria and a cost function, choose the best link from the identified possibilities.

Link detection can be used at any transmission layer, from the physical connectivity to any transport layer (PDH, SDH, ATM, IP, etc.). Once the network graph is defined, the connectivity between any two nodes can be checked regardless of the underlying technology. However, step 3 depends on several factors, for example:

- Distance.
- Line of Sight (LOS) in case of wireless links, and duct availability in case of wired links. They are assessed to check if two nodes can be connected directly.
- Occupation of existing links.
- Port availability in the nodes.

An example on those factors will be shown in Section 4.7.5.

In summary, this algorithm will ensure network connectivity and will propose the optimal modification in case this is not achieved. As stated previously, this algorithm will be more efficient if working in a proactive way, based on forecasted network topologies and traffic, since it allows ordering the required extra links in advance.

4.7.4.2. Capacity Expansion

As stated before, the main advantages of reactive protocols described in Section 4.7.1 is that they are able to quickly react to network changes or link overload by updating the routing tables. However, those protocols are not able to provide a solution when all the possible routes are blocked due to a high traffic increase. In this case, a capacity expansion is required.

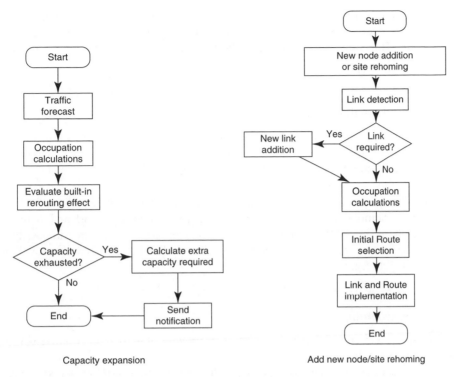

Figure 4.5   Algorithm flowchart for capacity expansion and node rehoming.

It is known that it could take weeks or even months since a link is detected to be overloaded until the new hardware is added to the network. The hardware purchase has to be approved, acquired and finally installed. This will have implications on network performance and budget provisions. In order to avoid this problem, an algorithm is required to evaluate the capacity expansions that are needed in the network. This algorithm mainly consists of traffic forecasting along with link occupation calculations (see Figure 4.5).

The traffic forecasting algorithm will be responsible of estimating the traffic offered to the network in the following months based mainly on historical network traffic and additional high-level marketing forecasts. Historical traffic will provide an initial estimate of traffic growth based on measured trends, and the marketing predictions will adjust the initial estimate and correct the expected traffic growth to include the impact of future network expansions, marketing campaigns, etc.

The link occupation calculations will estimate the load of the links in the network based on the predicted traffic. That would identify which links will be overloaded and when they will be overloaded, allowing engineers to predict the extra needed capacity in advance. For example, it may be required to expand a PDH connection from 1 to 2 E1 links.

It is important to highlight that capacity expansion algorithms must also consider any auto-configuration protocol being used in the network, to avoid adding capacity in situations when those protocols are able to cope with the extra traffic. For example, if OSPF or EIGRP protocols are active, they will distribute the load through different routes when the initial ones are getting overloaded, which affects the system positively by reducing the need for extra capacity. This step is represented in the flowchart with the label *Evaluate Built-In Rerouting Effect*.

4.7.4.3.   Path Protection Planning

Path protection planning consists of evaluating the impact of failure of certain links. The typical objective of this kind of planning is to secure appropriate network reliability. Path protection planning should be performed before the actual network failure, so that the network is protected should such event take place.

The core of the path protection analysis is the link occupation calculations described above. The link failure is emulated by virtually removing the link from the network and, in that way, its impact can be evaluated by counting the number of overloaded links. Once all possible link failures are considered, they are weighted by the failure probability of each link, providing the network availability figure.

4.7.4.4.   Site Rehoming

Site rehoming refers to migrating a site from a network controller to another one (e.g. moving a NodeB from one RNC to another).

In order to perform this change, the route through the transmission network must be obtained for the connection from the site to the new RNC. This can be achieved by using the algorithm in Figure 4.5.

The first step involves running the link detection algorithm described previously. With this algorithm, the connectivity between both nodes is evaluated. If it is not possible to find a path between them, new links need to be added to the network.

The second step describes the capacity expansion algorithm, which determines whether the network will be overloaded with the new logical relation, and whether additional bandwidth is required. In this step, a route between the nodes can also be calculated by the algorithm so the impact to the network load is minimized.

Finally, the changes should be implemented. This latest step may be done automatically or manually, depending on each specific case, as described before.

4.7.4.5.   Adding a New Node

Adding a new node can be seen as a special case of the site rehoming process. The difference is that in the link detection phase, a new link is always required: the first hop between the new node and the network. Therefore, in this case, the target of the link detection algorithm is not to determine if additional links are required, but where to add the new link.

Once the first hop is determined, the algorithm follows the same process as the site rehoming: it executes the capacity expansion algorithm to determine additional bandwidth that may be required for the new connection, together with an initial route proposal.

The node addition algorithm is described in Figure 4.5.

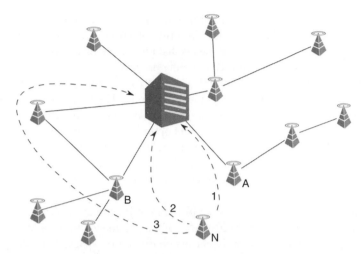

Figure 4.6   Transmission planning example.

4.7.5. Practical Example

Consider the network in Figure 4.6. It will be used to illustrate the process of adding a new node. Take the following assumptions:

- All nodes generate traffic equals to 1 unit (Erlang for example).
- All links have a capacity of 3 units.
- A new node N is added to the network.
- The cost of connecting N to A is 100 units and the cost of connecting N to B is 110 units.
- The cost of adding extra capacity to a link is 30 units.
- For existing nodes, shortest-path routing is assumed.

Assume the simplest algorithm (path 1). It will connect to A since the cost of the first hop is the smallest one. However, after connecting the node, there will be a lack of capacity and an expansion will be required. The cost will sum up to a total of 130.

Consider the option of connecting the new node to B, which has a higher cost for the first hop. There are two possible routes (path 2 and 3). A basic algorithm based on the number of hops will choose path 2. However, it does have the same problem as option 1: a new capacity expansion is required, increasing the cost by an extra 30 and summing up to 140, therefore making this option even worse than path 1.

Finally, with the third option (path 3), the cost of the first hop is higher and the number of hops is higher than in path 2. However, as the route has less traffic, no extra capacity is required. Thus, the total cost of path 3 is the smallest one.

With this example, we can see how using an automated transmission planning solution considering all the variables will positively impact in the overall cost of the network.

4.8.   Automated Site Selection and RF Planning

Planning site locations and RF settings of the network infrastructure is a fundamental problem for wireless operators. Apart from site locations, the specific RF settings that need to be planned and optimized include base station transmission powers and antenna parameters like azimuth, tilt and height. The goal is to minimize the amount of deployed infrastructure while maintaining appropriate levels of service, in terms of coverage, capacity and quality. The optimization can be achieved by several means that include manual and automatic processes. This section focuses on an automated RF planning process.

Automated site selection and RF planning are Self-Planning functions that represent an essential part of the SON network management paradigm. The selection of the optimum site location and the optimum RF configuration is challenging due to the inherent complexity of the requirements concerning radio modeling and optimization, particularly for Code Division Multiple Access (CDMA) or Orthogonal Frequency Division Multiplexing (OFDM) systems[2]. Generally speaking, automatic optimization[3] of any engineering system, including site configuration in mobile wireless networks, consists of three domains:

1. Configuration space: each possible state of the mobile wireless network, where a state is defined as a set of values for all the instances of the parameters to be optimized, is mapped to one of the points in a conveniently chosen configuration space. This mapping provides the necessary level of abstraction that allows implementation of more sophisticated search algorithms.
2. Evaluation engine: the evaluation engine is used to evaluate a point in the configuration space with respect to a given set of objectives. Typically this is done via a cost function (also known as objective function). The cost function maps every point in the configuration space into a non-negative number referred to as the cost. This number represents a measure of goodness for a given point when evaluated relative to a defined set of optimization criteria.
3. Optimization engine: while searching for the optimum solution, the algorithm evaluates many points in the configuration space. These evaluations are performed in a systematic manner dictated by the optimization algorithm. The method used to propose a new solution on the basis of the ones that have been previously examined constitutes the optimization engine.

The inputs to the automated site selection and RF planning process include propagation predictions, performance objectives (and prioritization between them), spatial traffic distribution, terrain data, clutter data, a list of site potential locations, monetary cost per item (CAPEX and OPEX per site, workload to change every type of RF setting in each specific site, etc.) and a set of restrictions (e.g. site locations that need to be included, allowed parameter values for every sector, maximum budget to be spent, etc.). The inputs may include measurements from the current network (if available) as will be explained in Chapter 5. The output of the process is an optimized set of site locations and antenna parameters for the chosen set of base stations that yield optimum performance, as defined by the user.

During optimization, base stations are placed in each of the candidate site locations, and initial values for powers and antenna parameters are assumed. The received signal values at each of the geographical pixels are computed as a function of the propagation model, the transmitting powers and the antenna orientations and patterns. During optimization, as the solution

---

[2] The selection of the optimum site locations and RF configuration is also critical for GSM. However, frequency planning can be used in GSM to mitigate the impact of interference caused by poor RF design.
[3] In this context, optimization refers to the optimization techniques used in planning or optimization processes.

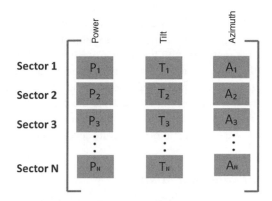

Figure 4.7   Format of the configuration state representation of a cellular system.

space is explored by the movement algorithm, the signal values are recomputed at each pixel based on the stored path-loss and the antenna settings of the configuration under assessment, and the overall cost is updated as a function of the new signal values at every pixel.

It is important to point out that the large number of parameters to optimize (due to the large number of sectors) and the interdependency between these parameters make the design process a complex optimization problem that belongs to the class of Nondeterministic Polynomial-time hard (NP-hard) problems [19, 20]. Therefore, the choice of movement algorithm is crucial. In addition, to achieve good results it is very important to define an evaluation model that provides an accurate representation of the network, which has to consider the particulars of the technologies being optimized, such as GSM, Wideband Code Division Multiple Access (WCDMA), Orthogonal Frequency Division Multiple Access (OFDMA), etc. However, the model must also be relatively simple to reduce the computational burden on the optimization engine, so that the optimization task can be completed within an operational timeframe.

The different aspects and trade-offs of the RF planning process will be investigated in the following sub-sections: The configuration space is described in Section 4.8.1. The evaluation model, its inputs and the general cost objective description are covered in Section 4.8.2. Section 4.8.3 describes the different approaches for the optimization search engine. Technology specific aspects of the RF planning are presented in Section 4.8.4.

4.8.1.   Solution Space

The RF planning process consists of selecting an optimal subset of sites among a list of site candidates and configuring the base stations to be installed at these sites in order to optimize network performance. The solution space is defined as the set of all possible combinations of sectors' parameters, where each parameter has an associated list of potential accepted values. For each sector, there is a parameter indicating whether it is included in the final design and there are other sector specific RF parameters, such as power configuration, antenna type, antenna azimuth, antenna height, mechanical tilt and electrical tilt. Each parameter type constitutes a dimension in the solution space. With this abstraction model, a simple way to represent the state of the system is by using the matrix presented in Figure 4.7.

As depicted in Figure 4.7, the matrix has a number of rows equal to the number of sectors in the system. Each parameter to be optimized introduces a column. For example, matrix entry *(i, j)* represents a value of the *jth* parameter at the *ith* sector in the cellular system.

Conceptually, all possible matrices with the format presented in Figure 4.7 form a configuration space over which the optimization algorithm searches for the optimum solution, which is the one that minimizes the cost function representing the set of requirements specified by the user. In practice, the ranges of acceptable values for the parameters are limited and hence, the search space is reduced. Moreover, there are other limitations to take into account, such as the fact that the activation of a certain site brings all its sectors on air.

4.8.2.   RF Planning Evaluation Model

The evaluation function is applied every time a new point of the solution space is explored by the optimization engine. Therefore, its execution must be both accurate and fast. There are several potential approaches to carry out such evaluation, from full-fledged network Monte Carlo (MC) simulations to network models with varying degrees of complexity. Unfortunately, the processing power required to run a comprehensive MC network simulation is remarkably intense and, with this approach, a complete network simulation would be required at every step in the exploration of the solution space. Therefore this approach, although the most accurate, becomes unfeasible since the number of evaluations to carry when optimizing a network with a few thousand sites is extremely large. As a consequence, in order to achieve operational optimization times, a technically sound network model based on a cost function is generally adopted.

With this model, the optimization process is driven by a set of KPIs that are evaluated during the optimization, with the aim to obtain the network configuration that achieves the best possible trade-off between these KPIs. The KPIs could be set to represent coverage, quality, capacity and/or other variables in any combination. There is a cost component that represents each KPI, and the combined cost metric can be computed by appropriately weighting the aforementioned components. The cost is then aggregated over a set of sectors and a set of bins (pixels) representing the geographic area being optimized, each pixel having associated signal level, interference, traffic, elevation and clutter type values. There are several methods to obtain signals with different levels of accuracy: pure measurements, field strength predictions (with or without calibrated propagation models) or measurement-weighted predictions. In the latter category operators can choose between predictions calibrated with drive test measurements or predictions scaled by different OSS statistics if available, like signal level distributions, traffic share with bad quality and Soft Handover (SHO) statistics.

One commonly used approach is to split the cost function into pixel cost and site cost. The pixel cost can be divided into pixel coverage and pixel quality costs. Given N measurement bins (representing a geographical area) and M sectors, the format of the cost function used by RF planning could be written as:

$$C_T = \sum_{b=1}^{N} C_b + \sum_{s=1}^{M} C_s, \tag{4.1}$$

where C_b is the cost contribution of the bth measurement bin, C_s is the cost contribution of sector s and C_T is the aggregated cost value associated with the point under evaluation in the configuration space. These cost contributions represent the degree of fulfillment of different KPIs, and a high corresponding cost would drive the optimizer to discard the design. Each cost component is weighted independently, and the weights can be adjusted by the user to steer the optimization in the desired direction. The main pixel components can be (but are not restricted to) coverage and quality (interference), and they can have different expressions for different technologies. Typically, the overall cost per sector is a weighted sum of DL and UL capacity cost components at sector level. More detailed expressions for the above components, for different technologies, will be given in the following subsections.

4.8.3.   RF Optimization Engine

As stated in Section 4.8.1, the dimensions in the solution space correspond to base station parameters, and each of these parameters has a range of values specified for each problem instance. The optimization engine is the agent in charge of exploring this space to determine the optimal parameter values. Exploring and scoring all possible solutions is not feasible. For example, to optimize 100 base stations that have 10 allowed tilt changes, five allowed power changes and four allowed azimuth changes, the size of the search space is $(10 \times 5 \times 4)^{100} = 1.27 \times 10^{230}$. Therefore, the choice of the optimization engine is crucial and the agent has to iterate over parameters in an efficient way for each parameter to converge to the appropriate value yielding a global optimum in the solution space. The methods typically used for RF optimization problem can include Greedy Search, Simulated Annealing (SA) [21], Genetic Algorithms [20] and [22, 23], Tabu Search [24], etc. In this section, two approaches are examined: Greedy Search and SA.

4.8.3.1.   Greedy Search Optimization Algorithm

A Greedy Search algorithm makes the locally optimal choice at each stage hoping to find the global optimum [25]. In RF planning, a greedy algorithm will iterate over all parameters (solution space dimensions), selecting for each parameter the value which will yield the best improvement in the optimization cost. Greedy Search has a disadvantage of being stuck in a local minimum in the search space. However, by its nature this algorithm can yield very good solutions when there is low intersector coupling in the network. Typically this can be the case when there are no overshooting cells in the network or extensive overlapping between sector serving areas.

The set of points that can be obtained in one perturbation step (one parameter changed) from a given point S is commonly referred to as the neighborhood of point S. In the optimization process, the local Greedy Search algorithm performs a simple loop:

1. Pick an arbitrary point S in the configuration space as an initial solution.
2. While there exists an untested neighbor of S, do the following:
    a. Let S' be a point in the untested neighbor of S.
    b. If cost(S') < cost(S), set S = S'.
    c. Go to 2.
3. Return S as the optimum solution to the problem.

4.8.3.2.   Simulated Annealing Optimization Algorithm

Simulated Annealing is a generic global optimization algorithm that can yield better solutions than the Greedy Search algorithm, by attempting to escape from local minimums. Its name and principle come from annealing in metallurgy, a technique that uses heating and controlled cooling to reduce the crystalline structural defects of the material. The heat frees the atoms of the crystal and the slow cooling allows them to find optimal configurations with lower energy. In the equivalent mathematical model used by SA optimization algorithms, at each step in the process, a new random solution is attempted in the neighborhood close to the current solution. If the new solution is better, it is chosen. If not, it is chosen with a probability that depends on the closeness of the current solution and a parameter called temperature (T) that is modified during the process using a gradual decreasing function. As T goes to zero, the solution ideally approaches the global minimum in the solution space. A typical application of the SA algorithm is outlined as follows:

1. Pick an arbitrary point S in the configuration space as an initial solution.
2. Set an initial value for the temperature parameter $T>0$.
3. While not frozen (i.e. $T\neq0$), perform the following steps:
   a. Let Δ=very large number.
   b. Perform the following loop until Δ<predefined error:
      i. Pick a random neighbor of S, say, S'.
      ii. Update Δ=cost $(S')-$cost(S).
      iii. If $\Delta\leq0$ (downhill move) set $S=S$'.
      iv. If $\Delta>0$ (uphill move) set $S=S$' with probability[4] equals to $e^{-\Delta/T}$.
   c. Reduce the temperature $T=rT$, $0<r<1$.
4. Return the best solution that has been evaluated as the optimum solution to the problem.

4.8.4.   Technology-Specific Aspects of RF Planning

4.8.4.1.   RF Planning for GSM Networks

In GSM/Time Division Multiple Access (TDMA) networks, RF planning is generally executed before Frequency Planning. Its mission is to improve the RF environment, as defined by KPIs such as coverage and quality (i.e. overlap between sectors), to make channel planning easier.

Bin Cost Function Component for GSM
For each individual bin, the format of the cost function is given as:

$$C_b = f(t_b)[f_{COV}(\mathbf{P}_b) + f_{INT}(\mathbf{P}_b)], \tag{4.2}$$

---

[4]This can be done by generating a random number between 0 and 1, and setting S = S' if the value of the random number is less than $e^{-\Delta/T}$.

Table 4.2 Typical GSM RSL thresholds for different environments

Environment	RSL threshold [dBm]
Dense urban	−66
Urban	−70
Suburban	−76
Rural	−84

where t_b is the amount of traffic served in bin b and $f(t_b)$ is a scaling factor associated with that traffic, and is used to provide weighting for the cost function, $f_{COV}(\mathbf{P}_b)$ is the function designed to penalize lack of coverage within the bin, $f_{INT}(\mathbf{P}_b)$ is the function designed to penalize the interference within the bin and $\mathbf{P}_b$ is the vector of received signals at that bin from all neighboring sectors.

Penalty for Lack of Coverage
To establish and maintain a call within any given geographic location, the received signal level on the DL needs to be above a certain threshold. It is usually assumed that density of man-made structures and their impact on the signal attenuation is highly correlated with the morphology classification of a given area [26]. Since Received Signal Level (RSL) predictions are typically computed outdoors, the coverage thresholds for the received signal are morphology dependent as well in order to take into account typical indoor penetration losses. Some typical values of the RSL thresholds for different environmental types are presented in Table 4.2.

In general, the RF planning algorithm assumes that a bin meets the coverage requirements if the RSL of the strongest server is above the threshold value specified for the bin's morphology.

Interference Penalty
Ideally, each sector in a cellular system should cover only its intended service area. If the actual coverage footprint is smaller than what was intended by design, coverage problems will arise. On the other hand, if the actual coverage area exceeds the desired one, intrasystem interference between sectors may occur. To maintain quality within a bin, the potential Carrier-to-Interference (C/I) ratio in the bin needs to be sufficiently high, where C denotes received signal level from the observed serving cell and I denotes interference (the sum of received signals from other cells). Typically, signals from several cells capable of servicing the call can be found within each bin. During the planning process, it is usually assumed that the principal serving cell is the one with the highest RSL. The rest of the cells represent interferers and their signal degrades the quality of service. Potential C/I (as opposed to actual C/I) does not take into account a specific channel plan, the assumption being that any potential interferer can be an actual one. Thus improving the potential C/I means cleaning the general RF environment and preparing the network to perform properly with any channel plan. The interference cost component is therefore used to penalize interference. Generally, there is a trade-off between coverage and quality, and the task of the RF planning algorithm is to find the optimum network configuration that balances both metrics.

GSM Sector Cost Function Component
In addition to coverage and quality, another important factor in GSM cell planning is capacity. Each base station has a certain number of TRXs capable of serving a predetermined amount of traffic (users). The sites in the network should be planned in such a way that two main capacity objectives are met with minimum hardware resources. Firstly, no users in the network should be blocked. Secondly, the load between the base stations needs to be as uniformly distributed as possible, thus achieving what is typically referred to as balanced load [26]. For each individual sector, the format of the sector component of the cost function is given as:

$$C_s = \max(\frac{E(s) - E_{MAX}(s)}{E_{MAX}(s)}, 0), \tag{4.3}$$

where C_s is the sector cost designed to penalize for exceeding capacity at sector s, $E(s)$ is the traffic served by sector s and $E_{MAX}(s)$ is the maximum traffic for sector s. Maximum traffic depends on number of TRXs assigned to sector s.

4.8.4.2. RF Planning for UMTS Networks

Running effective RF planning in UMTS networks is a challenging task. The main difficulty in UMTS network planning is the strong coupling among the WCDMA base stations, as well as the deep coupling between capacity, coverage and quality metrics. Modification of base station settings, such as antenna parameters, common channel power or admission control thresholds, can impact elsewhere in the network. The origin of this coupling is the sharing of bandwidth for all base stations so that for a given UE, all sectors (other than the serving one) are perceived as interferers.

The most accurate way to model these effects is running a Monte Carlo Simulation, which should take into account the amount and distribution of traffic in the network and take into account mechanisms like SHO, Power Control, etc. However, as mentioned in Section 4.8.2, the MC simulation is a computationally intensive process and is generally not practical as an evaluation mechanism for optimization search engines like the ones used in automated RF planning. Therefore, it is imperative to find fast evaluation models, at the same time ensuring that the technology-specific aspects of UMTS are properly considered. For example, simple *C/I* or dominance models are not adequate in UMTS, since other quality measures such as pilot *Ec/Io* and *Eb/No* are used. Pilot *Ec/Io* represents the ratio of pilot energy per chip (*Ec*) to the total wideband received spectral density of noise and signal (*Io*), and *Eb/No* represents the ratio of energy per bit (*Eb*) to the noise spectral density (*No*).

Bin Cost Function Component for UMTS
For each individual bin, the format of the cost function is given as:

$$C_b = f(t_b)[f_{COV}(\mathbf{P}_b) + f_{Pilot\ INT}(\mathbf{P}_b) + f_{Traffic\ INT}(\mathbf{P}_b)], \tag{4.4}$$

where t_b is the amount of traffic served in bin b and $f(t_b)$ is a scaling factor associated with that traffic, $f_{COV}(\mathbf{P}_b)$ is the function designed to penalize lack of coverage within the bin, $f_{Pilot\ INT}(\mathbf{P}_b)$ is the function designed to penalize the pilot interference within the bin, $f_{Traffic\ INT}(\mathbf{P}_b)$ is the function designed to penalize the traffic interference within the bin and $\mathbf{P}_b$ is the vector of signals representing the received power of pilots present in the bin.

Table 4.3 Typical UMTS RSCP thresholds for different environments

Environment	RSCP threshold [dBm]
Dense urban	−80
Urban	−84
Suburban	−90
Rural	−98

Penalty for Lack of Coverage

To establish and maintain a call within any given geographic location, the RSCP of the pilot channel needs to be above a certain threshold, which, as in GSM, will be different for different morphologies, and the RF planning algorithm assumes that a bin meets the coverage requirements if the RSCP of the strongest server is above the threshold value specified for the bin's morphology. Typical RSCP thresholds for different environmental types are presented in Table 4.3.

Pilot Interference Penalty

To correctly decode the pilot signal within a certain bin, the pilot Ec/Io of the serving cell in the bin needs to be sufficiently high, where pilot Ec/Io is the quality indicator of the pilot, which is also equal to the ratio between pilot RSCP and Received Signal Strength Indicator (RSSI) [26]. Typically, within each bin, pilot signals from several cells can be measured and, during the planning process, it is usually assumed that the main serving cell is the one with the highest pilot RSCP. The rest of the cells represent interferers and their signal degrades the quality of service unless they are part of the SHO Active Set. The pilot interference cost component is therefore used to penalize for pilot interference or pilot pollution.

Traffic Interference Penalty

In UMTS, another component is required in the cost function in order to take into account Eb/No of the traffic channels. This component is referred to as the traffic interference penalty and serves to ensure that a call is maintained with sufficient quality [26].

For both Ec/Io and Eb/No, it is essential that the components Io and No are updated properly during the optimization process. As the base station configuration is updated, the traffic is moved from one sector to another and/or the SHO Active Set changes. Moreover, the Io and No terms change in most pixels, and this needs to be properly captured by means of simple mathematical models that allow fast and acceptable estimation of how these variables will evolve when the RF environment changes.

UMTS Sector Cost Function Component

To a larger extent than in GSM, capacity plays a very important part in the RF planning process in UMTS/CDMA networks. In CDMA-based networks, capacity refers to DL and UL capacity. For each individual sector, the format of the sector component of the cost function is given as:

$$C_s = f_{CAP}^{DL}(s) + f_{CAP}^{UL}(s),$$

(4.5)

which represents the sum of DL and UL capacity components. The DL capacity component is defined on a per-sector basis, as a function of the power distribution at the base station:

$$f_{CAP}^{DL}\left(s\right) = \max(\frac{P_T(s) - P_T^{Max}(s)}{P_T^{Max}(s)}, 0), \tag{4.6}$$

where $P_T^{Max}(s)$ is the maximum total power of sector s and $P_T(s)$ is the total power currently transmitted at sector s. The UL capacity is a function of the Noise Rise at the base station. Noise Rise is defined as the total wideband received power divided by the thermal noise power. The UL capacity is defined as:

$$f_{CAP}^{UL}\left(s\right) = \max(\frac{NR(s) - NR^{Max}(s)}{NR^{Max}(s)}, 0), \tag{4.7}$$

where $NR^{Max}(s)$ is the maximum allowed UL Noise Rise in sector s and $NR(s)$ is the estimated UL Noise Rise in that sector.

Modeling of HSPA
HSPA is a set of protocols that extends and improves the performance of the UMTS network. It has two components, HSDPA (High Speed Packet Downlink Access) and HSUPA (High Speed Packet Uplink Access). The RF planning model can take HSDPA into account by considering the utilization of the HSDPA physical channels in the calculation of Io, which will impact the pilot and the traffic interference penalties of the bin costs. In addition, a new sector capacity component can be added to the DL sector capacity described in Equation (4.6) as follows:

$$f_{CAP}^{DL}\left(s\right) = \max(\frac{P_T(s) - P_T^{Max}(s)}{P_T^{Max}(s)}, 0) + \max(\frac{P_{HS-PDSCH}(s) - P_{HS-PDSCH}^{Max}(s)}{P_{HS-PDSCH}^{Max}(s)}, 0), \tag{4.8}$$

where $P_T^{Max}(s)$ is the maximum total power of sector s excluding the control and HSDPA channels, $P_T(s)$ is the total non-HSDPA power currently transmitted at sector s, $P_{HS-PDSCH}^{Max}(s)$ is the maximum power allocated for data in the High Speed-Physical Downlink Shared Channel (HS-PDSCH) and $P_{HS-PDSCH}(s)$ is the estimated HS-PDSCH power of that sector.

4.8.4.3.   RF Planning for LTE Networks

For LTE RF planning, the main technology-specific characteristics to consider are the following: (i) the use of OFDMA in the DL; (ii) the use of Single Carrier FDMA (SC-FDMA) in the UL; (iii) the use of Multiple Input Multiple Output (MIMO) antenna schemes to improve throughput; (iv) the potential application of interference coordination methodologies; and (v) the support for Frequency Division Duplex (FDD) and Time Division Duplex (TDD).

Bin Cost Function Component for LTE
For each individual bin, the format of the cost function is given as:

$$C_b = f(t_b)[f_{COV}(\mathbf{P}_b) + f_{Pilot\ INT}(\mathbf{P}_b) + f_{Traffic\ INT}(\mathbf{P}_b)], \tag{4.9}$$

where t_b is the amount of traffic served in bin b and $f(t_b)$ is a scaling factor associated with that traffic, $f_{COV}(\mathbf{P}_b)$ is the function designed to penalize lack of coverage within the bin, $f_{Pilot\ INT}(\mathbf{P}_b)$

Table 4.4   Typical LTE RSRP thresholds for different environments

Environment	RSRP threshold [dBm]
Dense urban	−84
Urban	−89
Suburban	−94
Rural	−108

is the function designed to penalize the lack of pilot quality represented by the Reference Signal Received Quality (RSRQ) within the bin, $f_{Traffic\ INT}(\mathbf{P}_b)$ is the function designed to penalize the traffic interference which represents lack of SNR to transmit data or insufficient throughput within the bin and $\mathbf{P}_b$ is the vector of signals representing the received power of pilots present in the bin.

Penalty for Lack of Coverage
In LTE, RSRP is a standard metric that captures the strength of the pilot signals received in the DL (it is conceptually equivalent to RSCP in UMTS). At every pixel, it is defined as the average DL received power on the resource elements that carry cell-specific Reference Signals (RSs) from the serving sector within the considered bandwidth [27]. To establish and maintain a call within any given geographic location, RSRP needs to be above a certain predefined threshold that is morphology dependant and, as in other technologies, it is assumed that a bin meets the coverage requirements if the RSRP of the strongest server is above the threshold value specified for the bin's morphology. Typical RSRP thresholds for different environmental types are presented in Table 4.4.

RSRQ Interference Penalty
In LTE, RSRQ is a standard metric that captures the quality of the pilot signals received in the DL (it is conceptually equivalent to *Ec/Io* in UMTS). At every pixel, it is defined as the ratio between RSRP and the wideband received signals from all base stations in the carrier bandwidth plus thermal noise [27].

To maintain the target quality within a bin, RSRQ of the serving pilot in the bin needs to be sufficiently high. During the planning process, it is assumed that, at every bin, the serving cell is the one with the highest RSRP.

Traffic Interference Penalty
In addition to the above KPIs, which will determine whether a mobile is able to get access to the LTE network, there is an additional KPI that has to be optimized. It determines the quality of the connection and the potential throughput achievable by the mobile unit. The coding type and rate should be computed or identified as a function of SNR. Improving and optimizing SNR is paramount and should be one of the objectives of the planning process. SNR is different from RSRQ, since it is impacted by the actual interference between sectors, taking into account the OFDM sector resource allocation, whereas RSRQ is affected by all the interference across the entire bandwidth of the utilized channel. To maintain the requested throughput throughout the network, a sufficiently high SNR needs to be achieved.

Moreover, the optimization needs to consider ICIC (if enabled). Due to the orthogonality provided by OFDM, at least in the ideal case, there should be no interference between transmissions within the same cell but only interference between different cells, especially for users at cell edge. Therefore, in order to reduce or control intercell interference and provide substantial benefits to LTE performance, particularly for the users at cell edge, ICIC mechanisms may be implemented, which wisely allocate fewer (and selected) resources (bandwidth) to users in the cell edge, which are accountable for the biggest share of interference in the network. If such functionality is active in the network, the Self-Planning function needs to take the impact of ICIC into account, i.e. its impact should be modeled when computing DL and UL SNR, which are then mapped to the potentially achievable throughput in both directions.

Apart from ICIC, there are other elements affecting the translation of SNR to throughput (for a given number of resource blocks). Among them, it is worth highlighting the performance of the applied MIMO schemes, which are translated into a modification in SNR. Note that, in the case of MIMO, the particularities of such system make it advisable to distinguish between those pixels with Line-of-Sight (LOS) from/to their serving antennas and those ones with Non-Line-of-Sight (NLOS), since the impact of MIMO will significantly differ in both cases.

LTE Sector Cost Function Component

As in GSM and UMTS systems, capacity is a critical component to optimize LTE. When running the network, the OFDM time-frequency resources at the base station have to be distributed among the served mobiles using a scheduler, typically a variation of the Round-Robin or Proportional Fair algorithms [27]. When the system becomes congested, some users will experience lower throughput than required, and these phenomena need to be taken into consideration when searching for the optimum system configuration. As in UMTS, the LTE capacity has a DL and UL component:

$$C_s = f_{CAP}^{DL}(s) + f_{CAP}^{UL}(s), \qquad (4.10)$$

and the DL and UL capacity components are defined as a function of the DL and UL utilization of the sector. Utilization is defined as the ratio between the average number of used RBs and the total number of available ones.

$$f_{CAP}^{DL}(s) = \max(\frac{U_{DL}(s) - U_{DL}^{Max}(s)}{U_{DL}^{Max}(s)}, 0), \qquad (4.11)$$

$$f_{CAP}^{UL}(s) = \max(\frac{U_{UL}(s) - U_{UL}^{Max}(s)}{U_{UL}^{Max}(s)}, 0), \qquad (4.12)$$

where $U_{DL}^{Max}(s)$ and $U_{UL}^{Max}(s)$ are the maximum total utilization of sector s for the DL and for the UL, respectively, and $U_{DL}(s)$ and $U_{UL}(s)$ are the total utilization of sector s for the DL and for the UL, respectively.

4.9.   Automated Neighbor Planning

The ability to maintain voice calls or packet sessions when moving from the coverage area of one cell to another has been one of the most fundamental enabling features of mobile communication networks. This feature, generally known as Handover (HO) or handoff, is transparent to the user, who remains unaware of serving cell changes during the call. Regardless of technology, the HO procedure is driven by mobile station measurements of received signal levels from candidate and serving cells, as well as optional cell traffic measurements from base stations. These measurements are processed, for example, in the BSC in GSM and the RNC in UMTS to ensure that mobile terminals are always served by cells offering the best possible quality and to ensure overall network efficiency. For each radio cell, the list of candidate HO counterparts to be measured is referred to as the neighbor (or adjacency) list. Since HO in GSM and UMTS is limited only to neighbor cells, the list must be wide enough to include all potential overlapping serving cells because dropped calls are frequently linked to missing neighbors. At the same time, the neighbor list should not be too long in order not to unnecessarily overload signaling trunks used to transport HO-related measurements. It means that cells with low probability of overlap should not be part of the neighbor list, but credible candidates must not be overlooked to avoid lack of HO paths when they are needed to improve quality or rescue a call from dropping.

The algorithms to process neighbor measurements and make HO decisions are proprietary to equipment vendors and are based on signal level thresholds and various hysteresis and timers [28, 29]. In GSM and UMTS, neighbor lists are typically limited to 32 neighbors with minor differences among equipment vendors (for example, recent extension to 64 neighbors of the same technology). In general, if no iRAT neighbors are defined, extension to 64 cosystem neighbors is supported by most vendors. If at least one neighbor from a different radio technology is defined, 32 cosystem neighbors are allowed. In other words, if a GSM cell has one or more UMTS neighbors, it cannot have more than 32 GSM neighbors defined.

While extending neighbor lists to the maximum of 32 (or 64) elements may appear as a robust and safe approach to minimize the danger of missing neighbors, such practice would affect the accuracy of the HO process due to the fact that shorter time would be dedicated to measuring and identifying each neighbor. To achieve completeness while avoiding excessively long neighbor lists, neighbors must be carefully chosen for each cell. In the past, before initial cell or cluster launch, neighbor lists have been created based on either proximity criteria or notoriously inaccurate propagation loss predictions. Following the launch, neighbor lists are tuned in a tedious manual process mostly based on drive test measurements. This approach is time consuming and depends heavily on the engineer's experience and capabilities.

Recently, automatic neighbor list generation and optimization methods have been formulated [30, 31]. These methods rely on measurements to detect both unnecessary and missing neighbors. Relying on measurements with properly defined criteria and thresholds eliminates the impact of varying engineers' experience and fits perfectly into the SON paradigm [1]. The following subsections will discuss various aspects of automatic neighbor list generation, which is becoming an important part of automated cell planning and optimization with the general goal of having as complete neighbor lists as possible, but at same time, as short as possible as well.

4.9.1. Technology-Specific Aspects of Neighbor Lists

4.9.1.1. Neighbor Lists in GSM

GSM neighbor lists can contain contributions from three domains:

1. Intraband: neighbors from the same band, for example 900 MHz/900 MHz or 1800 MHz/1800 MHz.
2. Interband: neighbors from different bands, for example 900 MHz/1800 MHz or 1800 MHz/ 900 MHz.
3. iRAT: handovers from GSM to UMTS.

Non-iRAT GSM neighbor lists contain BCCH frequencies to be scanned by the mobile terminals during idle timeslots and reported back to the BSC for HO decisions. The mobile's reporting role is commonly associated with the Mobile Assisted HandOver (MAHO). As mentioned, GSM neighbor lists should be complete enough to avoid missing important neighbors, potentially leading to dropped calls. At the same time, they should only include necessary BCCHs in order to avoid difficulties in BSIC decoding and slow HO execution. The number of GSM neighbor BCCH frequencies is typically limited to 32 (or 64 if no UMTS cells are configured as iRAT neighbors). If a cell does not require all 32 BCCH slots, free ones can be temporarily assigned to so-called dummy (non-HO) neighbors. Adding dummy neighbors creates an extended neighbor list where HOs to these dummy neighbors are disabled by setting prohibitive thresholds, but valuable mobile measurements on these BCCH frequencies are still reported back to the BSC despite disabled HOs. This neighbor list extension will temporarily increase signaling and may affect HO execution but, in practice, the impact has been found to be minimal and justified by the benefit of reliable data for automated network planning and audit. Besides network audit and neighbor planning, such extended neighbor measurements are used for Interference Matrix (IM) generation, thereby providing a valuable input for accurate automatic frequency planning, which at the same time can be used to calibrate coverage predictions and feed automatic antenna optimization algorithms.

If a network contains less than 32 BCCH frequencies, each cell can include all of them in its extended neighbor list. This enables accurate and complete estimation of interference in the network. An unavoidable limitation is the standard GSM characteristic by which only the six strongest BCCH frequencies are reported approximately every 0.48 s [32], so the potential for missed interference still remains, especially in dense site clusters with more than six strong overlapping cells. If a network deploys more than 32 BCCHs, dummy neighbor frequencies can be rotated in available slots or intelligent scan lists can be generated. Intelligent scan lists include the most likely interfering frequencies and are an important part of automated Self-Optimization algorithms because they ensure that interference measurements are concentrated on the most likely sources. These extended GSM neighbor list measurements are used in automated optimization, ranging from frequency and neighbor planning and optimization to sophisticated OSS-based automatic cell planning. However, the collection of extended neighbor list measurements should be limited to time periods that are just long enough to gather sufficient amount of data in order to prevent negative effect on HO performance when stretching the sizes of neighbor lists. Details and complexity of the extended neighbor list setup vary widely among GSM equipment vendors.

In addition to 32 GSM neighbors (that can include frequencies from multiple bands, for example 900 MHz and 1800 MHz), a GSM cell can also configure up to 32 UMTS cells for HO. UMTS neighbors of a GSM cell are configured through carrier center frequency and PSC. To monitor these neighbors, dual-mode mobile terminals must change their modulators and scan a different bandwidth. This is enabled by retuning to preferred UMTS carrier center frequency and increasing the receiver's pass-band from 200 kHz to 5 MHz during idle timeslots.

Although the impact of sub-optimal GSM neighbor lists on network performance can be somewhat hidden by good cell and frequency planning, it is always recommended to also properly select neighbor lists. Radio and frequency planning cannot overcome all neighbor problems (for example, missing neighbors), and proper neighbor lists solve problems at the source rather than patching them elsewhere.

4.9.1.2. Neighbor Lists in UMTS

UMTS neighbor lists may contain up to 96 neighbors from three distinct domains:

1. Intrafrequency (maximum 32 neighbors at the same UMTS carrier).
2. Interfrequency (maximum 32 neighbors at different UMTS carriers).
3. iRAT (maximum 32 GSM neighbors).

Most neighbors contributors are typically not at different carrier frequencies, but rather correspond to intrafrequency neighbor cells identified by PSCs reported to the User Equipment (UE) through the Primary Common Control Physical Channel (PCCPCH). As a fixed-power code channel in UMTS, CPICH serves as the reference for HO measurements and most automated cell planning and optimization tasks. Allocation of 512 available PSCs will be discussed in Section 4.11.

As in GSM, all UMTS HOs are MAHOs, with mobile terminals reporting received CPICH levels. However, reporting is not continuous by default as in GSM, but triggered when certain indicators exceed some configured thresholds [33], unless periodic reporting is activated. Moreover, unlike GSM, the WCDMA air interface, around which UMTS has been built, allows UEs to be simultaneously served by multiple cells within the same carrier. This is commonly referred to as Intrafrequency or Soft/Softer HO. Soft refers to cells from multiple NodeBs and Softer indicates cells from the same NodeB serving the call. The main benefit of SHO is the seamless (make-before-break) nature of the HO, where the connection is first established with the new cell and then, after a transient period with multiple servers, broken with the original server. Another benefit is increased Signal-to-Interference-Ratio (SIR) at the cell edge where multiple cells contribute to the service and thus may preserve a call that would otherwise drop or suffer from bad quality. Cells involved in Soft or Softer HO are referred to as the Active Set. Each time a UE updates its Active Set, it receives a new neighbor list aggregated from neighbors of each cell in the set. The neighbor list in UMTS is referred to as the Monitored Set. All remaining cells whose PSCs are detected by the UE on a best-effort basis are sorted into the so-called Detected Set and can be used to automatically adjust neighbor lists in case a strong pilot enters this set.

To be considered as an Active Set candidate, the cell's PSC must be part of the Monitored Set. The UE frequently scans the CPICHs from the Monitored Set and reports their quality back to UTRAN when certain conditions are met (or when periodic reporting is activated). Based on the mobile's measurements, UTRAN decides the HO combination for the mobile. However, as opposed to GSM, where these measurements are continuous with 0.48 s period, in UMTS reporting of neighbor measurements is event-based and is triggered when a neighbor's received pilot level or quality drops below a certain threshold [34]. UMTS, as already hinted, also supports periodic measurements and, to do so, UEs must be specifically instructed to provide such reports. Details on the scope and activation of these periodic measurements depend on the equipment vendor. For example, some UMTS vendors only allow collection of periodic neighbor reports through tracing specifically targeted UEs.

Problems occur when the neighbor list is incomplete and some of the strongest servers are not included in it. Since the UE only scans pilots from the neighbor list very frequently, pilots outside the list will rarely be scanned and not all the necessary MAHO measurements will be provided back to UTRAN. However, those signals are still present and, since they are not used within the Active Set, they will become interferers. The situation may become worse if the pilot outside the neighbor list becomes the strongest signal. The created interference may cause the UE to drop the call. After the call is dropped, the UE subsequently reacquires the connection and this time is served by the previously interfering cell. This prolongs the problem since the pilot that was serving the UE before the drop now acts as an interferer. Therefore, UMTS is vulnerable to erroneous neighbor lists.

Based on the above scenario, forming large neighbor lists may be tempting in order to include all possible PSCs present in an area. Such approach has negative implications, since the UE scans all neighbors from the Monitored Set in sequence and, if the list is long, the measurements of each individual CPICH are less frequent and hence the HO process is slowed down. In urban and dense urban environments, where the channel changes rapidly, this may lead to serious performance issues.

Another important consideration is the inclusion of GSM BCCH channels in the iRAT neighbor list of a UMTS cell. Upon command from the RNC, UMTS UEs enter the so-called compressed mode and temporarily disable WCDMA reception to scan BCCH frequencies included in the iRAT neighbor list [26]. These frequencies (channels) must be added carefully because entering compressed mode to measure GSM BCCHs injects additional interference in the UMTS network and may lead to throughput degradation if used excessively [31].

4.9.1.3. Neighbor Lists in LTE

While LTE brings significant enhancements over existing technologies, it also ensures coexistence with existing infrastructure by fully standardizing HOs with GSM, UMTS and even 3GPP2 CDMA2000 cells. The potential of LTE standardized HOs with legacy technologies will preserve enormous investment and allow a gradual migration path from earlier generations of wireless access standards. In addition to fully standardized iRAT HOs to legacy 2G and 3G technologies, LTE introduces significant enhancements into neighbor list maintenance and monitoring, both within LTE and between LTE and 2G/3G. These changes were motivated by the need to avoid the complexity and difficulties associated with maintaining neighbor lists in GSM and UMTS networks. To overcome these difficulties, LTE can even operate without predefined neighbor lists. This advanced mode is accommodated through the Automatic Neighbor Relation (ANR) function, where neighbor lists are dynamically created with cells detected from UE reports. LTE neighbor candidates are evaluated at the UE based on RSRP and RSRQ measurements with 200 ms periodicity. Measurements include serving cell strength and quality as well as all detected PCIs and corresponding signal levels. Thus, without a predefined neighbor list, after the UE detects the PCI of a suitable neighbor candidate as part of ANR, the Cell Global Identifier (CGI) of the candidate cell must be decoded. Once CGI is decoded and known, the Transport Network Layer (TNL) is set up and the X2 interface is established for handover to the newly detected neighbor. ANR is closely related to another LTE SON feature: the automatic Self-Optimization of PCIs. PCIs can be updated following automatic neighbor updates so that PCIs are adapted to the detected neighbors in order to avoid collisions. ANR is one of the most attractive standardized SON features in LTE (and indeed the entire cellular industry until now) and will be discussed more in Chapter 5.

ANR is an optional alternative to classic preconfigured neighbor lists that are still supported in LTE. In addition to such classic (and static) neighbor lists, LTE maintains two alternative cell lists [35]:

- White List: contains cells considered for reselection/HOs.
- Black List: contains cells that will not be considered for reselection/HOs.

Information about LTE neighbors is delivered as part of System Information Blocks (SIBs) carried by the BCCH in LTE. Relevant blocks containing neighbor list information are SIB types 4-8:

- SIB 4: information about intrafrequency LTE neighbors.
- SIB 5: information about interfrequency LTE neighbors.
- SIB 6: information about iRAT UMTS neighbors.
- SIB 7: information about iRAT GSM neighbors.
- SIB 8: information about iRAT 3GPP2 (CDMA2000) neighbors.

This list illustrates the rich support for HO from LTE to all legacy technologies. Regardless of the neighbor technology and frequency, neighbor cell measurements are based on the same principles. Since LTE UL transmission is not continuous, compressed mode is not necessary and other frequencies and radio technologies may be measured in idle timeslots. These neighbor measurements are not periodic but triggered by one of seven defined triggering events: five intra-LTE monitoring triggers (events A1-A5) and two iRAT monitoring triggers (events B1-B2):

- A1: serving cell becomes better than an absolute threshold.
- A2: serving cell becomes worse than an absolute threshold.
- A3: neighbor becomes better than serving cell by an offset.
- A4: neighbor becomes better than an absolute threshold.
- A5: serving cell becomes worse than an absolute threshold 1 and neighbor cell becomes better than an absolute threshold 2.
- B1: iRAT neighbor becomes better than an absolute threshold.
- B2: serving cell becomes worse than an absolute threshold 1 and iRAT neighbor becomes better than an absolute threshold 2.

Thresholds for all these events, together with corresponding timers, are inputs to the HO algorithm. If any of the triggering criteria is satisfied for sufficient time (exceeding a timer), periodic neighbor reporting is activated. Neighbors are then periodically monitored until either the serving cell quality improves beyond the triggering thresholds or HO is executed. If the serving cell quality improves, periodic neighbor monitoring is terminated.

4.9.2.   Principles of Automated Neighbor List Planning

As in every other aspect of automated cell planning and optimization, neighbor list tuning is most powerful when based on measurements. However, measurements are not available for greenfield designs or new sites/clusters being added to an existing network. Excluding trivial proximity criteria, propagation predictions are still the only available input to identify the cells that are suitable for inclusion into the neighbor list of cells that are not active yet. Typically, predicted overlapping coverage is used to generate neighbor lists and to provide ranking and weighting of potential neighbors. The following algorithm illustrates the neighbor list generation for a given cell s using predicted overlapping coverage:

- Perform the following steps for all potentially interfering cells associated with serving cell s:
    - For interfering cell v, set counter $C(v)=0$.
        - Perform the following step for all bins (pixels) in the coverage area of serving cell s:
            - For bin k: if $((C/I)$ at bin $k \leq \mathrm{Th}_{dB})$, $C(v)=C(v)+1$,

where C denotes received signal level from cell s and I denotes received signal level from cell v. Th_{dB} is a C/I threshold used to determine strong neighbor candidates. A reasonable choice of Th_{dB}, based on field experience, is 6 dB.

The output of the above algorithm is a set of counters $\{C(v), v=1, 2,..., N\}$, where N is the number of interfering cells, and each counter representing the number of bins at which the received signal from the corresponding interferer (neighbor candidate) was observed within a predefined threshold from the serving signal. Those counters, along with the total number of bins in the coverage area of serving cell s, can be used to compute the percentage of area overlap for all neighbor candidates and to rank them accordingly. An alternative method is to compute the amount of traffic in the bins where the received signal from the corresponding neighbor candidate was observed within Th_{dB}. In this case, a percentage of traffic overlap can be computed.

The main drawback of using propagation predictions is the low accuracy of the available computation models, especially in dense urban environments, which jeopardizes the quality of prediction-based neighbor lists generation. This can be enhanced by automatically tuning propagation models based on drive measurements or geo-located mobile measurements around the new site. With these approximate methods, propagation models can be derived from drive/mobile measurements and extrapolated to new cells ready to be activated in a particular morphology or area type.

In addition to predicted overlapping coverage, the following constraints are typically used to generate neighbor lists:

- Minimum predicted serving signal level to be considered in the computation area: only geographical areas[5] where signals from both the serving cell and its potential neighbors exceed a predefined threshold are considered for evaluating the percentage of overlap.
- Minimum percentage of overlapping served area/traffic required to qualify as a neighbor: coverage footprint of each serving cell is compared with those of all potential neighbors using the previously described algorithm. If the percentage of area or traffic overlap exceeds a predefined threshold, the potential neighbor is preliminary selected.
- Maximum number of neighbors per cell: any number of cells can qualify as neighbors depending on the site density and the selected overlapping criteria. For example, in dense site deployment scenarios, more than 32 neighbors may be selected according to the overlapping criteria. Therefore, it is necessary to rank all qualified candidate neighbors. Only up to M candidates will be finally proposed as neighbors, where M is the desired maximum size of the neighbor list.
- Other constraints such as maximum distance between a cell and its neighbors, enforcement policy for neighbor relationships between co-sited sectors, enforcement policy for neighbor reciprocity, etc.

An example of neighbor list planning based on overlapped predicted serving areas is shown in Figure 4.8. Referent serving cell is encircled and its proposed neighbors are highlighted in grey. This particular result was obtained by setting the minimum overlapping area level to 10% and the screenshot has been taken from a commercial RF planning tool.

After the new sites are turned on, mobile measurements and available OSS Performance Management (PM) and Configuration Management (CM) data can be used to optimize and tune the neighbor lists without using propagation predictions. Please refer to Section 5.7.2 for more details.

---

[5] If digital maps are used, geographical areas are represented by groups of pixels or bins.

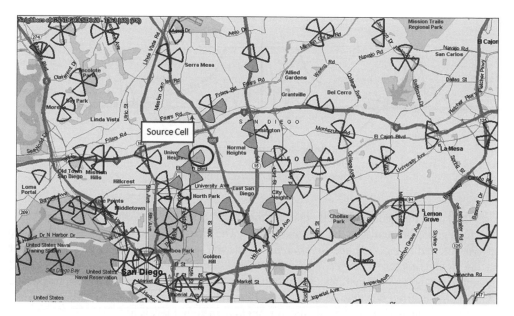

Figure 4.8 Example of neighbor list proposed by a commercial planning tool.

4.10. Automated Spectrum Planning for GSM/GPRS/EDGE

The function of frequency planning in wireless mobile communication systems is to divide the total number of available channels into subsets that can be assigned to radio cells [36]. Since valuable and limited portions of frequency spectrum are allocated to operators to build their mobile networks, channels must be reused among cells. Channel reuse introduces interference between co- and adjacent channels propagating into the same areas. With increased density and traffic demands of modern cellular networks, coverage overlap among cells grows, making frequency planning more challenging. Although dynamic channel allocation provides significantly more flexibility, current GSM/GPRS/EDGE systems support fixed channel assignments. Therefore, each GSM cell operates with a fixed set of preassigned nonoverlapping 200 kHz channels. The frequency planning process is responsible for assigning channels to GSM cells in such a way that guarantees certain availability and QoS. This means that the probability of finding an idle channel for an incoming call request must be sufficiently high (i.e. enough channels must be assigned to the cell). Simultaneously, the probability of interference between cells using the same or adjacent frequencies must be sufficiently low (i.e. assigned channels must be clear). In practice, compatibility constraints that determine whether two radio cells can or cannot deploy the same or adjacent channels are the RF propagation characteristics of a particular environment, the spatial separation between both transmitters, the orientation (azimuth) of the antennas and the system hardware properties. On the other hand, traffic demand determines the number of channels that need to be assigned to a particular radio cell. This section will not discuss the capacity aspects of channel planning and will concentrate on defining the principles of allocating an already defined number of channels per cell.

Reuse of frequencies is fundamental to the concept of cellular communications. To achieve high system capacity, frequencies have to be reused as tightly as possible. On the other hand,

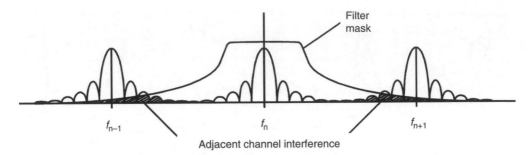

Figure 4.9   Adjacent channel interference.

using the same frequencies in cells that are too close to each other causes interference and tends to degrade the quality of the communication. The interference associated with frequency reuse is referred to as *cochannel interference* while the interference associated with channel spacing is called *adjacent channel interference*. Intrasystem interference (cochannel and adjacent channel) is the most significant capacity-limiting factor of GSM mobile communication systems.

Cochannel interference may be defined as undesirable signal energy attributed to the reuse of the same frequency. It can be calculated from known signal levels as:

$$I_{co-channel} = 10\log\left(\sum_{i=1}^{N} P_i\right)[dBm],$$
(4.13)

where P_i is the average received power of the ith cochannel cell (in milliwatts) and N is the number of cochannel cells.

Like cochannel interference, the adjacent channel interference represents a form of undesirable signal energy. It is attributed to the spillover from frequency components near the channel of interest. The radio receivers have tunable filters that reject a portion of the adjacent channel interference, but not all of it. The extent to which the receiver's filter is able to isolate the center frequency of interest from adjacent frequencies is referred to as the Adjacent Channel Rejection (ACR). This phenomenon is illustrated in Figure 4.9. Adjacent channel interference also occurs between neighboring portions of spectrum allocated to other operators or other technologies.

The task of frequency planning is categorized in mathematical theory as a combinatorial optimization problem where channels must be assigned without violating any imposed constraints. This makes the frequency planning problem closely related to the classic graph coloring problem known to be NP-complete. In essence, NP-completeness means that the complexity of the problem grows exponentially with its dimension without any closed-form method to obtain the optimal solution. In such cases, finding the exact optimal solution becomes practically impossible, even for problems of a moderate size. In existing literature, several methods have been proposed as viable solutions to the frequency planning problem. The exploited ideas encompass the classical graph coloring approach [37, 38], linear programming [39], neural networks [40, 41, 42], SA [43, 44], genetic computing [45, 46] and various heuristic methods [47, 48]. Among published work, SA methods are the most popular.

Since mid-1990s, these methods have been applied in various forms in commercial software solutions from vendors providing Automatic Frequency Planning (AFP) modules and services to operators. AFP is a crucial component of GSM/GPRS/EDGE Self-Planning.

4.10.1. Spectrum Planning Objectives

Frequency planning can be defined as a set of operations and procedures performed to efficiently allocate available channels to the radio transmitters in a certain network. In its fundamental form, the immediate objectives of GSM/GPRS/EDGE frequency planning are:

- Defining frequency planning strategies (such as mixed or segregated BCCH and TCH spectrum blocks), traffic allocation priority, Mobile Allocation List (MAL) size, etc.
- Assigning BCCH and TCH frequencies to control and traffic TRXs, respectively.
- Assigning BSIC values to all cells.
- Assigning HSN to hopping cells (both baseband- and synthesized-hopping cells).
- Assigning MAIO to synthesized-hopping cells.

The overall goal of these spectrum-planning deliverables is to ensure satisfactory KPI values. Such KPIs, together with CM data, can be used as inputs to the automated spectrum planning algorithms (i.e. the Self-Planning or Self-Optimizing algorithms). The main GSM KPIs affected by the spectrum plan are:

- DL and UL Received signal Quality (RxQual): RxQual is a quantized mapping of the raw channel Bit Error Rate (BER), and ranges between 0 (best quality, low BER) and 7 (worst quality, high BER) [32]. When monitoring performance, particular focus is given to RxQual5-7. Low RxQual5-7 percentage indicates a good frequency plan.
- Quality HO percentage: a quality HO occurs when RxQual is poor despite strong signal level (high RxLev). This is one of the best interference indicators. Reduction of the quality HO percentage is one of the primary indicators of successful spectrum planning. The quality HO percentage can be defined as the ratio between number of DL quality HOs and the total number of HO attempts.
- Dropped call rate: percentage of dropped calls relative to successful call establishments. The dropped call rate is one of the primary user satisfaction metrics, and may be driven by interference or other problems. Proper spectrum planning can reduce the number of dropped calls due to interference.
- Frame Erasure Rate (FER): percentage of erased 20 ms speech frames based on the outcome of Cyclic Redundancy Check (CRC) [32]. This KPI indicates user quality experience more accurately than raw RxQual, but is typically not available on the DL because reporting has been shown to impact the performance of the call. It is available on the UL where collection has no negative impact on performance.
- Standalone Dedicated Control Channel (SDCCH) dropped call rate: an important KPI describing system accessibility in GSM. Any interference among channels carrying SDCCH will reflect in higher SDCCH drop rate.
- Radio Link Control (RLC) layer throughput: average GPRS/EDGE throughput per timeslot during active transmissions (i.e. excluding queuing). This KPI represents the average RLC throughput based on the selected coding schemes and thus reflects the interference on the channels carrying traffic. Note the difference between RLC throughput and Link Level Control (LLC) layer throughput. LLC also includes delays due to lack of resources and thus combines both RF quality effects and queuing.

- Coding scheme distribution: GPRS uses coding schemes CS1 to CS4 and EGPRS uses modulation and coding schemes MCS1 to MCS9 [32]. Higher coding schemes lead to higher RLC throughput per timeslot. Therefore, a low percentage of samples with high coding scheme indicates interference on (E)GPRS channels and inadequate frequency plan.

Spectrum planning is typically triggered by one of the following:

- Periodic optimization of the network to reduce interference and improve performance.
- Ad-hoc optimization triggered by the observation of degradation in one or more of the aforementioned KPIs.
- Periodic maintenance to plan for traffic growth (addressing transceiver unit additions and integration of new sites).
- Spectrum addition, e.g. acquiring and activating new channels.
- Spectrum carve/reduction, e.g. migration from GSM to UMTS.

As illustrated when introducing the typical triggers for spectrum planning activities, frequency plans are dynamic. They are changed to allow for new base stations to satisfy additional spectrum requirements in heavy traffic areas, to eliminate problems resulting from unacceptable interference levels, or even to accommodate for seasonal changes in the propagation environment. A good frequency plan will increase capacity and improve QoS. Depending on the stage in the lifecycle of the sectors under consideration, different approaches for determining the optimal spectrum allocation will be appropriate and feasible. Before the commercial launch of a sector, OSS statistics are typically unavailable unless the network has been partially launched for friendly customers. Therefore, this case falls within the Self-Planning scope and the optimization engine will be naturally fed with an IM based on propagation predictions. However, when the sectors are fully operational, a much more accurate IM can be build with OSS statistics, thereby entering the Self-Optimization domain. Although throughout this book Self-Planning is covered in Chapter 4 and Self-Optimization is described in Chapter 5, for the sake of clarity, all the aforementioned aspects of frequency planning and optimization have been addressed together in Chapter 4 in a combined, integrated manner.

4.10.2.   Inputs to Spectrum Planning

Regardless of the approach to frequency planning, the IM is the most critical input. The role of the IM is to estimate the *C/I* relationship between each pair of sectors in a system if they were to use the same channel. *C/I* between any two cells varies widely depending on the observing location, from high values close to the server to low values further from the server and closer to the interferer. Thus, the values populating an IM may be averages, medians or any other relevant percentiles (fifth, tenth, etc.) of all *C/I* observations. For every location bin where the source transmitter is the best server, the ratio between the received signal level from the source transmitter and that from the interferer is recorded. This results in a set of *C/I* values from every bin in the source transmitter serving area. The histogram is calculated from individual bin values and IM entries are reported as desired (fifth or any other percentile) as illustrated in Figure 4.10.

In a network with *N* cells, the IM is an *N-by-N* matrix that contains interference relationship values for every active sector into every other active sector (see the example in Table 4.5). Serving cells are lined as matrix rows and again in columns when assuming the role of an

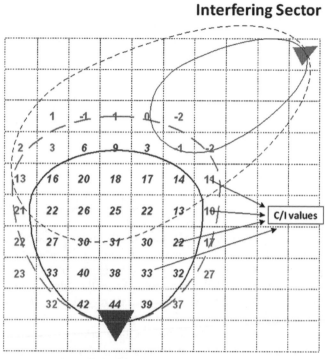

Serving Sector

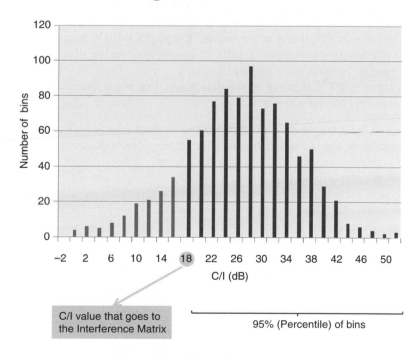

Figure 4.10   IM calculation principles.

Table 4.5   Example of an Interference Matrix

Sector	AA	BB	CC	DD	EE	FF	GG	HH	II	JJ
AA	×	4	127	5	−2	−4	14	32	10	48
BB	13	×	−2	−1	−4	2	23	12	−7	18
CC	33	−1	×	7	15	19	19	−2	−2	4
DD	6	1	14	×	26	−5	−5	6	32	11
EE	−6	3	28	19	×	11	45	38	−5	44
FF	−6	3	37	−4	12	×	6	17	27	34
GG	14	7	25	−6	35	5	×	−2	35	12
HH	45	6	−2	5	29	16	−1	×	22	3
II	11	2	5	19	−4	21	35	23	×	32
JJ	43	4	−3	−3	28	18	13	−5	24	×

interferer. The value of every entry in the IM represents the interference relationship value, i.e. a *C/I* percentile value (in dB) between the two sectors intersecting at that point in the matrix. Thus, for example in Table 4.5, when cell AA is serving, it will be on average (if the IM reflects average values) 4 dB above BB if they are to reuse the same channel. That is insufficient for acceptable performance in GSM. Cells with no recorded overlap may be assigned a high value (100, for example), indicating safe or penalty-free reuse. The diagonal of an IM is obviously not relevant as it represents the serving cell's overlap with itself.

While single-valued IMs have been traditionally used in frequency planning, additional improvement can be achieved if algorithms are based on full histograms instead of single values. This means that instead of approximating *C/I* distributions and calculating median and percentiles based on the Gaussian distribution or other assumptions, the IM stores entire histograms rather than calculated single values. Keeping a full *C/I* histogram to calculate reuse penalties allows the avoidance of artificial assumptions on the distribution of widely-varying *C/I* random variables for each pair of cells.

Different types of IM, depending on the data source that is used for *C/I* calculation, and including any hybrid combination, are:

1. IMs based on local knowledge: this type is used in manual frequency planning based on market experience and is not suitable for large scale spectrum-planning. More details are provided in the next section.
2. IMs based on propagation predictions: propagation models are notoriously inaccurate, especially in dense urban environments. The typical standard deviation of the error may be as high as 10 dB. Therefore, these IMs are not suitable to develop reliable frequency plans with today's demands in high capacity areas.
3. IMs based on drive measurements: these IMs are more accurate than the ones based on predictions because signal levels from each cell are measured instead of predicted. Errors related to propagation model inaccuracy, model classification assignment for each cell, antenna patterns and site database errors are no longer a problem. Requirements for drive measurements are:
    o Drive test routes need to be defined to ensure that the entire serving area from each cell is measured and sufficient samples are collected.
    o Special scanners should be used to measure all (or most) interfering BCCH channels.

    o BSIC assignment for each cell should be optimized before the drive test in order to max-imize the co-BCCH-BSIC reuse distance and thus make cell identification more reliable. Drive test measurements provide more accurate inputs to generate an IM as compared with local knowledge and propagation predictions. However, drive testing is time consuming and RxLev measurements are limited to measured areas, thus not reflecting actual traffic patterns nor indoor propagation.

4. IMs based on Mobile Measurement Recordings (MMR): these measurements are collected by mobiles in active mode and report the difference between signals from the serving cell and its interferers[6], identified by means of their BCCH and BSIC. The signal difference (in dB) represents the C/I between server and interferer if they were to use the same frequency. The initial purpose of MMRs was to identify missing neighbors. Different GSM equipment vendors report MMRs differently, from full C/I histograms to counters describing the percentage of MMRs with C/I below selected thresholds.

Properly processed MMRs are the most powerful input to the automated spectrum planning/optimization process. The advantages of the MMR-based approach for creating the IM are the following:

- MMRs require no drive test time and resources.
- MMRs provide statistics that are representative of the entire network and are not biased towards the measured areas, i.e. interference inside buildings is available.
- The generated IM considers the actual serving area of each cell, taking into account Hierarchical Cell Layout (HCL) priorities, HO parameters that control cell boundaries and all other complexities.
- Data processing is limited to matching BCCH/BSIC to a cell. No additional processing of every sample is needed as in drive testing.
- MMRs take into account the actual location where users place calls and the actual signal levels the mobiles experience.

The entries from an IM are used in frequency planning to indicate several important factors:

- Identify channels that can be assigned to a particular transmitter without violating the specified co- and adjacent channel C/I requirements.
- Identify channels that, if assigned to the selected transmitter, would violate the specified co- and adjacent channel C/I requirements (conflicts).
- Show current violations of the specified Co- and adjacent channel C/I requirements.

Besides the IM, other factors also need to be considered, such as:

- Traffic: the number of required channels depends on the cell's traffic demand. Traffic is also used to weight the reuse so that cells carrying less traffic will be more likely to reuse channels than high-traffic cells. In this way, the overall impact of interference is lower.
- Neighbor-lists: cochannel BCCH frequencies cannot be assigned to neighbor cells.

---

[6]The top six interferers are reported in every measurement report every 0.48 seconds.

- Combiner constraints: maintaining minimum frequency separation between channels assigned to combined TRXs.
- Frequency separation between cells belonging to the same site, etc.

Satisfying these rules in order to derive acceptable spectrum plans without automatic frequency planning and optimization solutions is time consuming and prone to errors, especially in large networks.

4.10.3.   Automatic Frequency Planning

Manual frequency planning today is mostly an abandoned practice. The main basis of manual spectrum planning is the planner's skill and experience as to which cells may deploy co- and adjacent channels while delivering acceptable network performance. Due to the need for frequency expansions and periodic network-wide retunes (every 2-3 months) and due to the limitations of manual frequency planning, especially in terms of cost of resources (OPEX) and inability to generate high quality frequency plans, AFP solutions became commercially available in the 1990s. A major advantage of AFP tools over traditional manual methods are the superior quality and the significant reduction in the time to generate and implement a frequency plan. That allows the engineering personnel to be more efficient and enables them to be consistent in the frequency planning and network optimization strategy.

Spectrum planning belongs to a mathematical class of constrained combinatorial optimization problems. The goal of the optimization algorithm is to automatically plan frequencies satisfying design requirements without violating imposed constraints. A formal way to mathematically define frequency plan requirements is to create an objective cost function. A proper definition of the cost function is crucial for the success of the AFP routine. However, this is not an easy task since precise mathematical expressions for requirements, such as minimizing interference subject to combiner restrictions on minimum frequency spacing, are not obvious. Assuming the ability to mathematically postulate requirements and constraints, an AFP can use the concept of the composite penalty function based on the simple idea of penalizing every violation of the constraints posed before the channel planning routine. Whenever a constraint is violated, the value of the cost function is increased. In other words, the AFP algorithm is not prevented from violating the constraints, but every time it happens, the cost associated with that particular channel assignment is increased and the chances of that plan being accepted are lowered. Thus, the constraints are soft in the sense that the objective function does not prevent, but rather penalizes their violation. This way, the channel plan obtained at the end may not satisfy all imposed constraints, but the violated constraints do not seriously degrade the performance.

Objective functions are usually broken into local constraints. The cost of each constraint is then calculated separately and added. The objective function integrated from individual constraint penalties is called the aggregated cost function. Such formulation allows algorithm upgrades once additional requirements are postulated. However, the aggregated approach is computationally inefficient because of possible redundancy in the objective function definition. The total aggregated sum is repeatedly recalculated in an iterative loop for each tested system configuration. Iterations repeat until no constraints are violated (zero cost function) or when a preselected maximum number of iterations is reached. At that point, the current parameter set (frequency plan) is frozen and proposed for implementation.

An appealing approach to this problem might be to take all available channels and test all possible frequency plan combinations. For each configuration, the resulting *C/I* is calculated and the configuration with the maximum C/I is selected. This approach avoids the need for complicated objective functions, but is prohibitively slow for practical implementation. Therefore, the definition of the objective function is a critical element of an AFP solution. This definition must accommodate all design requirements and constraints imposed on a cellular system. A technically sound frequency planning solution should provide enough flexibility to accommodate for many design requirements. Some of them are essential from the standpoint of the service quality provided to the customer (interference avoidance, ability to serve offered traffic), while others are important from the standpoint of cellular system hardware restrictions (combiner constraints). Finally, some constraints express the preferences of the frequency planner in terms of following certain regularity in the channel assignment pattern. Furthermore, an AFP module should also allow the spectrum planning engineer to make decisions on which requirements are more critical in each particular situation.

In addition to the requirements, constraints must also be accommodated in the objective function. The quality of the automatically generated frequency plan still depends on the engineer's ability to prioritize those constraints and assign penalties for their violations. Typical constraints are related to frequency reuse (co- and adjacent channel limitations per cell/site), combiner characteristics, consideration of neighbor lists, permission matrix and frequency coordination.

AFP solutions are considered system optimization tools because they use the existing frequency plan as an input and try to generate a more efficient channel allocation plan through the minimization of the cost function. The accuracy and completeness of the input data (both for the planned area and the surrounding buffer zones) is a critical factor in preparing a frequency plan. Input data to the AFP process include:

- Site database: information about all cells in the network together with their existing frequency assignment.
- IM: this input is used by the optimization algorithm to score and evaluate candidate frequency assignments.
- Neighbor lists: the algorithm prevents the assignment of the same or adjacent channels to the neighbor cells.
- HO attempts: this input is used to perform a neighbor audit, i.e. identify neighbors with few HO attempts that can be deleted. In addition, this input also allows the algorithm to minimize cochannel reuse between neighbor cells with higher HO attempt count, which allows reduction of interference in the HO areas where the mobile is more vulnerable to drop a call or initiate a quality HO to a less dominant cell.
- Traffic: used to minimize total interfered traffic by penalizing reuses between cells carrying more traffic.
- Exceptions lists, e.g. separation requirements for different cell relations and illegal frequencies for certain sectors.

This input data is used to assign the constraint violation penalties. This step requires advanced engineering skills and experience to properly weight the relative importance of each cost function penalty. The model can be perfect, but without realistic and constructive penalties, the result will be less than adequate. Usually, default penalties (that work on average) can be assigned although they may not give the best possible results. Currently there is not a formal scientific approach for assigning penalties that would provide the optimal configuration and it is left to the RF planner to recommend the best settings. This seems like an arbitrary manual technique, but once these penalties are assigned for a given network, the process is fully automatic and well defined. Besides finite penalties assigned to soft constraints, hard constraints are identified as those that are prevented from being violated. A configuration with hard constraints that are not respected will not be considered.

4.10.4. Spectrum Self-Planning for GSM/GPRS/EDGE

Since one of the goals of SON is to minimize human involvement in the process of network planning and optimization, the GSM spectrum Self-Planning process needs to be designed along the same philosophy. Objectives of SON spectrum planning can be summarized as follows:

- Autonomously generate network-wide frequency plans to produce the best possible performance based on actual network statistics. Automation of the entire frequency planning process is very important especially when the focus of the operator is on new technologies (3G and 4G) but significant traffic is still present in GSM.
- Routinely retune (automatically) only subareas and groups of cells where the KPIs are below acceptable levels.
- Self-Planning of spectrum for new sites.
- Minimize the amount of time required to generate frequency plans as compared to manual planning.
- Allow automatic selection of the frequency planning strategy (segregated versus mixed BCCH/TCH spectra, MAL size, traffic allocation considerations, etc.). Generating multiple plans using different strategies and comparing them allows the selection of the best plan and strategy.

Before generating a new spectrum plan, a number of important steps must be followed to ensure that the new plan will be optimum. Each step can be considered as a function or component in a spectrum Self-Planning process. Full automation can be applied in some of these steps:

1. BSIC cleanup before activating MMR: in order to improve the reliability of the BCCH/ BSIC-to-Cell matching algorithm, the BSIC plan needs to be cleaned prior to MMR recordings. The purpose of the BSIC cleanup is to maximize the distance between co-BCCH/ BSIC pairs in order to properly identify the most likely interferer in case multiple cells reuse the same BCCH/BSIC.
2. Data consistency and integrity checks: if the input data is not accurate and complete the automatic spectrum plan will be suboptimal. Various tests and consistency checks can be automated to determine whether the input data is valid to proceed with optimization. Examples include:
   o Check completeness of physical, network, traffic, HO and IM data.
   o Identify cells with abnormal average neighbor distance to detect wrong relationships.
   o Identify cells with outlying low and high utilization.
   o Identify inconsistencies between number of channels and number of TRXs.
3. BCCH Allocation List (BAL) evaluation: the integrity of the MMR-based IM depends on the ability to measure all potential interferers. The BCCH of a potential interferer needs to

be scanned for a sufficient period of time. To scan a channel, it has to be part of cell's BAL, which normally contains only neighbors. The process of adding channels to BALs can be automated keeping in mind the BAL size limit of 32 channels (extended to 64 if there are no UMTS neighbors).

4. BAL generation: as part of the fully automated process, scripts can be prepared and downloaded to the OSS to update BALs of cells that will be activating MMR.
5. Daily automatic MMR checks to ensure smooth data collection: any problems in data collection must be detected early enough to ensure that the collected data set is sufficient for proper spectrum planning.
6. Neighbor list deletions: daily HO attempt counters should be collected for 2-3 weeks and should be neighbors with low counts can be removed to avoid unnecessary constraints and increase the potential BAL size.
7. TRX count reduction: the integration of new sites is sometimes not followed by a potentially feasible TRX count reduction in the surrounding sites. Ensuring that only necessary TRXs are kept reduces interference and makes planning easier.
8. Spectrum plan generation and verification: the final step in the spectrum Self-Planning process following all integrity checks and measurement collection.

4.10.5. Trade-Offs and Spectrum Plan Evaluation

As in most optimization activities, spectrum Self-Planning algorithms must incorporate certain trade-offs:

- Performance versus capacity: as in any mobile communication technology, QoS in GSM can be improved at the expense of capacity. Conversely, traffic growth will increase interference and negatively impact QoS.
- Simplicity versus performance in different planning strategies:
  - o 1×1: simple use of the entire TCH pool in large MALs without TCH planning. This planning strategy requires MAIO and HSN planning, i.e. it does not require any TCH planning.
  - o 1×3: simple use of three different MALs per site. This planning strategy requires MAIO and HSN planning only, i.e. it does not require any TCH planning.
  - o Ad-Hoc: complex method deploying custom MALs for each cell. Frequency plans associated with this strategy need to be created using an AFP solution and they usually show significant performance and Effective Frequency Load (EFL) improvements as compared to 1×1 and 1×3 reuses.
- Accessibility and HO performance versus call quality performance: balancing the number of BCCH channels (handling SDCCH accessibility and HOs) and the number of TCH channels (carrying voice traffic).
- Underlay/Overlay: the Underlay/Overlay concept, available in different forms in main GSM vendors, is to divide a cell into two serving areas. The Overlay is the area close to the cell with high signal level, and the Underlay covers the remaining area, including cell edge. In order to increase the utilization on the BCCH layer without degrading performance, the BCCH layer is allocated to the Overlay and the TCH layer to the Underlay where benefits of frequency hopping on TCH are important. In doing so, the cell has been effectively divided into two subcells, each with its own TRXs. The trunking efficiency of the cell decreases.

- Traffic allocation priority on BCCH versus TCH: in general, the optimum setting should be decided at cell level, i.e. the layer that provides best performance should be selected to serve the traffic first. However, even though the RxQual on the BCCH layer might be better than that of the TCH layer, the TCH layer might perform better because of the frequency diversity gain. Thus a simple practical rule may be to compare the percentage of samples with BCCH RxQual4-7 versus the percentage of samples with TCH RxQual5-7, since an additional RxQual4 bin is needed for BCCH (nonhopping) to provide equivalent FER performance figures as hopping TCH layer. For GPRS/EDGE, vendors report RLC throughput per ayer, thereby enabling decisions on which layer should be the preferred.
- DL versus UL performance: in general, the GSM performance is DL-limited because the mobile station receivers are simpler, with a higher noise figure and no antenna diversity as compared to the Base Transceiver Station (BTS) receiver. DL is also subject to continuous full power transmission on the BCCH. However, certain cells may be limited in UL, such as high sites suffering from interference generated by a large number of mobiles served by other cells. UL performance can be enhanced by mixing (partially or fully) BCCH and TCH channel pools to increase the number of potentially useful frequencies in moderately-loaded networks with TCH channel allocation priority. Improvement on UL comes from under-utilization of UL BCCH duplex-counterpart channels that may otherwise not be used much in UL in the segregated spectrum case.

Understanding these and other trade-offs is important in creating a technically sound GSM/ GPRS/EDGE self spectrum planning solution. The quality of a frequency plan generated by a self spectrum planning solution needs to be evaluated pre- and post implementation. While KPIs are available for post implementation evaluation, simulations are executed prior to rolling out new spectrum plans into live networks. The purpose of simulations is to rank potentially multiple AFP scenarios and choose the most promising candidate plan for implementation. These scenarios may vary in spectrum planning strategy or relative weights assigned to various constraints. For example, neighbor or site reuses may be relatively balanced versus the IM cost component. Because of the importance of choosing the optimal plan, it is crucial for an AFP solution to embed a QoS simulator enabling automatic and convenient way to rank optimized scenarios. Typically, percentage of RxQual5-7 is the main simulated comparison metric. Another useful metric is the *EFL*, describing network efficiency and defined as the average time duration for which the channel is loaded:

$$EFL = \frac{E_{Tot}(1-0.5HR)}{N_{TSL}N_f N_s}, \qquad (4.14)$$

where:
E_{Tot}: total subscriber Erlang count, typically evaluated during the network busy hour.
HR: ratio of half-rate usage [32].
N_{TSL}: average number of timeslots per TRX.
N_f: number of frequency channels.
N_s: number of active sectors.

EFL needs to be evaluated together with QoS because a network can achieve better QoS with lower *EFL* (i.e. higher infrastructure cost). Therefore, when comparing QoS it is fair to specify also the *EFL* to avoid favoring expensive networks with more under-utilized spectrum versus tighter and more efficient spectrum utilization. While RxQual and *EFL* may be simulated and

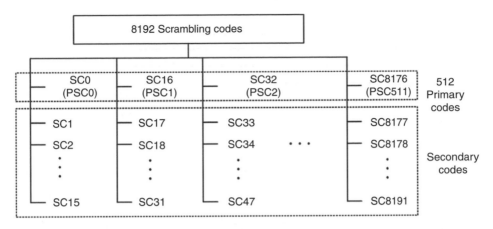

Figure 4.11   Structure of the UMTS-FDD scrambling codes.

calculated analytically, more direct evaluation methods are mostly available through measurements following the implementation of the plan. Evaluation KPIs have been listed in Section 4.10.1, and the expected outcome of the spectrum replan is the improvement of all these KPIs.

4.11.   Automated Planning of 3G Scrambling Codes

Scrambling is a process used in UMTS to distinguish and decode signals from different sources (cells or UEs). It is used on top of channelization in both DL and UL. Without knowing the exact scrambling sequence applied at the source, the original user data sequence cannot be decoded at the receiver. Scrambling is used in the DL to uniquely distinguish signals from different cells (sectors) and, in the UL, to uniquely distinguish signals from different UEs.

4.11.1.   Scrambling Codes in UMTS-FDD

4.11.1.1.   DL Scrambling Codes

UMTS-FDD uses 8192 scrambling codes. Those codes are divided into 512 groups of 16 codes each. Each group consists of one primary and 15 secondary codes. The 512 PSCs numbered from 0-511 are further divided into 64 groups of eight codes each. The structure of the scrambling codes is illustrated in Figure 4.11 and Figure 4.12.

Each cell within a UMTS-FDD system is allocated one PSC that identifies that cell in the network, following a philosophy that is similar to the way in which a BSIC/BCCH combination is used in GSM networks to identify a cell.

The cell's PSC does not only uniquely identify the UMTS cell, but also dictates the value of the Secondary Synchronization Code (SSC) for the Secondary Synchronization Channel (S-SCH). The S-SCH is used to help the mobile to acquire frame synchronization with the serving cell. The UMTS-FDD specifications define 64 different coding groups that can be

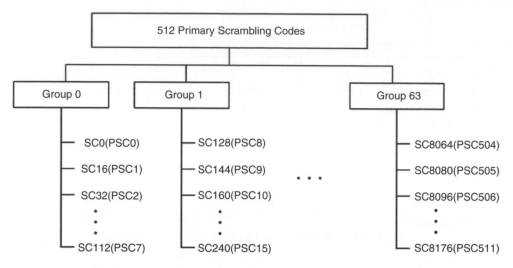

Figure 4.12   Structure of the UMTS-FDD primary scrambling codes.

assigned to the S-SCH. The coding groups consist of 15 code words and are designed so that they remain unique under cyclic shifts smaller than 15. There is a one-to-one correspondence between the S-SCH group and the PSC group. Therefore, the S-SCH is used by the base station as a pointer to the appropriate group of eight PSCs, which helps the mobile to determine the scrambling code used at the cell. Thus PSC assignment for a cell and its neighbors will impact cell search, synchronization time and battery life of the UE [49]. Improper PSC assignment will have negative impact on call quality, HO and mobility in the network.

4.11.1.2.   UL Scrambling Codes

UL channels are scrambled with complex scrambling sequences. UMTS-FDD defines two kinds of scrambling code sequences, long scrambling codes and short scrambling codes. There are 2^{24} long UL scrambling codes and 2^{24} short UL scrambling codes. The long codes are Gold codes that have low cross-correlation properties so that as many users as possible can use the radio channel with minimum mutual interference, and they are used if the base station implements a Rake receiver. If advanced multiuser detectors or interference cancellation receivers are used in the base station, short scrambling codes can be used to make the implementation of the advanced receiver structures easier [50].

Millions of scrambling codes are available for the UL, and they are assigned by the RNC on a per call basis in the connection establishment phase. The 2^{24} long and 2^{24} short scrambling codes are divided between RNCs during the planning process. Each RNC thus has its own planned range. In the UL, the scrambling codes are used to distinguish the signals from different mobiles as well as a means to provide encryption and privacy. No sophisticated planning is needed for UL scrambling codes as there are millions of them and they can be split easily between RNCs.

4.11.2.   Primary Scrambling Code Planning

PSCs are used to uniquely identify cells in UMTS as introduced in Section 4.11.1.1, similar to BSIC-BCCH utilization in GSM systems in order to uniquely identify a GSM cell. The 512 PSCs use different Gold Code sequences and, therefore, they are not constrained by some of the planning rules used in other CDMA systems[7]. UMTS is a direct sequence spread spectrum CDMA system, and all sectors use the same carrier frequency. PSC planning must take into consideration the distance between cells, their coupling in terms of propagation/interference and their neighbor relationships.

PSC interference causes either code interference or code confusion. Code interference happens when two cells that have strong signal overlap use the same PSC. Code interference makes it hard for the UE to decode the signal from any of the cells that use the same PSC, degrades call quality and might cause call drops. Note that assigning adjacent PSCs to cells with strong overlap does not create any code or signal interference, since PSCs represent separate codes rather than separate channels or shifts in a given code.

A neighboring cell cannot be uniquely identified during handover if it shares the same PSC with another cell. Code confusion will occur when the same PSC is used by two or more neighbors of the serving cell or by a non neighbor and a neighbor of that cell. Cells that use the same PSC will create code confusion to mobiles connected to the serving cell. Code confusion will trigger event 1A[8] that may result in the addition of a neighbor to the UE's Active Set based on PSC measurements made on a different cell with the same PSC. Once a neighbor is added based on false measurements, events 1B[9] and 1C[10] will be triggered to replace or remove that neighbor. Moreover, the serving cell should not reuse the same PSC with either direct neighbors or composite neighbors, where the composite neighbor list is the combined neighbor list from all cells in the Active Set. Furthermore, in order to simplify system optimization it is also important to ensure that PSCs are only reused between sectors that are far from each other. This allows accurate identification of cells based on the reported PSCs, and can help in identifying missing neighbors and detecting overshooting cells.

In the following, PSC planning rules and constraints are presented in detail:

1. Neighbors: PSC planning has to ensure that cells do not reuse the same PSC with either direct or composite neighbors, and that composite neighbors have unique PSCs. Neighbors using the same PSC will cause confusion to the network and the system will execute a HO to the wrong cell, which will degrade performance. The UMTS system supports SHO, where one UE can communicate with more than one cell at the same time. The neighbor

---

[7] For example, IS-95 uses 512 PN offsets (or codes), each representing a different time shift of the same code. Therefore, a code assigned to a given sector can appear to belong to another sector if the pilot signal travels a certain distance. PN offset planning has to take into account code aliasing and there are restrictions on code assignment based on sector orientation.

[8] Event 1A, also known as radio link addition request, is the event of reporting a neighboring cell to be added to the Active Set when the pilot of a neighboring cell that is not in the Active Set is measured with quality that fulfils a predefined criterion.

[9] Event 1B, also known as radio link removal request, is the event of requesting the removal of a cell from the Active Set when the pilot of that cell is measured with quality that fulfils a predefined criterion.

[10] Event 1C, also known as combined radio link addition and removal (replacement), is the event of requesting swapping of cells between the Active Set and the Monitored Set when the Active Set is full and the pilot of a neighboring cell that is not in the Active Set is measured with quality that fulfils a predefined criterion.

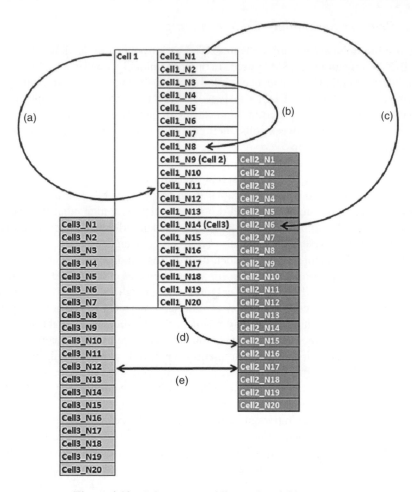

Figure 4.13 Primary scrambling code neighbor reuse.

list for a UE in SHO is the combined neighbor list from all servers in the Active Set. Thus not only is it important to avoid reuse of PSCs with and among immediate neighbors, but it is also important to avoid reuse of PSCs with and among neighbors of defined neighbors (composite neighbor list). The following five neighbor constraints illustrated in Figure 4.13 have to be taken into account for during planning in order of importance.

a) Serving cell reuse with a direct neighbor: the serving cell should not use the same PSC as any of its primary neighbors. This is the most severe PSC neighbor reuse. SHO execution will fail between serving and neighbor cells, and the neighbor cell will cause code interference to the serving cell, especially at cell edge where the serving cell signal strength is weak.

b) Direct neighbors PSC reuse: two or more primary neighbors of the serving cell should not use the same PSC. This is the second most severe neighbor reuse. Neighbor confusion will exist between neighbors that use the same PSC due to the fact that the serving

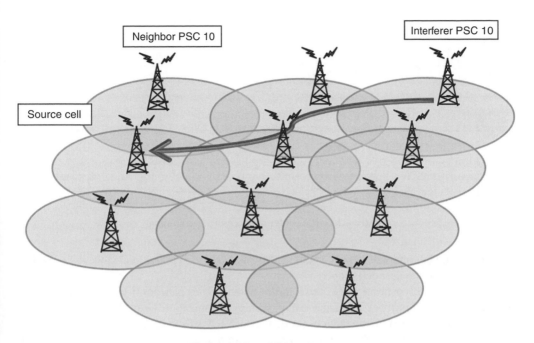

Figure 4.14   PSC interference.

cell has no way to uniquely identify the cell associated with the reported PSC, since this can correspond to any of the neighbors that share the same PSC. Thus, SHO might be executed with the wrong neighbor.

c) Serving cell reuse with a neighbor of a direct neighbor: the UE has the potential to be in SHO with any member of the composite neighbor list, which includes the primary neighbors of the serving cell, as well as the neighbors of other cells in the Active Set that may not be direct neighbors of the serving cell. If a composite neighbor uses the same PSC as the serving cell, it will cause code interference if it is received with strong signal level and most likely SHO with that composite neighbor will not be successfully executed.

d) Direct neighbors and non-direct neighbors PSC reuse: direct or primary neighbors and non-direct neighbors should not share the same PSC. Neighbor confusion will happen between cells in the composite neighbor list using the same PSC.

e) Non-direct neighbors PSC reuse: non-direct neighbors should not share the same PSC with each other.

2. PSC interference: in UMTS, this phenomenon manifests itself in the form of code interference or code confusion. Cells not defined as neighbors but overshooting into the serving cell coverage area should not use the same PSC as the serving cell, any of its primary neighbors or any of its composite neighbors. An overshooting cell with the same PSC as the serving cell will cause code interference and might impact reception. Moreover, an overshooting cell with the same PSC as one of the neighbors of the serving cell will cause code confusion and SHO will be triggered to that neighbor cell based on measurements from the overshooting cell as shown in Figure 4.14. Furthermore, an overshooting cell that has strong signal overlap with a neighbor sharing the same PSC will cause code interference

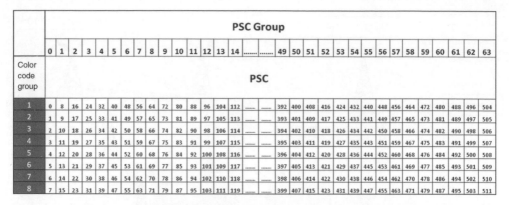

Figure 4.15 PSC color code groups partitioning.

to that neighbor and will impact signal quality; as a result, SHO will fail since the neighbor signal will not be decoded properly.

3. Reuse distance: ensuring large reuse radio distance (i.e. path-loss) between cells using the same PSC improves the accuracy of system optimization tasks that rely on mobile measurements. Maximizing the reuse distance between cells using the same PSC will make it easier to associate a PSC reported by mobiles to a unique cell in the network with high confidence. This will help in RF optimization as well as automatic neighbor optimization.

PSC planning and optimization can further apply restrictions and partitioning of PSCs based on operators' best practices. For example, overshooting cells and high sites can be assigned a PSC from a reserved set of codes so that their interference can be detected easily without ambiguity based on PSCs reported by mobiles. In-building sites and femto cells can have their own reserved set of codes. Another set of codes can be reserved for expansion and new sites. It is also a common practice to ensure that PSCs of neighboring cells are not using the same PSC group in order to save battery life, although this approach impacts system acquisition time negatively [49]. One approach to ensure that neighbors of a cell belong to different PSC groups is to divide the 512 available PSCs into eight color code groups with 64 PSCs per color code group belonging to different PSC groups as illustrated in Figure 4.15.

After color code group partitioning, cells in the network are divided into geographical clusters (with a maximum of 512 cells per cluster). Each cluster is then subdivided into eight regions and each region is assigned one of the eight color code groups as shown in Figure 4.16. This ensures that, within each region, PSCs assigned to cells and their neighbors do not belong to the same PSC group. This approach reduces the number of cells that have neighbors sharing the same PSC group. However, cells on the edge of the region will still have neighbors that belong to same PSC group.

4.11.3. PSC Planning and Optimization in SON

UMTS provides 512 PSCs, which makes PSC planning much easier than frequency planning for TDMA/FDMA systems. Nonetheless, for networks with high site density or aggressive expansion, PSC planning becomes a challenge since network operators have to continuously update and optimize current PSC plan, and they also need to derive new PSCs for new sites.

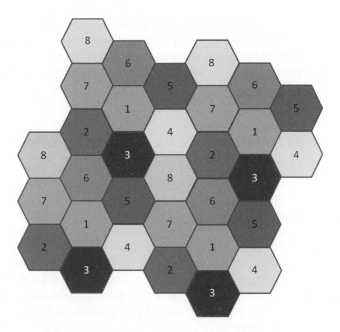

Figure 4.16 PSC color code to region assignment.

PSC assignment belongs to the class of NP-hard problems [19][20]. Therefore, the optimum PSC assignment can be found by using optimization search techniques such as Greedy Search, SA [21], Genetic Algorithms, etc. The configuration space consists of all possible PSC assignments to all cells to be optimized. A cost function can be used to map every possible PSC assignment to a nonnegative number or cost. If the cost function models the PSC planning rules and constraints presented earlier in this section (distance, interference and neighbors), with all cost components weighted appropriately, the assignment that minimizes the total cost will be the optimum plan.

Automated PSC optimization/planning can be implemented as a SON function or a 3G SON Use Case. This function can be used to optimize the PSC assignment of new sites and existing ones. Typical triggers for PSC optimization include the addition of new sites, antenna configuration changes (e.g. electrical tilt, mechanical tilt and azimuth) and changes in transmit power. Such changes will have an impact on neighbor lists, inter-cell interference and overshooting cells.

To optimize live and operational sites, PSC optimization can be run periodically (daily/weekly) or it can be event triggered. For example, when the neighbor list of a given sector is updated, a request can be made to the PSC optimization function to tune the PSC plan for that sector and its neighbors (since updating the neighbor list may introduce new PSC collisions and reuses).

Both CM and PM data can be used for PSC optimization. CM data provides information on network topology, including neighbor lists and current PSC plan. PSC conflicts between neighbors and problematic reuses with other sectors are identified by analyzing CM data. Neighbor list changes will also be reflected in CM data and may trigger PSC optimization. Interference information can be captured through PM data. Moreover, the PSC optimization function can use

PM counters and call traces to take into account missing neighbors or overshooting cells, thereby allowing the avoidance of PSC reuse with sectors that can be mistaken with existing neighbors. Missing neighbors can be identified using mobile measurements reporting detected cells that are neither defined nor monitored, and overshooting cells can be identified using propagation delay information also reported by mobile measurements.

4.12. Automated Planning of LTE Physical Cell Identifiers

The planning of the LTE PCI is parallel to the UMTS scrambling code planning process described in Section 4.11, as there are similarities between both concepts for the two technologies.

The LTE PCI is used to identify the different cells in the system within the scope of several physical layer procedures [51], becoming an essential cell configuration parameter. The main purpose of assigning the PCI is to enable the UE to identify the source of the received signal. Besides, the PCI determines the sequence to generate the DL RS, which the UE needs in order to demodulate the received signal. It is also used in a number of procedures requiring cell specific settings such as the bit scrambling carried out prior to modulation, which should be different for neighbor cells to ensure interference randomization and full processing gain of the channel coding. Other examples of PCI usage are the selection of the hopping pattern of the sequence that is used to generate the UL RS and the mapping to resource elements of certain channels.

4.12.1. The LTE Physical Cell ID

Similar to the WCDMA/HSPA scrambling code, the LTE RS sequence can be seen as a PCI indicator. There are 504 RS sequences, corresponding to 504 different cell identities. Each RS sequence is built through the product of a pseudorandom sequence and an orthogonal sequence. There are three different two-dimensional orthogonal sequences and 168 different two-dimensional pseudorandom sequences. Each cell identity corresponds to a unique combination of an orthogonal sequence and a pseudorandom sequence, thus allowing for 504 unique cell identities (168 cell identity groups with three cell identities in each group). The PCI can be expressed as shown in Equation (4.15), being defined by a number $N_{ID}^{(1)}$ in the range of 0-167, which represents the cell identity group, and a number $N_{ID}^{(2)}$ in the range of 0–2, further defining the exact cell identity within the cell identity group.

$$N_{ID}^{cell} = 3N_{ID}^{(1)} + N_{ID}^{(2)} \qquad (4.15)$$

The PCI also determines the sequences to generate the primary and secondary synchronization signals, which help the UE to acquire the time and frequency synchronization during the cell search procedure [27]. From these signals, the UE can detect the PCI and then obtain knowledge about the sequences used for generating the RS in the cell. The PCI in LTE is a very short identifier that mobile devices can read without having to decode the full broadcast channel. With only 504 values available, the identities need to be reused across the network. Thus, the PCI assignment needs to be carefully planned to avoid conflicts and undesirable interference.

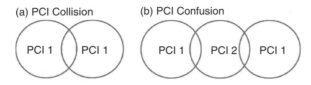

Figure 4.17   PCI collision and confusion.

4.12.2.   Planning LTE Physical Cell IDs

The main objective of this planning is to assign a PCI to each cell so that mobile terminals can identify neighboring cells without ambiguity. To achieve that objective, two criteria must be fulfilled: the assignment has to be collision and confusion free. The former means that two neighbor cells may not have the same PCI; the latter implies that a cell may not have two neighbors with the same PCI. Both concepts are illustrated in Figure 4.17.

Despite the existence of 504 different values, the actual available identities to choose from may be limited to a smaller subset in order to overcome various planning constraints; for example, to make subgroups for macro-, micro- and femto cells in order to simplify the introduction of new cells in one layer without impacting other layers; or to make subgroups to handle planning constraints at network borders (e.g. between two countries), where coordination between network operators is difficult.

The PCI planning process should take into account the reuse distance. An efficient plan should maximize the radio distance between sectors using the same PCI. By doing so, it will be easier to avoid collision and confusion, and the mobile terminals will be able to identify the different measured cells without ambiguity. In order to consider this criterion, physical information (at least location and basic orientation) of the different cells is needed.

There are some other best practices like assigning PCIs within the same cell identity group to cells belonging to the same eNodeB so that their RS are based on the same pseudorandom sequence but use different orthogonal sequences, thereby minimizing interference between them.

Operators can use planning tools and other automated solutions to develop a network PCI plan or to derive the PCI for a new eNodeB. Different algorithms can be used to derive the optimum PCI assignment. The election of the algorithm presents a typical trade-off between optimality of the results, cost and complexity.

PCI assignment belongs to the class of NP-hard problems [19] and, therefore, the optimum PCI assignment can be found using optimization search techniques such as Greedy Search, SA [21], etc. The configuration space consists of all possible PCI assignments to all cells to be optimized. An objective function or cost function can be used to map every possible PCI assignment to a nonnegative number or cost. If the objective function is accurate and contains all necessary cost components weighted appropriately, the assignment that minimizes the total cost will be the optimum plan. The cost function must take into account the distance between sectors, the signal level received from every sector and neighbor relations. The collisions penalizing the PCI reuse in the cost function are essentially similar to those described in Section 4.11.2 except for those related to the composite neighbor list since LTE does not support SHO.

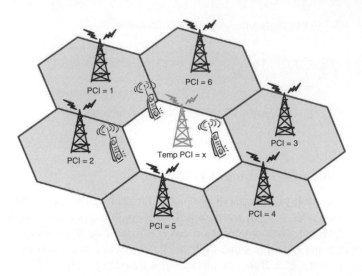

Figure 4.18   Example of deployment with a new introduced cell.

A different approach is presented in [52], where the PCI assignment is mapped to the well-known problem of graph coloring. The cells are depicted as vertices, and the vertices representing neighbor cells are connected by an edge; the nodes in the resulting graph must be colored in such a way that two connected nodes are not assigned the same color. The proposed method provides a solution for initial deployments and also adapts to network growth with new cells being added. This algorithm is also suitable for automated PCI assignment within a network auto configuration process, which is presented in the following subsection.

4.12.3.   Automated Planning of PCI in SON

The automatic planning of PCI is included in the Self-Configuration Use Cases defined by the Next Generation Mobile Networks (NGMN) Alliance [1]. The PCI is one of the radio parameters that need to be self-planned in a new eNodeB. Automating PCI planning as part of SON has many advantages over traditional planning techniques. It replaces a tedious and complex manual process with an automatic one and avoids human errors.

The SON Use Case Automatic Configuration of Physical Cell Identity [53] proposed by 3GPP aims at configuring the PCI of a newly introduced cell automatically, fulfilling the requirements of collision and confusion free, but no specific mechanisms have been standardized yet. Different proposals are discussed in [54, 55], in which some vendors defend a distributed solution, allowing eNodeBs to auto-configure their PCIs, and others prefer centralized control, making the Operation And Maintenance (O&M) system responsible for making the final decision.

One of the proposed distributed methods [55] employs a scheme in which the new cell choosing a temporary PCI randomly from a predefined set, then using the ANR function during a certain period of time to obtain neighbors' PCI information, and finally adjusting its PCI accordingly to avoid collision and confusion. Figure 4.18 illustrates the explained scenario, where the cell in the center is the newly introduced one.

An alternative option based on a centralized solution is proposed in [54]. This method relies on a central function that stores information about eNodeBs, including their location, supported cells and related configuration such as the PCI assignment. Assuming that the location of the eNodeB is available, a centralized entity can then rely on simple distance calculations to automatically provide an optimum PCI assignment to the eNodeB fulfilling the corresponding requirements. The location does not need to be very accurate and can be based on other types of information, e.g. some neighbor relations. Compared to this method, the distributed approach does not guarantee a collision and confusion free selection and needs repair mechanisms to solve suboptimal or faulty configurations.

The lack of processes specified in the standards will result in different vendor-specific distributed functions which will enable infrastructure vendors to innovate and differentiate their solutions. Meanwhile, there is also additional room for differentiation based on the O&M capabilities, which can take an important role in the PCI planning process by providing a global view of the network, thereby making the assignment more efficient and more robust to avoid collision and confusion. Solutions that used to optimize BSIC/BCCH assignments for GSM and scrambling code assignments for UMTS, which have been proven to work in the field, can also be utilized to optimize LTE PCIs.

4.13. References

[1] Next Generation Mobile Networks (NGMN) Alliance, Deliverable (2008) *NGMN Use Cases Related to Self Organising Network, Overall Description*, Version 2.02, December 2008, www.ngmn.org (accessed 3 June 2011).

[2] Hata, M. and Nagatsu, T. (1980) Mobile Location Using Signal Strength Measurements in Cellular Systems, *IEEE Transactions on Vehicular Technology*, **29**, pp. 245–352.

[3] COST 231 TD(91)109 (1991) *1800 MHz Mobile Net Planning Based on 900 MHz Measurements*.

[4] Lee, W.C.Y. (1993) *Mobile Communications Design Fundamentals*, John Wiley & Sons, Inc., New York.

[5] Kostanic, I., Guerra, I., Faour, N. Zec, J. and Susanj, M. (2003) Optimization and Application of W.C.Y Lee Micro-cell Propagation Model in 850 MHz Frequency Band, *Proceedings of Wireless Networking Symposium*, Austin, TX.

[6] The Nielsen Company (2010) *Quantifying the Mobile Data Tsunami and its Implications*, 30 June 2010, http://blog.nielsen.com (accessed accessed 3 June 2011).

[7] Tsang, D. and Ross, K. (1990) Algorithms to determine exact blocking probabilities for multirate tree networks, *IEEE Transactions on Communications*, **38**(2), pp. 1226–1271.

[8] Moungnoul, P., Laipat, N., Hung, T. and Paungma, T. (2005) GSM Traffic Forecast by Combining Forecasting Technique, *IEEE Fifth International Conference on Information, Communications and Signal Processing*, pp. 429–433.

[9] Tikunov, D. and Nishimura, T. (2007) Traffic prediction for mobile network using Holt-Winter's exponential smoothing, *IEEE 15th International Conference on Software, Telecommunications and Computer Networks*, SoftCOM 2007, pp. 1–5, September 2007.

[10] Ramos-Escano, G. and Pedraza, S. (2004), *A-bis Interface Dimensioning for EGPRS Technology*, *IEEE Vehicular Technology Conference*, **6**, pp. 4471–4475, Fall/September 2004.

[11] Plummer, D. (1982) *An Ethernet Address Resolution Protocol*, RFC 826, MIT-LCS, November 1982, http://www.rfc-editor.org/rfc/rfc826.txt (accessed accessed 3 June 2011).

[12] Droms, R. (1997) *Dynamic Host Configuration Protocol*, RFC 2131, March 1997, http://www.rfc-editor.org/rfc/rfc2131.txt (accessed accessed 3 June 2011).

[13] Moy, J. (1998) *OSPF Version 2*, RFC 2328, April 1998, http://www.rfc-editor.org/rfc/rfc2328.txt (accessed accessed 3 June 2011).

[14] Cisco Systems Inc. (2005) *Enhanced Interior Gateway Routing Protocol*, September 2005, http://www.cisco. com/en/US/tech/tk365/technologies_white_paper09186a0080094cb7.shtml (accessed accessed 3 June 2011).

[15] Dijkstra, E. (1959) A note on two problems in connexion with graphs, *Numerische Mathematik,* **1**, pp. 269–271.

[16] ATM Forum Technical Committee (2002) *ATM Inter-Network Interface (AINI) Specification*, (af-cs-0.125.001), http://www.broadband-forum.org (accessed accessed 3 June 2011).

[17] Cisco Systems Inc. (2004) *Introduction to PNNI*, April 2004, http://www.cisco.com/en/US/docs/switches/wan/ mgx/software/mgx_r5.0/data/pnni/network/planning/guide/pintro.pdf (accessed accessed 3 June 2011).

[18] Cisco Systems Inc. (2007) *Cisco ONS 15454 SDH Reference Manual*, Release 4.0., 23 October 2007, http://www. cisco.com/en/US/docs/optical/15000r4_0/15454/sdh/reference/guide/sdh40ref.html (accessed 3 June 2011).

[19] Evans, J. R. and Minieka, E. (1992) *Optimization Algorithms For Networks And Graphs*, Marcel Dekker Inc.

[20] Jamaa, S., Altman, Z., Picard, J.M. and Fourestie, B. (2003) *Optimisation des reseaux mobiles utilisant les Algorithms Genetiques*, Chapter 8 in 'Metaheuristiques pour L'optimisation Dificile', J. Dreo et al. (eds.) Eyrolles.

[21] Hurley, S. (2000) Automatic base station selection and configuration in mobile networks, *52nd IEEE Vehicular Technology Conference*, **6**, pp. 2585–2592, September 2000.

[22] Goldberg, D.E. (1989) *Genetic Algorithms in Search, Optimization and Machine Learning*, Addison-Wesley Publishing Comp. Inc., New York.

[23] Altman, Z. Picard, J.M., Ben Jamaa, S. et al. (2002) New Challenges in Automatic Cell Planning of UMTS Networks, *IEEE Vehicular Technology Conference*, **2**, pp. 951–954, Fall/September 2002.

[24] Lee, C. Y. and Kang, H. G. (2000) Cell Planning with Capacity Expansion in Mobile Communications: A Tabu Search Approach, *IEEE Transactions on Vehicular Technology*, **49**, pp. 1678–1691, September 2000.

[25] Cormen, T. H. Leiserson, C. E. Rivest, R. L. and Setin, C. (1990) *Introduction to Algorithms*, MIT Press and McGraw-Hill.

[26] Holma, H. and Toskala, A. (eds.) (2004) *WCDMA for UMTS: Radio Access for Third Generation Mobile Communications*, John Wiley & Sons, Ltd, Chichester.

[27] Holma, H. and Toskala, A. (eds.) (2009) *LTE for UMTS - OFDMA and SC-FDMA Based Radio Access*, John Wiley & Sons, Ltd, Chichester.

[28] 3GPP, Technical Specification, Technical Specification Group GSM/EDGE Radio Access Network (2006) *Physical Layer on the Radio Path; General Description*, 3GPP TS 45.001 Version 7.3.0, Release 7, 17 October 2006, http://www.3gpp.org/ftp/Specs/archive/45_series/45.001/45001-730.zip (accessed 3 June 2011).

[29] 3GPP, Technical Specification, Technical Specification Group Radio Access Network (1999) *RF Parameters in Support of Radio Resource Management*, 3GPP TS 25.103 Version 2.0.0, Release 99, 13 October 1999, http:// www.3gpp.org/ftp/Specs/archive/25_series/25.103/25103-200.zip (accessed 3 June 2011).

[30] Magnusson, S. and Olofsson, H. (1997) Dynamic neighbor cell list planning in a microcellular network, *Proceedings of the 1997 IEEE 6th International Conference on Universal Personal Communications*, **1**, pp. 223–227, October 1997.

[31] Parodi, F., Kylvaja, M., Alford, G., Li, J. and Pradas, J. (2007) An Automatic Procedure for Neighbor Cell List Definition in Cellular Networks, *IEEE International Symposium on a World of Wireless, Mobile and Multimedia Network*, pp. 1–6, June 2007.

[32] Halonen, T., Romero, J. and Melero, J. (eds.) (2003) *GSM, GPRS and EDGE Performance*, John Wiley & Sons, Ltd, Chichester.

[33] 3GPP, Technical Specification, Technical Specification Group Radio Access Network (2010), *Radio Resource Control (RRC), Protocol Specifications*, 3GPP TS 25.331 Version 8.12.0, Release 8, 8 October 2010, http:// www.3gpp.org/ftp/Specs/archive/25_series/25.331/25331-8c0.zip (accessed 3 June 2011).

[34] Hiltunen, K., Binucci, N. and Bergstrom, J. (2000) Comparison between the periodic and event-triggered intra-frequency handover measurement reporting in WCDMA, *Wireless Communications and Networking Conference*, **2**, pp.471–475, September 2000.

[35] 3GPP, Technical Specification, Technical Specification Group Service and System Aspects (2009) *Automatic Neighbour Relation (ANR) Management; Concepts and Requirements*, 3GPP TS 32.511 Version 9.0.0, Release 9, December 2009, www.3gpp.org (accessed 3 June 2011).

[36] Lee, W. (1995) *Mobile Cellular Telecommunications*, McGraw Hill.

[37] Hale, W. (1980) Frequency Assignment: Theory and Applications, *Proceedings of IEEE*, **68**(12), pp. 1497–1514.

[38] Box, F. (1978) A heuristic technique for assigning frequencies to mobile radio nets", *IEEE Transactions on Vehicular Technology*, **27**(2), pp. 57–64.

[39] Giortzis, A.I. and Turner, L.F. (1996) A Mathematical Programming Approach to the Channel Assignment Problem in Radio Networks, *IEEE Vehicular Technology Conference*, **2**, pp. 736–740, April 1996.

[40] Funabiki, N. and Takefuji, Y. (1992) A Neural Network Parallel Algorithm for Channel Assignment Problems in Cellular Radio Networks, *IEEE Transactions on Vehicular Technology*, **41**(4), pp. 430–437.

[41] Kunz, D. (1991) Channel Assignment for Cellular Radio Using Neural Networks, *IEEE Transaction on Vehicular Technology*, **40**(1), pp. 188–193.

[42] Kim, J., Park, S., Dowd, P. and Nasrabadi, N. (1997) Cellular radio channel assignment using a modified Hopfield network", *IEEE Transactions on Vehicular Technology*, **46**(4), pp. 957–967.

[43] Duque-Anton, M., Kunz, D. and Ruber, B. (1993) Channel Assignment for Cellular Radio Using Simulated Annealing, *IEEE Transaction on Vehicular Technology*, **42**(1), pp. 14–21.

[44] Mathar, R. and J. Mattfeldt, J. (1993) Channel assignment in cellular radio networks, *IEEE Transaction on Vehicular Technology*, **42**(4), pp. 647–656.

[45] Ngo, C. and Li, V. (1998) Fixed channel assignment in cellular radio networks using a modified genetic algorithm, *IEEE Transactions on Vehicular Technology*, **47** (1), pp. 163–172.

[46] Lai, W. and Coghill, G. (1996) Channel assignment through evolutionary optimization, *IEEE Transaction on Vehicular Technology*, **45** (1), pp. 91–96.

[47] Leese, R. (1997) A unified approach to the assignment of radio channels on a regular hexagonal grid, *IEEE Transaction on Vehicular Technology*, **46**(4), pp. 968-880.

[48] Wei, W. and Rushforth, C. (1996) An adaptive local-search algorithm for the channel-assignment problem (CAP), *IEEE Transaction on Vehicular Technology*, **45** (3), pp. 459–466.

[49] Kourtis, S. (2000) Code planning strategy for UMTS-FDD networks, *IEEE Vehicular Technology Conference*, **2**, pp. 815–819, May 2000.

[50] 3GPP RAN WG1 Meeting #4, R1-99333 (1999) *Short scrambling codes for the UTRA/FDD uplink*, April 1999, www.3gpp.org (accessed 3 June 2011).

[51] 3GPP, Technical Specification, Technical Specification Group Radio Access Network (2010) *E-UTRAN; Physical Channels and Modulation*, 3GPP TS 36.211 Version 9.1.0, Release 9, 30 March 2010, http://www.3gpp.org/ftp/Specs/archive/36_series/36.211/36211-910.zip (accessed 3 June 2011).

[52] Bandh, T., Carle, G. and Sanneck, H. (2009) Graph Coloring Based Physical-Cell-ID Assignment for LTE Networks, *The 5th International Wireless Communications and Mobile Computing Conference (IWCMC 2009)*, June 2009.

[53] 3GPP, Technical Specification, Technical Specification Group Radio Access Network (2010) *E-UTRAN; Self-Configuring and Self-Optimizing Network (SON) Use Cases and Solutions*, 3GPP TS 36.902 Version 9.1.0, Release 9, 6 April 2010, http://www.3gpp.org/ftp/Specs/archive/36_series/36.902/36902-910.zip (accessed 3 June 2011).

[54] 3GPP RAN WG3 Meeting #59, R3-080812 (2008) *Solution(s) to the 36.902s Automated Configuration of Physical Cell Identity Use Case*, April 2008, www.3gpp.org (accessed 3 June 2011).

[55] 3GPP RAN WG3 Meeting #59, R3-080376 (2008) *SON Use Case: Cell Phy_ID Automated Configuration*, February 2008, www.3gpp.org (accessed 3 June 2011).

5

Multi-Technology Self-Optimization

Juan Ramiro, Rubén Cruz, Juan Carlos del Río, Miguel A. Regueira Caumel, Carlos Úbeda, Szymon Stefański, Andreas Lobinger, Juanjo Guerrero, Salvador Pedraza, José Outes, Josko Zec and Khalid Hamied

5.1.   Self-Optimization Requirements for 2G, 3G and LTE

Self-Optimization is defined as the utilization of measurements and performance indicators collected by the User Equipments (UEs) and the base stations in order to auto-tune the network settings. This process is performed in the operational state, which is defined as the state where the Radio Frequency (RF) interface is commercially active (i.e. when the cell is neither barred nor reserved). Ideally, Self-Optimization solutions should portray the following characteristics:

- Measurement-based. Network adjustments based on theoretical propagation prediction models should be replaced with decisions driven by network statistics in order to take into account the real system status in terms of propagation environment, indoor penetration losses, traffic profile and spatial distribution of the users.
- Sector- and adjacency-based. Experience has shown that homogeneous network configurations are in general suboptimal, and therefore require optimization at the finest possible resolution, i.e. at sector or even adjacency level.
- Multi-vendor. Able to: (i) access and interpret the performance statistics that are provided by each vendor; (ii) derive qualified adjustments of vendor-specific parameters; and (iii) provide automatic or autonomous mechanisms to implement the derived changes onto the network. The difference between automatic and autonomous processes was explained in Chapter 2.

Self-Organizing Networks: Self-Planning, Self-Optimization and Self-Healing for GSM, UMTS and LTE,
First Edition. Edited by Juan Ramiro and Khalid Hamied.
© 2012 John Wiley & Sons, Ltd. Published 2012 by John Wiley & Sons, Ltd.

- Multi-technology. Able to: (i) deal effectively with all relevant technologies, both separately and in a combined way; (ii) carry out inter-Radio Access Technology (iRAT) load balance in accordance with the operator's traffic management policies; and (ii) take into account all cross-technology coverage, capacity and quality implications associated with the proposed adjustments of technology-specific or shared parameter settings.
- Multi-source. Able to leverage all the pieces of information that can be collected autonomously, such as performance counters from the Operations Support System (OSS), Charging Data Records (CDRs), call traces, etc.
- Configurable. Able to operate according to high-level objectives, trade-offs and policies set by operators in terms of capacity, coverage, quality or a weighted combination of them.

5.2. Cross-Technology Constraints for Self-Optimization

Self-Optimization algorithms modifying settings in a given radio access technology need to be aware of the impact of such changes on other technology layers. Observation of these mechanisms goes beyond the availability of independent technology-specific functions in the different layers and calls for the concurrent application of expert knowledge concerning multiple technologies within the automated decision processes. Some examples of such cross-technology constraints are the following:

- The cases in which antennas are shared among different radio access technologies. In these situations, it is not appropriate to carry out an automated optimization that only considers one technology, since the performance in one layer might be improved at the expense of degrading the rest.
- All capacity/coverage optimization exercises that influence the footprint of a technology, indirectly pushing nonserved traffic to other layers. If these decisions are carried out without taking all the affected layers into account, excessive traffic could be offloaded to other layers and this would degrade their capacity, coverage and/or quality beyond acceptable levels.
- All explicit iRAT load balance policies that are based on biasing iRAT Handover (HO) processes by means of offsets. These algorithms need to put performance of all involved layers into perspective in order to avoid improving one performance dimension in a certain layer at the expense of the rest.

These mechanisms will be further illustrated throughout this chapter, especially in Section 5.11.

5.3. Optimization Technologies

5.3.1. Control Engineering Techniques for Optimization

Historically, automatic control has played a crucial role in advanced engineering and science, using different approaches to drive the system output as close as possible to the desired behavior. A few examples are electricity distribution systems, airplanes, medical instruments, robotics, etc. Modern control engineering relies on electronics and computers. The latter have become cheaper and more powerful over the years. Many approaches have been devised to solve linear and nonlinear control problems, but no general method exists to control complex systems, such as the one subject to discussion in this chapter.

Radio Access Network (RAN) performance in a mobile cellular network can be modeled as a system in which the inputs are the configuration parameters and the outputs are performance counters, call traces and alarms. In general, performance counters are often used to build Key Performance Indicators (KPIs), although these can also be built with other information elements, such as postprocessed call traces. Additionally, there are other noncontrollable input variables such as propagation behavior, profile of customers using the system, etc.

The target of an optimization system is to improve RAN performance. The inputs of such optimization system are, on one hand, the aforementioned performance counters, call traces and alarms, as well as the operator's requirements and KPI targets. On the other hand, the outputs of the optimization system are the new proposed values for the radio configuration parameters. The operator's requirements include relative prioritization between different KPIs, allowed KPI margins, allowed parameter ranges and step values, and criteria to accept data as statistically relevant.

The adopted solution must lower the required operational effort and improve the performance of the network. The process is typically automated by following three steps: collecting information from the network, executing an optimization algorithm based on the gathered information and implementing the required changes in the network [1].

A common feature in most automatic control cases is the use of feedback information, characterizing the so-called closed-loop systems (versus open-loop systems). In closed-loop schemes, an adaptive, iterative process is adopted, in which the optimization system modifies the network parameters and the resulting performance indicators are subsequently fed as input for the optimization system to derive the next set of parameter settings. The advantage is the ability to control unstable systems and compensate for perturbations and modeling errors. The main disadvantage is the need for continuous monitoring of the system output, which is also subject to errors. It is worth pointing out that the periodicity of this feedback loop does not need to be necessarily constant, since the adaptive adjustment can be also event-driven or scheduled to happen at predetermined times.

In open-loop control systems, the output has no effect on the control action, i.e. they do not use feedback information. This means that the optimized parameter settings are provided in one single step, without taking the performance indicators resulting from the system modification into account for the execution of additional iterations. They do not have the benefits of closed-loop systems (they cannot control unstable systems nor compensate for perturbation and modeling errors), so they are more suitable if the impact of the parameters under optimization on the system performance is very well known and no major disturbances affect it.

In general, closed-loop optimization is the most suitable way to proceed in order to optimize soft parameters, e.g. those governing Radio Resource Management (RRM) algorithms. However, the nature of other parameters (e.g. some RF settings, such as antenna azimuths or mechanical tilts, which require costly manual intervention to implement modifications) makes it advisable to adopt open-loop optimization solutions in which the final optimized design is generated in one single step.

Depending on the complexity of the system to optimize and the availability of information, several control techniques can be applied:

- Analytical solutions. If the system to control/optimize is known (or at least subject to be modeled) with linear or simple nonlinear behavior, the solution can be analytically derived. Unfortunately, few real systems are so simple and mobile networks are not among them. Examples of these types of systems are oven temperature control or satellite trajectory tracking.
- Searching algorithms. When the system involves highly nonlinear dynamics but can be modeled and simulated offline (i.e. disconnected from the real system), the solution can be reached by means of searching algorithms. The solution is derived using the model and then applied to the real system. A good match between the real system and the model is crucial to obtain efficient solutions. Real system outputs can be used to tune the model. Examples of these types of problems are Automatic Frequency Planning (AFP) or Physical Cell ID (PCI) planning.
- Expert systems. When no model can be obtained, only expert knowledge based algorithms are possible. In this case, the expert knowledge is modeled and included in the control/optimization system. Most RRM optimization problems fit in this category.

Next, different alternatives based on these techniques will be presented as solution candidates for the formulated control problem.

5.3.1.1.   Metaheuristic Searching Algorithms

Metaheuristic algorithms constitute a class of optimization methods belonging to the category of searching algorithms, and are based on iteratively trying to find better candidate solutions based on given cost/objective functions (or quality measures). The cost/objective functions are derived from a model of the problem. This type of algorithm makes few or no assumptions about the problem subject to optimization and can search very large spaces of candidate solutions. However, metaheuristic algorithms do not guarantee that an optimal solution is ever found. An example of suitability for this type of algorithms is the travelling salesman problem [2], where the search-space of candidate solutions grows exponentially as the size of the problem increases, which makes an exhaustive search for the optimal solution unfeasible. Relevant examples of metaheuristic searching algorithms for combinatorial problems are Greedy Search, Simulated Annealing and Genetic Algorithms:

- Greedy Search [3]. This is the simplest searching algorithm. Starting from an initial solution, in every iteration, the current solution is replaced by the best adjacent solution. The main drawback of Greedy algorithms is the impossibility to continue improving once a local optimal solution has been reached.
- Simulated Annealing [4]. The idea comes from annealing in metallurgy, a technique involving heating and controlled cooling of a material to increase the size of its crystals and reduce their defects. The atoms are initially static, but the increase of the temperature moves them from their initial positions (a local minimum of the internal energy) towards random states of higher energy. Then, slowly decreasing the temperature gives them more probability to reach states with lower internal energy than the initial one. Similarly, each step of the simulated annealing algorithm replaces the current solution by a randomly perturbed solution, chosen with a probability that depends on the difference

between the corresponding cost function values and a global parameter that simulates the temperature, which is gradually decreased during the process. The dependency is such that the current solution changes almost randomly when the temperature is large, but improves as the temperature approaches zero. The allowance for deterioration of the solution saves the method from becoming stuck at a local optimum.

- Genetic Algorithms. This technique is based on a heuristic search that emulates the process of natural evolution [5]. In a Genetic Algorithm, a population of strings (called chromosomes), which encode candidate solutions to an optimization problem (called individuals), evolves towards better solutions. In every iteration, the population evolves with the creation of new individuals that are the result of the combination of good previously existing individuals, as well as the removal of some of the bad individuals. The population is also modified by applying mutation to some of the individuals, which basically introduces random changes in the solutions.

5.3.1.2.   Artificial Neural Networks

Artificial neural networks are a computational method that tries to simulate the structure and/ or functional aspects of biological neural networks [6], and they are classified as an expert system. A neural network consists of an interconnected group of artificial neurons, which process information using a connectionist approach to computation. In most cases a neural network is an adaptive system that changes its structure based on external or internal information that flows through the network during the learning phase. Adaptive control involves modifying the applied control laws used by a controller to cope with the fact that the system being controlled is slowly time-varying or uncertain. In an artificial neural network, the expert knowledge is applied during the learning phase.

5.3.1.3.   Bayesian Networks

Bayesian networks are a probabilistic graphical model that represents a set of random variables and their conditional independencies via a directed acyclic graph [7]. They model the probabilistic relationships between causes and symptoms. Given symptoms, the network can be used to compute the probabilities of the presence of various causes. In order to fully specify the Bayesian network and thus fully represent the joint probability distribution, it is necessary to specify, for each node, the probability distribution conditioned on all its parents. A clear drawback of Bayesian networks is that building a model requires the definition of all the conditional probabilities, which makes it very challenging for complex system such as wireless networks. An automatic control system based on Bayesian networks is also an expert system.

5.3.1.4.   Fuzzy Logic

This expert system is an evolution and enlargement of formal multi-value logic that deals with approximate facts instead of precise ones [8]. In contrast with pure conditional logic where binary sets have binary logic, fuzzy logic variables may have a truth value that ranges between

0 and 1. Fuzzy logic is typically adopted when the complexity of the process is high and no accurate mathematical models are available, in highly nonlinear processes or with imprecise definition of the impacting variables. Expert rules are evaluated in a fuzzy way to obtain control/optimization actions. Several rules can be triggered simultaneously and then aggregated to get a combined decision about the system output.

5.3.2. Technology Discussion for Optimizing Cellular Communication Systems

Based on the different alternatives introduced in the previous subsection, the suitability of the different technologies for the optimization of a cellular communication system is discussed.

Metaheuristic searching algorithms represent an attractive choice for RF parameter optimization. Experience has shown that for the particular case of RF parameters, e.g. frequencies, physical/electrical antenna parameters and transmission powers in the Global System for Mobile Communications (GSM), the use of search algorithms based on analytical cost-functions provides remarkable gains, not only in theoretical planning tools, but also in the field. Due to the nature of the problem, the state of the network in terms of RF metrics can be computed by means of simple cost functions, easy and fast to evaluate, which makes metaheuristic searching solutions appropriate.

The use of neural networks to optimize wireless network parameters would imply an initial learning process which should be carried out offline, with the help of expert information. An example of this expert information in wireless systems is a set of input-output pairs, from which the neural network would construct outputs for any new inputs based on internal interpolation. This technique is useful to optimize RRM parameters when an exhaustive set of input-output pairs is available so that the control system does not need to extrapolate. However, in practice, this is typically unfeasible. In the case of Bayesian networks, similar difficulties can be found, since extensive offline observation of system performance under different circumstances is required, at least at the beginning, and the derived models are likely to be highly dependent on each specific network.

Fuzzy logic fits the requirements to optimize RRM parameters of a wireless network due to the high complexity of the system and the difficulties in characterizing the network model. With this solution, it is necessary to embed human expertise in the defined rules. Knowing all the implicit trade-offs in the system to be optimized is therefore crucial when designing the optimization algorithms. On the other hand, a small degree of memory and/or learning process should not be discarded to compensate for unpredicted behaviors that are impossible to consider during the definition phase of the optimization system (e.g. through algorithm adjustments based on previous inputs and outputs). However, this needs to be included carefully in order to properly filter out noisy information elements.

5.4.   Sources for Automated Optimization of Cellular Networks

A clear requirement for Self-Optimization systems is to be measurement-based. However, not all traditional automated optimization schemes have always complied with this requirement, and the nature of the considered inputs has followed an evolutionary process that started with simple propagation predictions based on theoretical models. This section provides a short overview of the different stages of that process.

5.4.1. Propagation Predictions

The most basic source for automated optimization is the use of propagation predictions carried out by means of theoretical models based on equations (e.g. Hata [9], COST231 [10], Lee [11], etc.). These models contain parameters that influence their behavior, and extensive research has been carried out in this field in order to obtain propagation predictions that model different generic environments accurately (dense urban, suburban, rural, etc.). However, in practical systems, accurate matching between predictions and reality is not straightforward unless the applied mathematical models are calibrated with measurements. This is a cumbersome and costly process to carry out, and still implies a standard deviation of the prediction error between 6-10 dB. Moreover, in general, most problems of wireless networks affect indoor users, and achieving acceptable prediction accuracy for these users would require extensive indoor measurement campaigns, which is even more challenging.

5.4.2. Drive Test Measurements

Drive testing campaigns can be done with two different purposes: (i) to benchmark the behavior of the wireless system in a wide range of geographical locations; and (ii) to gather input information that allows calibration of propagation models. In order to automate optimization processes, the second use case is the most relevant one and, compared with the use of purely theoretical models, the combination of drive test measurements and propagation prediction models provides remarkable improvements in terms of accuracy. In general, this combination can be carried out in two ways:

1. Propagation model tuning. In this case, an optimization algorithm is applied in order to find the optimum set of values for the parameters in the propagation model, so that the difference between measurements and predictions is minimized [12].
2. Drive test smart interpolation. With this method, measurements are given absolute precedence for those locations in which they are available. For the rest, a smart interpolation algorithm is applied, taking into account the relative spatial trends suggested by the propagation models and some other measured quantities, such as the correlation distance.

In a normal operational environment, the utilization of drive tests presents three main drawbacks:

1. They involve an expensive and cumbersome process.
2. They do not provide sufficient insight about the indoor user experience.
3. They only characterize the system at the measurement time, which can be insufficient if load-dependent metrics need to be considered.

5.4.3.   Performance Counters Measured at the OSS

Performance counters contain extremely insightful information about network performance with detailed granularity and have the following key advantages:

- Their utilization for optimization is extremely cost effective, since this information is already available at the OSS.
- They are collected continuously. Unlike drive tests, the insight they provide is not limited to the time period in which an active measurement campaign has been scheduled. They can be collected on a 24/7 basis.
- OSS counters consider all users, taking into account their real experience in terms of radio conditions, indoor losses and traffic distribution.
- They are not limited to standard radio measurements or statistics related to protocol events. Additionally, they describe system performance in many other domains, such as resource utilization, internal system events, RRM statistics, etc.

Although they do not provide user location, they can be used for statistical calibration of the input data for RF optimization, providing remarkable performance improvements in the field as will be illustrated in Section 5.7. Moreover, as will be seen in this chapter, the application of optimization algorithms based on OSS performance counters yields outstanding results in the field when fine-tuning RRM soft parameters.

5.4.4.   Call Traces

In this context, call traces are defined as recordings of the detailed message flow of a certain set of calls for a selected number of protocols in different interfaces (e.g. Iub, Iu, etc.). Traditionally, it has been possible to collect them by means of Hardware (HW) probes, although the capability to collect them by means of embedded Software (SW) functionality in the network nodes has become increasingly popular in the recent years.

Call traces provide massive amounts of detailed information that needs to be processed efficiently in order to carry out interesting tasks, for example:

- Derive specialized counters that are not available at the OSS.
- Compute counters only for a subset of calls that follow a configurable pattern.
- Detect outstanding RF problems such as missing neighbors.
- Geo-locate calls and their associated performance reports, being able to segment all these deliverables depending on multiple attributes, such as subscriber type/group, terminal type/group, service class, etc. Some examples of the geo-located maps that can be generated for performance surveillance (i.e. virtual drive testing) and optimization support are the following: traffic maps, coverage and signal quality maps, throughput maps, dropped call maps per drop reason, intersystem HO maps, etc.

In general, postprocessed call traces constitute a rich set of insightful information that can be used as input for automated optimization of soft and physical parameters. Appendix A provides more details about geo-location techniques for the Universal Mobile Telecommunications System (UMTS) based on call traces, and Appendix B focuses on the estimation of X-maps for Long Term Evolution (LTE).

5.5.   Self-Planning versus Open-Loop Self-Optimization

By nature, Self-Planning schemes are open-loop since calculations are carried out offline before the Network Elements (NEs) under consideration are commercially launched. Therefore, it is not feasible to extract any performance indicators in order to include them in any feedback loop.

On the other hand, practical Self-Optimization schemes can follow open-loop or closed-loop approaches, depending on the characteristics of the problem to be solved. In this respect, several considerations need to be taken into account when selecting the optimization technology:

- A requirement for open-loop schemes to be feasible is that it must be possible to model the dependency of the system performance on the parameters under optimization by means of a relatively simple mathematical cost function, which can then be used for optimization by, for example, applying metaheuristic algorithms. Examples of this are pilot power optimization, Remote Electrical Tilt (RET) optimization and frequency plan allocation.
- If the aforementioned cost function is continuous on the variables under optimization, then the problem can be, in principle, solved with either open-loop or closed-loop iterative schemes. Otherwise, open-loop schemes are the only choice and the application of closed-loop iterative solutions is either problematic or unfeasible. In this respect, pilot power can be optimized with both techniques, but frequency plan allocation requires an open-loop optimization scheme due to the lack of continuity of the cost function on the variables under optimization.
- A requirement for closed-loop iterative schemes to be feasible is that it must be possible to modify the parameters under optimization at virtually no cost, e.g. by means of software scripts. Otherwise, if costly manual intervention is required to implement modifications, iterative schemes become economically unfeasible. This is the case of antenna azimuths, which are typically optimized with open-loop schemes that provide the final parameter values in a single step. However, since adjusting a RET involves marginal incremental cost, this can be done by means of both open-loop and closed-loop iterative techniques.

At a high, conceptual level, some Self-Optimization Use Cases can also be seen as part of the Self-Planning category, since they involve the derivation of parameter values that can be computed during the planning and/or optimization phase. Moreover, when open-loop schemes are applied, the same basic optimization technology is utilized (e.g. the same combination of cost function plus metaheuristic searching algorithms is recommended for both Self-Planning and open-loop Self-Optimization). Note that, in essence, any Self-Planning exercise is an optimization problem, meaning that the best parameter values must be selected among the available alternatives, according to a set of predefined criteria. However, there are two key practical differences between Self-Planning and Self-Optimization:

1. During the Self-Planning phase, the purpose is to derive the settings for a NE (or a set of NEs) that have not been activated yet, whereas Self-Optimization is applicable to NEs that are already being operated.
2. As a consequence, the available input information is totally different for both cases (see Section 5.4). For example, a Self-Planning exercise concerning the RF design of new sites will be typically based on propagation predictions, possibly fine-tuned to match the characteristics of the environment in a generic way. However, when the same problem is addressed from a Self-Optimization perspective, live statistics and/or call traces from the OSS are available,

which allows their incorporation as inputs to the optimization process. In this case, the utilized propagation information will be massively fine-tuned to match the statistics in every particular sector, or even replaced (totally or partially) with geo-located information derived from call traces. However, at the end, the same optimization technology (cost function plus metaheuristic search) will be applied. The only difference is that the availability of measurements allows smart modification of the inputs for the mathematical optimization process.

5.5.1. Minimizing Human Intervention in Open-Loop Automated Optimization Systems

Self-Planning exercises are in general automated, but not autonomous, which means that the decision to plan and deploy new NEs necessarily involves human intervention, even though the bulk of the calculations is assisted to a large extent by automated processes run by computers.

When using the same technologies (e.g. to generate new frequency plans) under the scope of open-loop Self-Optimization, it is debatable whether this process should remain automated or become fully autonomous. In the example of frequency planning, there is nothing that prevents the transition to a fully autonomous operational scheme. However, in practice, many operators still wish to review different high-level strategies and the implications of different trade-offs. Therefore, in many cases some human intervention is still required, unless it is considered acceptable to apply predetermined, fixed strategies in an autonomous way without revisiting them periodically.

Even if open-loop Self-Optimization processes are to be kept automated (i.e. not autonomous), this does not prevent suppliers of Self-Organizing Networks (SON) solutions from making some of the phases in this process fully autonomous. More precisely, the most tedious and time-consuming parts of any open-loop automated Self-Optimization process (i.e. data collection, validation and preprocessing) can be made fully autonomous, in such a way that this part of the process is always running in the background and, at any time, engineers have the latest input information already available to smoothly initiate the value-added part of the automated process (e.g. to evaluate different high-level strategies, etc.).

5.6. Architectures for Automated and Autonomous Optimization

This section is focused on three different architectures for Self-Optimization. Section 5.6.1 is devoted to the centralized version of open-loop Self-Optimization systems, which tend to be automated instead of autonomous although, as discussed in Section 5.5.1, some stages of the process (e.g. data gathering, preprocessing and validation) are subject to be made fully autonomous. Section 5.6.2 covers the centralized version of closed-loop Self-Optimization schemes, and distributed Self-Optimization is described in Section 5.6.3. Both of them are envisioned to be autonomous.

5.6.1. Centralized, Open-Loop Automated Self-Optimization

As already stated, open-loop Self-Optimization schemes do not involve iterative, adaptive processes in which system performance after a change in the network is fed back as input for the derivation of the following configuration change. However, as depicted in Figure 5.1, even though the output of the process is derived in a single step, it considers the network

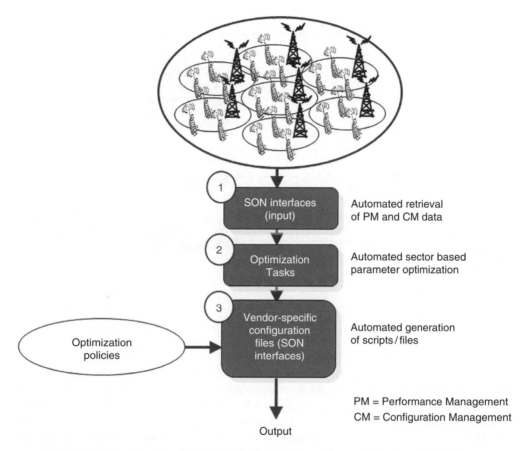

Figure 5.1   Open-loop Self-Optimization scheme with a centralized architecture.

performance before the optimization as a valuable input that makes the process robust against modeling errors that are normally inherent to theoretical models, such as propagation predictions.

5.6.2.   Centralized, Closed-Loop Autonomous Self-Optimization

In the remainder of this chapter, when referring to a centralized, closed-loop and autonomous Self-Optimization scheme, the high level architecture and process in Figure 5.2 will be assumed[1]. Since 2nd Generation (2G) and 3rd Generation (3G) networks were specified and generally built before the SON paradigm was officially formulated by the Next Generation Mobile

---

[1] In this context, there is no specific consideration about the possibility to run centralized Self-Optimization algorithms in the BSC and the RNC for 2G and 3G networks, respectively. However, in practice, those scenarios are similar to the centralized deployments under analysis, with three main differences: (i) when running closer to the base stations, algorithms can react faster to network changes, although quick reactions are not always advisable and statistical reliability of the input information must be ensured in any case; (ii) when running at lower level in the network architecture, visibility is limited to a lower number of network nodes, which is likely to lead to alignment problems in borders between adjacent BSCs/RNCs; and (iii) the implementation of SON algorithms in the BSC/RNC makes it significantly harder to deploy third-party SON solutions.

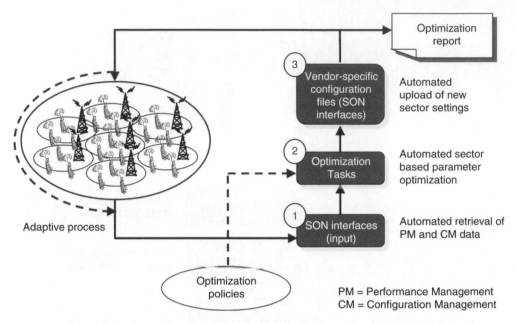

Figure 5.2   Closed-loop Self-Optimization scheme with a centralized architecture.

Networks (NGMN) Alliance, available solutions are dependent on the support provided by equipment vendors. In this chapter, the discussed autonomous Self-Optimization technologies for 2G and 3G RRM parameters (see Sections 5.8, 5.9 and 5.11) are based on a centralized, closed-loop architecture. For LTE, some SON features, such as Automatic Neighbor Relation (ANR), naturally fit a distributed approach, whereas others are powerful when implemented with a centralized architecture. The difference between centralized and distributed SON is explained in Chapter 3. The two architectures can be combined in a hybrid approach to take advantage of the benefits of both.

As depicted in Figure 5.2, different key steps are identified within the Self-Optimization process:

1. Gathering detailed information about the network area to be optimized, with sector or even adjacency resolution. As shown in Figure 5.2, the SON interfaces centralize the access to the different data sources available for each infrastructure vendor and radio access technology.
2. Execution of the optimization algorithm. In this phase, the intelligence embedded in the algorithm is leveraged in order to update the sectors' configuration, emulating the way in which a human expert with no time/effort constraints would reason.
3. Closing the loop, i.e. autonomous implementation of the suggestions (parameter changes) that are proposed by the optimization algorithm. This is done autonomously, in other words, without human intervention. Note that, only by doing this, it is possible to claim full support for the SON requirements formulated by the NGMN Alliance.

This process is assumed to be conducted iteratively to enable closed-loop, adaptive Self-Optimization that ensures the best possible system configuration with sector- and

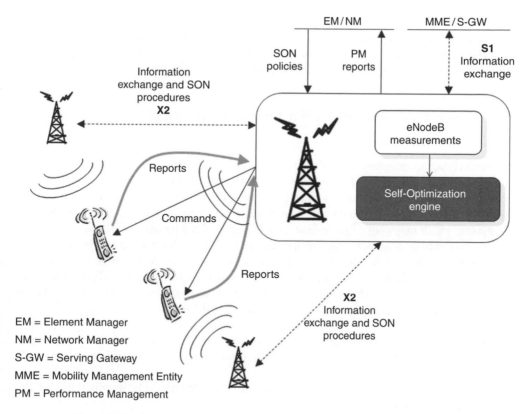

Figure 5.3 General Self-Optimization scheme with a distributed architecture.

adjacency-level granularity. Again, note that the execution of the recurrent adaptive adjustments can be event-driven or scheduled to happen at predetermined times.

5.6.3. Distributed, Autonomous Self-Optimization

In a distributed architecture, Self-Optimization algorithms are executed at the NEs, which are located at a very low level in the architecture. For LTE, multiple mechanisms have been standardized in order to support distributed SON for a selected number of Use Cases. To avoid purely local decisions, information exchange and negotiation processes have been specified (mainly over the X2 interface [13]) in order to allow eNodeBs to rely on information coming from its neighbors for Self-Optimization decisions, which can be coordinated among neighboring eNodeBs. In order to cover Use Cases that have not been specifically addressed during standardization, the necessary standard coordination mechanisms may not exist at a distributed level, which would mean that, for the time being, vendor-specific solutions or centralized SON are the only option to make such functionality available until the necessary functionality is standardized. The distributed optimization paradigm is sketched at a high level in Figure 5.3.

Since 2G and 3G systems were specified before the SON concept was formulated, no standardized support for distributed SON is currently available for those technologies, since the necessary functionalities need to be embedded in the base stations. When no support is available from the equipment vendor, the only alternative is a centralized architecture.

5.7. Open-Loop, Automated Self-Optimization of Cellular Networks

5.7.1. Antenna Settings

The underlying optimization technology for open-loop Self-Optimization of RF settings is the same one that was described for Self-Planning in Chapter 4. However, when applied to live, operational networks, there is the possibility to: (i) extract configuration information directly from the OSS, reflecting reality in a much more accurate way as compared with other databases, which may contain inconsistencies in terms of key parameter values, set of active sectors and current neighbor relationships; and (ii) include OSS performance statistics that allow the utilization of intelligent techniques that calibrate basic propagation predictions in order to fully align them with the network evaluation that results from the analysis of the OSS statistics (see Section 5.4.3) and eventually call traces. Such consistency must hold in different dimensions. Among others, the most important ones are the following:

- Received signal level distribution per sector, e.g. distribution of the Received Signal Code Power (RSCP) of the Common Pilot Chanel (CPICH) and RSCP-driven iRAT HO statistics for UMTS.
- Received signal quality distribution per sector, e.g. CPICH Ec/Io distribution and Ec/Io-driven iRAT HO statistics for UMTS.
- Relative radio coupling (and mutual interference) between pairs of sectors, e.g. Soft Handover (SHO) statistics in UMTS and interference matrix in GSM.
- Propagation delay for each sector, measured by means of timing advance or random access statistics, depending on the technology. This is useful to determine the actual cell footprint, as well as the existence of potential overshooting cells.
- Technology-specific load indicators per sector, e.g. Downlink (DL) transmit power or Uplink (UL) noise rise in UMTS.
- Absolute and relative traffic amount per sector.

Note that, by using networks statistics in this way, the generated RF design truly takes into account the real end user experience in terms of radio conditions, indoor losses and traffic distribution. Additionally, counters related to relative load in the transport domain can be considered during the process in order to avoid RF settings that, although technically sound from a pure radio perspective, may cause transport congestion in some of the base stations.

Moreover, as explained in Section 5.4.4, call traces can be imported (if available at a reasonable cost in terms of investment and operational complexity) in order to further enrich the network view by means of, for example, geo-located traffic, signal level or interference maps. Additionally, call traces can be used to rebuild network KPIs in such a way that they are segmented by different criteria, such as user group and terminal type, thereby allowing prioritization among them when conducting the RF optimization process.

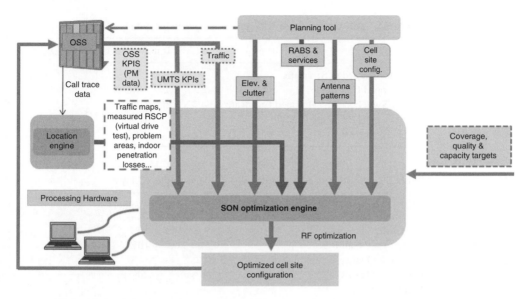

Figure 5.4   Open-loop, automated Self-Optimization of RF settings (UMTS example).

In order to generate a detailed plan for Self-Optimization, the process to collect, validate and preprocess all the required input information can be set to run autonomously so that the optimization can be initiated at any time, always with the latest information available (see Section 5.5.1). A functional layout of the system in charge of carrying out this open-loop, Self-Optimization process in an automated way is depicted in Figure 5.4.

This approach has been used in many commercial networks. A dual-technology example is provided in the following case study, which describes the application of this methodology in a European city to address a joint optimization of the 2G and 3G layers physically sharing antennas. Measured with OSS counters, the results showed remarkable improvements in both 2G and 3G (see Table 5.1 and Table 5.2) after jointly optimizing antenna azimuths and mechanical tilts for both layers (due to antenna sharing) and electrical tilts for 3G.

Another advantage of a measurement-driven approach is that this scheme allows operators to detect inconsistencies in the input information: for example, when the observed propagation delay is totally misaligned with the nominal cell radius reflected in the operator's database. Note that, except for RETs and maybe remote azimuth steering, other RF settings (such as traditional antenna azimuth) can be wrongly stored in the databases, whereas this is not the case for soft parameters, such as RRM settings. Another example is the presence of cross-feeder problems, where two sectors are swapped in the databases and therefore radiate towards totally unexpected directions, which can be detected by comparing cell coupling indications from the OSS with simple geometrical considerations. In general, the detection of outstanding errors should raise a flag and call for manual verification before continuing with the optimization.

Table 5.1 Self-Optimization KPI improvements in the 3G layer

KPI	Improvement
Circuit Switched (CS) Traffic	5%
CS Dropped Call Rate	8%
Packet Switched (PS) Dropped Call Rate	40%
Call Setup Failure Rate (CSFR)	20%
SHO Overhead	10%

Table 5.2 Self-Optimization KPI improvements in the 2G layer

KPI	Improvement
CS Traffic	9%
CS Dropped Call Rate	16%
Good DL Quality	14%
Good UL Quality	20%
Bad DL Quality	14%
Bad UL Quality	23%
DL Quality-HO	22%
UL Quality-HO	15%

5.7.2. Neighbor Lists

In Section 4.9 of Chapter 4, the principles of automated neighbor list planning were presented. Prediction-based neighbor list generation has been traditionally utilized for new sites before activation, based on predicted overlapping coverage. However, in live networks, measurements can be used to audit and automatically tune initial neighbor lists without the need to tune prediction models. As a first step, for all cellular technologies, HO statistics can be accumulated over a period of time recording HO attempts at all neighbor relationships of each cell. Relationships not reporting any or below-threshold number/percentage of HO attempts can be erased from the neighbor list. Enough time should be allowed for accumulation of HO counters, ensuring not to remove relationships relevant to special events that might not have been covered by the accumulated statistics. For example, if HO counters were collected during weekdays and an area is subject to weekend sporting or trade events, counters will be misleading as different mobility patterns and significant increase in HO counts may occur during these special events. This has to be taken into account whenever collecting neighbor statistics.

Lack of HO attempts, as the obvious criterion applicable to neighbor deletion, may be enhanced by processing mobile measurement reports for GSM, and activated periodic measurements (or pilot scanner recordings) for UMTS. These measurements are used to evaluate the probability of interference between cells. The probability of interference is typically expressed through certain percentile C/I ratio, where C denotes the received signal level from the observed serving cell and I denotes the received signal level from a cell considered as a neighbor candidate. Typical percentiles for neighbor candidate C/I ranking are fiftieth, tenth or fifth percentiles. Candidates with a percentile C/I ratio below the chosen threshold are overlapping and should be automatically added to the neighbor list. The percentile *C/I* ratio can be combined with HO counters as an additional criterion for neighbor removal, whereby not only lack of recorded HO attempts, but also a high *C/I*, triggers safe deletion of a neighbor relationship. For example, the criterion for adding/deleting a neighbor relationship may be stated as:

- For each (server *s*, interferer *v*) pair with a percentile *C/I* ratio denoted by *CIR*:
    - If *v* is an existing neighbor:
        - Delete the relationship if (HO attempts $< \mathrm{Th}_{HO}$) and ($CIR > \mathrm{Th}_{dB1}$).
    - Else (*v* is not an existing neighbor):
        - Add the relationship if ($CIR \leq \mathrm{Th}_{dB2}$).

A reasonable choice of parameters, based on field experience, for this simple algorithm might be:

- *CIR*: The value of *C/I* below which a certain percentage of observations fall. For example, if the fifth percentile is adopted, this refers to the value in dB below which *C/I* observations or samples occur with a probability of 5%.
- Th_{HO}: HO attempt threshold, for example 1% of all the HO attempts reported by a given server.
- Th_{dB1}: *C/I* threshold to determine weak neighbor candidates, for example 15 dB.
- Th_{dB2}: *C/I* threshold to determine strong neighbor candidates, for example 12 dB.

In addition to the above algorithm, neighbor lists may be tuned with various additional constraints, including:

- Consider neighbor deletion beyond (or addition within) certain distance, which can be absolute or relative to the average neighbor distance.
- Consider enforcing symmetry when removing/adding neighbors, so that neighbor lists are changed only if the criterion is satisfied in both directions.

An example of automated neighbor list optimization is given in Figure 5.5 with the serving cell circled on the map. This figure identifies three cell classes:

1. Keep: cells that were part of the original planned neighbor list and were confirmed as proper neighbors after deploying measurement-based automatic neighbor optimization. This group is shown in the figure with grey fill.
2. Remove: cells that were part of the original planned neighbor list but were flagged by the automatic neighbor optimization as unnecessary and not justified. This group is shown in the figure with light grey fill.
3. Add: cells that were not part of the original planned neighbor list but were suggested by the automatic neighbor optimization as suitable based on the estimated overlap. This group is shown in the figure with black fill.

As mentioned in Section 4.9 of Chapter 4, to enable this automated measurement-based concept, GSM neighbor lists may be temporarily extended to include more than just current neighbor BCCH frequencies, and periodic reporting for UMTS must be activated.

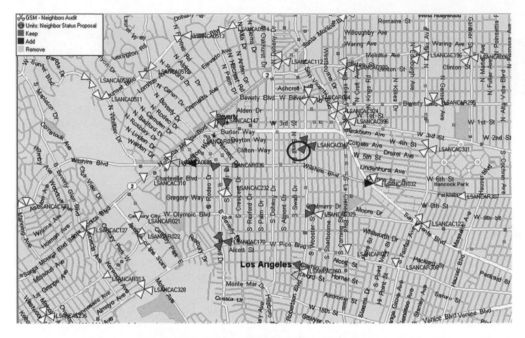

Figure 5.5   Automatic neighbor list optimization.

5.7.3. Frequency Plans

In general, to derive the optimal frequency assignment of NEs, the most accurate option is to apply open-loop Self-Optimization, feeding the optimization algorithms with an interference matrix that has been generated with OSS statistics. However, when NEs have not been launched yet, the only option is to apply a Self-Planning function using an interference matrix that has been generated based on propagation predictions. Moreover, the frequency assignment of NEs that are already on-air may need to be reconfigured due to the introduction of new neighboring base stations. For on-air NEs, the Self-Planning function can use existing OSS statistics to characterize the corresponding victim-interferer relationships in the interference matrix. In other words, the interference matrix used to derive the optimum frequency assignment of the new NEs and the existing neighboring elements should contain a combination of prediction based relationships and OSS measurements based relationships. For the sake of clarity, all the aforementioned aspects of frequency planning and optimization have been addressed in a combined, integrated manner in Section 4.10 of Chapter 4.

5.8.   Closed-Loop, Autonomous Self-Optimization of 2G Networks

Extensive field experience has shown that there are two main Self-Optimization Use Cases that consistently offer noticeable performance improvement when applied in live, operational 2G networks: Mobility Load Balance (MLB) and Mobility Robustness Optimization (MRO). Since 2G networks were specified and generally built before the SON paradigm was defined,

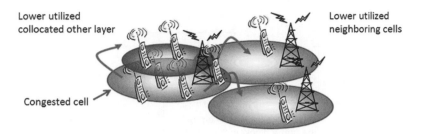

Lower utilized collocated other layer

Lower utilized neighboring cells

Congested cell

Figure 5.6   Traffic balance for congestion relief and CAPEX rationalization.

autonomous Self-Optimization schemes for RRM parameters are centralized by nature (see Section 5.6.2, where the reference closed-loop architecture for autonomous Self-Optimization is presented), although there are increasing initiatives from vendors to offer distributed solutions in this area.

5.8.1.   Mobility Load Balance for Multi-Layer 2G Networks

5.8.1.1.   Description

In live networks, traffic is typically distributed among sectors in a nonuniform way, which results in suboptimal utilization of the available resources. Whereas some sectors may be congested, therefore blocking traffic and calling for capacity expansions, others may have spare capacity at the same time. By applying MLB, some traffic is shifted from the highest loaded sectors to those with lower resource utilization, taking into account the installed capacity in each sector (see Figure 5.6). By doing this, the utilization of the resources across the network becomes more equalized, which results in higher trunking efficiency and therefore lower need for capacity expansions and Capital Expenditure (CAPEX).

For the successful application of this capacity-enhancing strategy, it is crucial to closely monitor and consider coverage and quality metrics, which impose some restrictions when shifting traffic between neighbor sectors, since a massive traffic offload may jeopardize the quality of the RF environment experienced by some users under their new serving cell. As will be seen, it is possible to achieve remarkable capacity improvements while at the same time maintaining or improving quality. In this respect, it must be noted that field results consistently show that, for networks with significant traffic growth, the application of this technology can provide savings in radio expansions of up to 0.15 radios per sector.

The scope of the MLB strategy covers both CS and PS traffic, and moves traffic between sectors within the same frequency band layer and also between (typically colocated) sectors belonging to different layers. During this process, quality metrics need to be continuously monitored and weighted in the decision process in order to preserve Quality of Service (QoS). Since the majority of the parameters governing this process are adjacency-specific, there is full freedom to tune the shapes and sizes of the different sectors, thereby adapting the footprint of the different sectors to the real traffic distribution and the capabilities of the currently installed equipment.

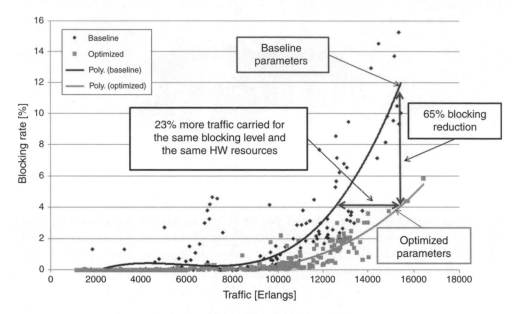

Figure 5.7   MLB results in the field.

Inputs

Among others, the decision process is driven by congestion and quality KPIs, such as Traffic
Channel (TCH) congestion rate, HO attempt/failure rate, Dropped Call Rate (DCR) per
reason where available, Enhanced General Packet Radio Service (EGPRS) KPIs (multiplexing, utilization, etc.), Received signal Quality (RxQual) and Received signal Level (RxLev)
distributions, interference matrix, etc.

Outputs

Based on this information, qualified decisions are made about how to modify HO-related
parameters at the finest possible granularity. Among the modified parameters, the following can
be highlighted: (i) best-cell HO thresholds and hysteresis; (ii) interband HO thresholds; and (iii)
quality- and level-driven HO thresholds. Moreover, cell reselection offsets are tuned in parallel
with the adjustments affecting users in dedicated mode in order to ensure cell border alignment.

5.8.1.2.   Field Results of Mobility Load Balance

In this case study, MLB was applied in a live dual-band GSM cluster with 500 sectors covering a
dense urban area with 10000-15000 Erlangs. Due to remarkable traffic growth, the cluster was
experiencing a TCH blocking rate of up to 15%. The measured results are described in Figure 5.7.

Each point in Figure 5.7 represents the overall blocking and traffic figures aggregated over the
entire cluster during a certain hour of the benchmarking period. There is a set of points that corresponds to the original configuration and another set characterizing the optimized one. As can be
seen, during those hours in which the network carries more traffic, the blocking rates are higher.

Moreover, it is clearly perceived that the blocking rates with the initial parameters are much
higher than the ones with the optimized configuration. For the same traffic level (approximately

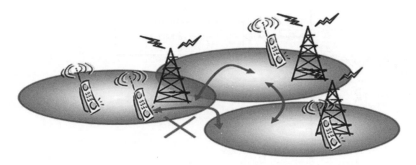

Figure 5.8   MRO for 2G: example of relationship with poor HO performance.

15400 Erlangs for the cluster under consideration), the optimized configuration presents a 65% reduction in the blocking rate (from 12% to 4%). Furthermore, for the same blocking rate (around 4%), the optimized configuration carries 23% more traffic (15400 Erlangs versus 12500 Erlangs for the cluster under analysis). This implies a clear capacity increase, since more traffic is carried with the same HW resources and the same blocking figures. As a result of the optimization, quality metrics slightly improved, mainly due to reduction of HO failures, since a large fraction of them were caused by the blocking of incoming HO calls due to congestion.

5.8.2.   *Mobility Robustness Optimization for Multi-Layer 2G Networks*

5.8.2.1.   Description

In its 2G version, this Use Case is focused on the continuous and autonomous Self-Optimization of several HO parameters in order to improve HO performance and handle network interference. The main objective of this process is quality improvement, materialized in improved HO performance and lower DCR.

The applied algorithm shall identify neighbor relations with high HO failure rate (see Figure 5.8), and analyze the reasons for this. Based on this diagnosis, which is built by cross-correlating all available quality related KPIs at the finest possible granularity, the algorithm shall modify the HO settings in order to reduce these failures. Among others, the algorithm will fine-tune HO margins, triggers for level- and quality-driven HOs, minimal Received Signal Strength Indicator (RSSI) level for a HO candidate, etc. As a result, the following KPIs are typically reduced: number of unnecessary HOs, HO failure rate, HO drop rate and, as a direct consequence, overall network DCR. Moreover, call quality shall improve in general due to timely HOs.

5.8.2.2.   Field Results of Mobility Robustness Optimization

In this case study, MRO was applied in a live GSM cluster with approximately 1000 sectors covering a dense urban area with around 10000 Erlangs, which had been identified to have poor HO performance (in terms of both high HO failure rate and HO drop rate),

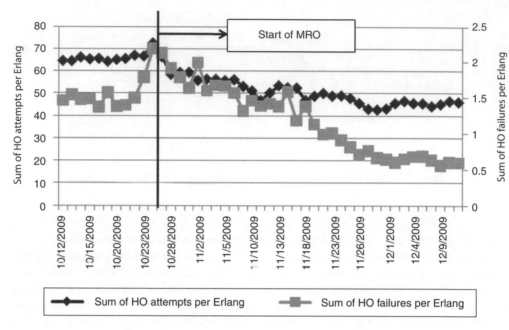

Figure 5.9   Field results from the application of MRO for 2G.

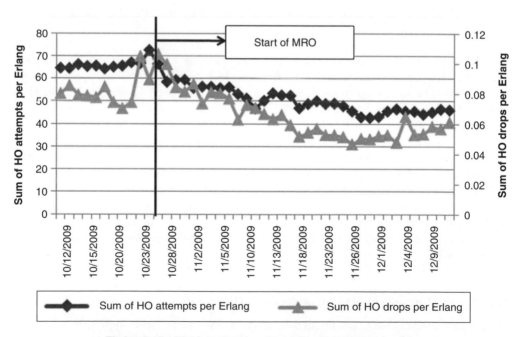

Figure 5.10   Field results from the application of MRO for 2G.

as well as interference problems (high RxQual with high RxLev). The measured results are depicted in Figure 5.9 and Figure 5.10, which represent the daily evolution of the number of HO attempts per Erlang, the number of HO failures per Erlang and the number of HO drops per Erlang. Moreover, a vertical line marks the date in which the MRO algorithm started being applied. From that day onwards, a clear reduction trend can be noticed in all three KPIs, which implies an overall improvement of the HO performance. In addition, and as a consequence of this improvement, the overall DCR was reduced by more than 10%.

5.9.   Closed-Loop, Autonomous Self-Optimization of 3G Networks

As in any Code Division Multiple Access (CDMA) based system, UMTS coverage, quality and capacity are tightly coupled to each other. Therefore, at least in the radio domain, it is not advisable to address the optimization of the settings for the different RRM functionalities separately, since that approach will lead to suboptimal configurations that may counteract each other's effects and even degrade network performance. Instead, the set of parameters to adjust, the implicit mechanisms and their trade-offs need to be considered in a holistic, unified manner, so that all decisions about parameter changes are made by a single algorithm that analyzes the network performance in all important dimensions and pulls all available levers (i.e. all key parameter settings, each one of them governing a certain RRM functionality) in order to pursue the desired goal. Such coordinated decision may include prioritization of one lever over the rest.

In real networks, there are many optimization levers that currently remain untouched or, at best case, fine-tuned at the network level, i.e. with the same settings for all the different cells. As a consequence, even though a UMTS network may be delivering acceptable KPIs, most likely there is still room for increasing its performance in the desired direction (capacity, coverage, quality or a weighted combination of them), just by carefully tuning the different settings on a sector-by-sector basis.

Since UMTS networks were specified and generally built before the SON paradigm was defined, autonomous Self-Optimization schemes in this domain are centralized by nature (see Section 5.6.2 for a description of the reference closed-loop architecture for autonomous Self-Optimization), although there are increasing initiatives from vendors to offer distributed solutions in this area.

5.9.1.   UMTS Optimization Dimensions

As already introduced, UMTS performance can be systematically analyzed by structuring performance information in three dimensions: coverage, capacity and quality. Each one of them is characterized by a set of KPIs that provides the optimizer (either a human expert or an intelligent software module) with insightful information about how the system is performing in each one of these important domains. The discussion in the following is kept at conceptual level, since detailed implementation will depend on the performance counters that are available for each specific infrastructure manufacturer.

5.9.1.1.   Pilot Coverage

In this context, a certain location is considered to have sufficient pilot coverage when the pilot signal and the rest of relevant common signaling channels of the serving sector are received with the signal strength and quality that allow proper decoding according to the Third Generation

Partnership Project (3GPP) specifications [14]. In practice, the coverage area of a sector is typically defined as the set of locations in which the sector is the best server, i.e. the one with the strongest pilot signal. Usual problems related to pilot coverage are the following:

- Poor CPICH related KPIs: presence of areas in which the pilot RSCP or the pilot Ec/Io of the best serving cell is too low.
- Cell overshooting: existence of spots in which the best serving cell is excessively far away.
- Pilot pollution: presence of significant pilot RSCP from an excessive number of cells.
- Cell umbrella: cells with very wide coverage range that overlap with their surrounding cells and create interference problems.

The main cell-specific KPIs that may indicate a coverage problem are the following:

- Share of iRAT HOs due to poor pilot RSCP and/or poor pilot Ec/Io.
- Accessibility indicators, such as Radio Resource Control (RRC) connection setup failure rate.
- Propagation delay, measured via statistics that are collected when accessing the cell via Random Access Channel (RACH).
- SHO statistics, to demonstrate the degree of overlapping between sectors and the severity of pilot pollution.

Of course, bad coverage will also affect other KPIs, but in a collateral way. As already mentioned, in UMTS all dimensions are tightly related to each other.

5.9.1.2. Traffic Quality

One way to define quality is the ability to convey the user plane information to and from the UE with the desired attributes in terms of data integrity, throughput, delay, etc. Bad quality normally results in dropped calls (mainly for voice services) or data calls with poor throughput.

Bad quality is tightly related to bad coverage, which means that spots with bad coverage are likely to suffer from bad quality as well. In those cases, it is recommended to solve coverage problems first. However, there may be quality-specific issues that are mainly related to the configuration of RRM processes that only affect user data transmission, but do not impact common signaling.

The main cell-specific KPIs that may indicate a quality problem are the following:

- DCR, which measures the share of calls that ended due to radio problems. This KPI is relevant for Radio Access Bearers (RABs) and also for RRC connections.
- Frequent power clipping, measured through the share of calls that are frequently close to the maximum allowed DL transmitted power.
- Share of iRAT HOs due to poor pilot Ec/Io.
- Share of calls with poor Channel Quality Indicator (CQI) for High Speed Downlink Packet Access (HSDPA) traffic.

- UL bad quality indicators, derived from RRM events related to high UE transmit power, high UL Block Error Rate (BLER) or high CQI reception failure rate.

5.9.1.3. Capacity

The capacity of a cell is defined as the maximum amount of traffic that it can carry with the desired coverage and quality level. Certain parameterization (e.g. increasing the pilot power significantly or allowing voice channels to transmit with very high power) can solve quality and coverage problems, but may have a very negative impact on capacity when applied blindly. Therefore, capacity considerations need to be factored in when deciding on changes in the parameter settings. The main cell-specific KPIs to consider when assessing potential capacity problems are the following:

- DL transmitted power.
- UL noise rise.
- Channelization codes utilization.
- Blocking statistics related to lack of resources.
- Statistics describing congestion and rejection of bit rate upgrades due to lack of resources.

Moreover, at NodeB and Radio Network Controller (RNC) level, there are other interesting capacity related KPIs:

- Utilization of the Iub bandwidth.
- Utilization of channel elements.
- Call processor load at the RNC.

5.9.2. Key UMTS Optimization Parameters

The key UMTS optimization parameters are conceptually summarized in Figure 5.11. In the following subsections, further comments about the main trade-offs affecting each one of these parameters are provided.

5.9.2.1. Neighbor Lists

Although neighbor lists were covered when presenting Self-Planning technologies in Chapter 4, they can be also treated as a Self-Optimization problem (see Section 5.5 and Section 5.7.2).

5.9.2.2. Scrambling Codes

Although scrambling codes were covered when presenting Self-Planning technologies in Chapter 4, they can be also treated as a Self-Optimization problem (see Section 5.5). All details about the implicit mechanisms and restrictions are covered in Chapter 4.

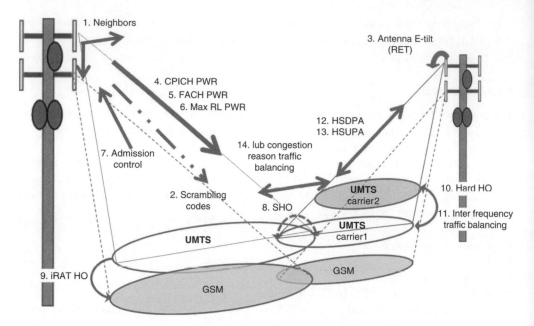

Figure 5.11   Key UMTS optimization parameters.

5.9.2.3.   Electrical Tilt

Although electrical tilt settings were covered when discussing open-loop Self-Optimization of antenna settings in Section 5.7.1, the autonomous closed-loop approach presented in Section 5.6.2 is also valid to fine-tune electrical tilts, provided that the network has RET capabilities. Otherwise, the proposed adaptive, continuous adaptation is not feasible. At a high level, the following trade-offs apply:

- When decreasing the electrical tilt (i.e. when uptilting the antenna), the coverage area of the cell is expanded, but problems can be created or aggravated, such as overshooting, cell umbrella, pilot pollution or capacity shortage.
- When increasing the electrical tilt (i.e. when downtilting the antenna), the coverage footprint of the cell is reduced, which can be useful to correct overshooting problems, to correct cell umbrella problems, to prevent the cell from polluting its neighbors or to offload traffic to other sectors due to capacity problems. However, coverage holes can be created if the tilt is excessive.

As can be seen, its direct impact on cell footprint makes this parameter a powerful lever to carry out traffic balance between sectors. However, unlike the 2G MLB case described in Section 5.8.1, the mechanisms here are less versatile, since the electrical tilt is a cell-specific parameter, not an adjacency-specific one. Moreover, the degrees of freedom to carry out traffic balance with this parameter are further reduced due to the many mechanisms and KPIs (e.g. coverage, pilot pollution, etc.) that also need to be carefully considered. However, in 3G, there are other parameters to support traffic balance such as adjacency-specific SHO offsets.

5.9.2.4.   CPICH Power

The CPICH channel is crucial for system performance and determines the cell coverage area. Therefore, it is important to ensure that it is received with the required signal strength (i.e. RSCP) and quality (i.e. Ec/Io) across the entire cell. If these conditions are not fulfilled, the UE cannot decode the CPICH, which is used for channel estimation in signal reception. Moreover, even though the CPICH RSCP and Ec/Io are above the minimum values required by 3GPP specifications, if there are locations in which they are below the established thresholds for iRAT HO, those calls will be handed over to 2G, which will have negative consequences for the users in terms of the set of accessible services and data rates.

However, having too much CPICH power is also negative and therefore the goal of any optimization strategy must be to minimize overall CPICH power consumption while still ensuring that the required coverage and quality constraints are fulfilled. Minimizing CPICH power is important, due to the following reasons:

- CPICH acts as a source of interference.
- The power that is allocated to the CPICH in a given cell constitutes the reference for the power allocation for the rest of the common control channels.
- Both the CPICH and the rest of common control channels consume transmission power from the amplifier. Since this power is finite, all the power that is allocated to the CPICH cannot be used for user data transmission.
- The CPICH is not power-controlled, which in most cases means that it constitutes an artificial noise floor that determines the absolute level of transmit power for power controlled channels (i.e. the amount of power required by some of the channels actually conveying user data), and indirectly causes an increase in the amount of consumed HSDPA resources per traffic unit.

From a purely algorithmic point of view, the following trade-offs apply:

- If the CPICH is increased, the coverage area of the cell is expanded, but problems can be created or aggravated, such as overshooting, cell umbrella, pilot pollution or capacity shortage.
- When decreasing the CPICH, the coverage footprint of the cell is reduced, which can be useful to correct overshooting problems, correct cell umbrella problems, prevent the cell from polluting its neighbors or offload traffic to other sectors due to capacity problems. However, pilot coverage problems can be created or aggravated.

At first sight, it can be noticed that the trade-offs affecting CPICH power and RET are very similar. In fact, there are strong similarities. However, there are two key differences:

1. When increasing the CPICH, the potential consequences in terms of capacity shortage (e.g. blocking of calls) are more severe than the ones experienced when uptilting the antenna. In both cases, these problems are likely to be worsened due to the fact that more traffic is captured. However, when increasing the CPICH, there are two additional capacity-damaging effects: (i) less power is left for user data transmission; and (ii) resource (power)

consumption per traffic unit is further increased by the additional interference coming from the CPICH and the common control channels.

2. Changes in the antenna tilt affect UL and DL in the same way. However, CPICH changes only improve DL coverage. As a consequence, problems related to UL/DL imbalance might appear and this is the reason why some engineering guidelines discourage the utilization of different CPICH values in different neighboring cells. However, real field experiences have shown that adjusting the CPICH on a per sector basis is a powerful technique when the potential effect of UL/DL imbalance is considered and closely monitored during the optimization process.

As in the case of electrical tilt, CPICH power is also a powerful lever to carry out traffic balance between sectors. However, the same limitations and considerations that were deemed as important in the case of electrical tilt (see Section 5.9.2.3) are applicable here.

5.9.2.5. FACH Power

The power of the Forward Access Channel (FACH) is normally automatically referenced to the CPICH power. Therefore, when tuning the CPICH power on a per sector basis, the absolute value of the FACH power is also adjusted. However, in very seldom cases, it may be desirable to tune the offset that relates the power of both channels, mainly to improve the performance of some call access procedures.

5.9.2.6. Maximum Radio Link Power

This parameter determines the maximum transmission power that is allowed for power-controlled channels, and it is typically defined as a function of the CPICH power. Note that power-controlled channels need to increase their transmission power considerably in certain occasions in order to counteract channel fading and maintain the desired Signal to Noise Ratio (SNR). In this case, the following trade-offs apply (see Figure 5.12):

- If the limit is very high, power clipping will be very seldom, which results in better quality. However, transmission power peaks associated to the compensation for deep fades will consume a significant share of resources and degrade capacity.
- If the limit is very restrictive, no capacity problems will be created, but users in bad radio conditions will often experience power-clipping, which results in poorer service quality and eventually higher DCR.

5.9.2.7. Admission Control and Congestion Control Thresholds

Admission Control (AC) is responsible for controlling the load of the system so that the available capacity can be exploited without compromising the system stability. Before admitting a new UE or modifying the connection of an already admitted UE, AC checks whether these actions will sacrifice the planned coverage area or the quality of the existing connections.

In general, it is well-known that power is a robust integral measure of the network load for Wideband Code Division Multiple Access (WCDMA) systems, supporting both speech and

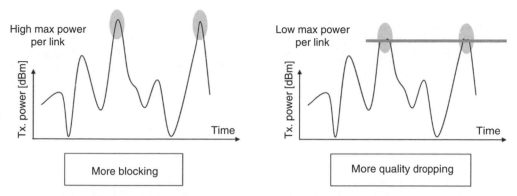

Figure 5.12   Fine tuning the maximum power per link.

variable bit rate data users (see [15, 16, 17]). In general, power based AC algorithms are attractive as they offer trade-offs between capacity and coverage, while taking advantage of the soft capacity offered by WCDMA systems. Therefore, most commercial UMTS systems measure the load in the cells as a function of the total wideband transmitted power for the DL case, and the total wideband received power for the UL case. Since the received power in the UL case is strongly related to the thermal noise power, it is common to measure the UL cell load by means of the so-called noise rise, which is defined as the total wideband received power divided by the thermal noise power.

Typically, AC needs to estimate how the cell load is going to grow after admitting a new call (which can be a completely new call in the network or an incoming HO) or after allowing a service quality upgrade for an existing call. Then, the call will be admitted if the estimated new load does not exceed the applicable AC threshold, which can be different for different service types. In most systems, such comparison is based on average magnitudes, e.g. on the comparison of the average cell load with a given threshold.

Let us focus on the DL case, where the main constraints are typically caused by the limitations of the base station power amplifier. In this case, the AC threshold must be set in order to make sure that the power amplifier can cope with all power demands of the users in the cell, including the power peaks due to fast power control. The maximum AC threshold depends on several factors which include:

- The service profile. A large number of low bit rate users will make the variations of the total transmit power lower, since there will be more independent stochastic processes contributing to the final composite transmit power.
- The propagation environment. In those scenarios with larger frequency diversity, the power variations due to fast power control are expected to be smaller, which allows the use of higher AC thresholds.

The two plots in Figure 5.13 illustrate two scenarios. Both of them present the time evolution of the instantaneous DL transmitted power. The one in the left corresponds to the case in which the variance of the transmitted power is high, while the one in the right illustrates the

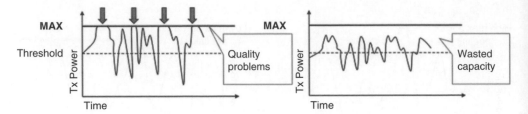

Figure 5.13   Adjusting the AC threshold for different environments.

case in which this variance is lower. Note that a large variance may be caused by a propagation scenario with little frequency diversity and a traffic mix with few high bit rate users. In both plots, the maximum transmit power of the base station amplifier is depicted, together with the AC threshold. Both values are the same in both plots.

In the scenario with large variance, the AC threshold is found to be too high, since the power amplifier cannot cope with the power peaks that the users request by using fast power control. Whenever, the transmitted signal cannot track the requested power variations, the quality of the received signals at the terminal side is lower than required, which causes quality problems that may even lead the system to drop calls.

On the other hand, the scenario with low variance shows that, for the same AC threshold, the power amplifier can perfectly cope with all the requested power variations. In fact, it could even cope with more users before experiencing any kind of limitation due to the power amplifier. Therefore, it can be concluded that, in this case, the AC threshold is too low.

Other mechanisms that affect the variability of the system load for a constant number of served users are the following: variability in the number of incoming HOs, user mobility (towards areas with less favorable propagation environments, such as indoor locations or cell edge) and delivery of services that are likely to experience sudden need for large amounts of additional bandwidth, e.g. when using Variable Bit Rate (VBR) services. The expected incidence of all these mechanisms in each particular cell need to be implicitly considered when proposing values for the AC thresholds, since it may not be acceptable to admit a remarkably large amount of users in a cell in which the load variability that is attributed to already existing users is likely to be very high.

In general, it can be stated that setting aggressive targets may lead to quality (and instability) problems, while a too conservative approach may lead to a waste of capacity. Therefore, the optimal operation point for each sector must be found by seeking the appropriate trade-off between the mechanisms that have been identified here.

Furthermore, it must be explicitly pointed out that analogue considerations can be made for the UL case, in which too aggressive settings may drive users into outage, or make the whole system unstable.

5.9.2.8.   Soft Handover Settings

A user is defined to be in SHO when it is served by multiple sectors at the same time. By means of this mechanism, users near the cell-edge benefit from macro- and microdiversity, which can be translated into capacity and/or quality improvement. The SHO margins control the proportion of radio links in SHO by setting the power imbalance that is allowed between different links within the Active Set. Higher SHO margins imply more users in SHO and therefore a higher average number of serving sectors (commonly referred to as SHO branches) per user.

When the SHO margins are too high, hard blocking may occur in DL due to channelization code shortage. Furthermore, the DL transmit power is used in a suboptimal way, which generates extra unnecessary interference and causes DL congestion problems. On the other hand, when the SHO margins are too low, the UL interference is increased and coverage, congestion and quality problems may become more frequent in the UL.

The benefit from SHO can be seen in two ways. On one hand, its net effect can be noticed by the fact that the call quality is improved (especially at the cell-edge), which also results in fewer dropped calls. On the other hand, in the DL case, the benefits from SHO can be exploited in such a way that the quality improvement is traded for an equivalent increase in the system capacity. Given a certain SHO configuration, the manner in which the benefits from SHO are exploited can be controlled by tuning the parameters that govern the maximum allowed power per link.

Therefore, the optimization of this feature comprises two deeply interrelated tasks: (i) finding the optimal SHO overhead (measured as the average number of radio links per user) given a certain set of performance criteria; and (ii) adjusting the parameters governing the maximum DL transmission power per link so that the SHO benefits are exploited in the desired way.

5.9.2.9. iRAT Handover Thresholds to Ensure Mobility

Within the context of iRAT HO optimization to ensure mobility, the discussion is focused on coverage-based HOs (based on RSCP or Ec/Io), which happen when no contiguous WCDMA coverage is available. In general, coverage based intersystem HOs are conducted when the quality in the current UMTS Terrestrial Radio Access Network (UTRAN) frequency is below a certain threshold and the estimated quality in GSM is above another specific threshold (for more details, see the description of event 3A in [18]). In the case depicted in Figure 5.14, the calls are originally handled by the WCDMA network, and they are handed over to the GSM network when coverage problems are detected.

A common optimization practice is to try to maximize the utilization of the 3G layer, provided that call retainability, smooth mobility and service quality are maintained. This means that the iRAT HO thresholds should be configured in such a way that:

- HOs are not executed too soon (i.e. when the 3G quality/coverage is still good), in order to leverage available 3G resources.
- HOs are not executed too late (i.e. when the 3G quality/coverage is already excessively degraded) since this will lead to poor call quality while in 3G (even dropping the call) and increased probability of iRAT HO failure (therefore damaging mobility as well).

The optimum pilot RSCP and Ec/Io thresholds triggering the iRAT HOs depend on many factors, which change as the network layout, the propagation environment and the traffic patterns evolve. Therefore, it is necessary to apply automated, sector-based optimization on a

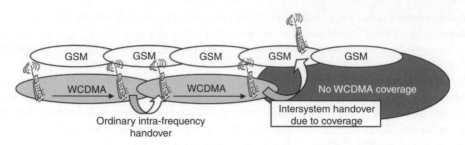

Figure 5.14   Coverage-based iRAT HOs from UMTS to GSM.

periodical basis. For the design of such automated optimization policies, other important factors need to be carefully considered:

- Situations in which too much traffic is offloaded to 2G must be avoided. Therefore, the optimization algorithms must monitor the appropriate 2G quality and congestion metrics very closely, which leads to the need for operational multi-technology support in the applied optimization solution.
- The optimization of thresholds must be carried out independently for each service type. In this way, it is possible to grant more priority to data users in the 3G layer, which may be desirable due to the fact that voice users can be seamlessly served in 2G and 3G, whereas data users suffer from a drastic QoS penalty when transferred to 2G.

5.9.2.10.   Interfrequency Handover Thresholds to Ensure Mobility

The same general ideas contained in the previous subsection (5.9.2.9) may be applicable, depending on the operator's traffic management strategy for multi-carrier UMTS networks.

5.9.2.11.   Interlayer Load Balance

Interlayer load balance is deemed beyond the scope of 3G RRM optimization and is therefore discussed independently in Section 5.11.

5.9.2.12.   HSDPA Settings

In general, HSDPA performance will benefit from a clean RF environment, which is typically obtained by means of careful, cell-by-cell CPICH and antenna tilt optimization. Apart from this, remarkable performance improvements can be obtained by adjusting the following parameters on a per-sector basis:

- Number of High Speed Physical Downlink Shared Channel (HS-PDSCH) codes. Although this allocation is generally dynamic and included as part of the RRM functionality, proper optimization of the initial values and the boundaries for the dynamic allocation scheme results in more suitable trade-offs between HSDPA and Release 99 (R99) traffic, depending on each sector's traffic and resource consumption profile.

- Number of High Speed Shared Control Channel (HS-SCCH) codes. Careful optimization of this setting will result in the proper equilibrium between efficient resource utilization and signaling congestion.
- Measurement Power Offsets. Fine-tuning this parameter on a per sector basis prevents CQI clipping at both upper and lower boundaries, thereby avoiding suboptimal capacity utilization and excessive BLER, respectively.
- HSDPA HO power hysteresis and time to trigger. Sector-based adjustment of these parameters facilitates the optimal trade-off between selection of unsuitable cells that will lead to low throughput (or drops) and excessive ping-pong HOs, which cause throughput degradation.

5.9.2.13.   HSUPA Settings

As in the HSDPA case (see Section 5.9.2.12), High Speed Uplink Packet Access (HSUPA) performance will also benefit noticeably from antenna tilt optimization. On top of this, another key parameter is the maximum UL Noise Rise that is allowed for Enhanced Dedicated Channels (E-DCHs), which limits the maximum interference that is permitted for HSUPA users. Resembling the trade-offs affecting AC thresholds, if this value is too high, quality will be impacted. On the other hand, too restrictive values will involve system operation below its maximum UL capacity.

5.9.2.14.   Parameters for Load Balance between Sites due to Iub Congestion

In this case traffic is balanced between sectors belonging to different sites in order to shift traffic from a congested Iub link to a less congested one. The mechanisms to carry out the actual traffic balance are the same ones that have been mentioned so far: CPICH adjustments, RET fine-tuning and optimization of adjacency-specific SHO offsets. The only particularity is given by the driving criteria, which are oriented to mitigate Iub congestion.

5.9.3.   Field Results of UMTS RRM Self-Optimization

This centralized approach for UMTS Self-Optimization has been proven in numerous deployments in the field. Since the formulation of the policies can be flexibly adapted to the operator's strategy, the different sets of field results can be classified according to the high-level objectives that were devised for the optimization process. Note that no further details are provided about network size and other conditions, since the main purpose is to simply illustrate that the proposed methodology provides tangible performance improvement in operational networks. The results in Table 5.3 correspond to networks in which there was no capacity limitation and the only objective was to improve quality. Results show relative gains

Table 5.3 Self-Optimization KPI improvements in quality-driven scenarios (N/A means that the corresponding KPI was not assessed in that deployment; ~ means that the KPI modification was negligible)

Operator Vendor	1 A	2 B	3 B	4 C	5 A	6 B	7 B
Traffic CS	9.1%	~	10.0%	~	2.7%	~	N/A
Traffic PS	~	9.7%	10.7%	3.0%	~	~	~
CS CSFR	30.7%	20.4%	22.4%	36.8%	35.9%	~	17.5%
PS CSFR	26.7%	59.0%	7.7%	34.4%	44.3%	46.6%	73.6%
3G CS DCR	39.3%	14.3%	35.3%	16.8%	27.3%	10.7%	32.2%
3G PS DCR	5.5%	N/A	12.5%	14.1%	23.8%	15.2%	N/A

Table 5.4 Self-Optimization KPI improvements in capacity-quality-driven scenarios (N/A means that the corresponding KPI was not assessed in that deployment; ~ means that the KPI modification was negligible)

Operator Vendor	8 A	9 B	10 B
Traffic CS	9.5%	9.4%	5.9%
Traffic PS	N/A	N/A	N/A
CS CSFR	N/A	~	30.0%
PS CSFR	N/A	~	N/A
3G CS DCR	~	1.1%	8.7%
iRAT HO CS	30.9%	19.3%	17.3%
HSDPA traffic	N/A	N/A	1.7%
HSDPA throughput	18.6%	19.3%	2.3%
Average CQI	28.2%	28.2%	10.2%
Avg. Tx Power	24.1%	33.6%	15.6%
CS capacity	25.0%	13.0%	9.0%
3G PS DCR	5.5%	N/A	12.5%

in the different KPIs. Positive gains mean improvement, e.g. reduction in DCR, increase in the amount of traffic, etc.

Moreover, Table 5.4 shows results from networks in which the high-level objective was to improve capacity while maintaining or improving quality. As can be seen, capacity was improved while maintaining and/or improving quality and/or coverage. Key quality metrics, such as DCR or HSDPA throughput, were maintained or improved; and the fact that coverage was improved can be noticed by looking at the increased amount of traffic and the reduced share of users being handed over to 2G.

5.10.   Closed-Loop, Autonomous Self-Optimization of LTE Networks

LTE is the first wireless system in which some Self-Optimization functionalities have been standardized, which means that measurement, reporting, negotiation and information exchange mechanisms are available to implement distributed Self-Optimization functionalities. As already introduced throughout this book, for some Use Cases, the distributed approach is the most suitable one, whereas others are better covered by means of a centralized Self-Optimization entity that looks at the network as a whole. Notwithstanding this discussion, a centralized Self-Optimization approach will be the only option for those cases in which the necessary functionality has not been standardized and no vendor-specific distributed functionality is available. In these cases, the modus operandi will be the one that has been successfully applied over the years in 2G and 3G networks (see Section 5.6.2 for the reference closed-loop architecture for autonomous Self-Optimization, although some Use Cases, e.g. those affecting antenna settings that cannot be modified remotely, might require the application of the open-loop architecture defined in Section 5.6.1). In the following, the key advantages of a centralized architecture are summarized for convenience[2]:

- Provided that a rich variety of performance counters is available at the OSS, powerful algorithms can be built without the need for standardization or implementation of specific functionality at enhanced NodeB (eNodeB) level. In this respect, support for detailed performance measurements in today's 2G and 3G systems is sufficient.
- Optimization algorithms can look at the network in a global, holistic way, since the information of the entire region is available in a single source: the OSS. Experience from 2G and 3G shows that, for some key algorithms, it is crucial to cross-correlate information from different cells, which adds complexity and calls for the standardization of new procedures for negotiation and information exchange in a distributed architecture. As a result, time to market with a centralized approach is shorter for new optimization ideas requiring standardization of new inter-eNodeB coordination processes when implemented through a distributed approach, unless nonstandard mechanisms are implemented to coordinate actions between eNodeBs at distributed level.
- Since distributed SON is not currently available for current 2G and 3G networks, all Multi-Technology Self-Optimization strategies need to be built around a centralized architecture (or at least around a hybrid one, allowing the execution of selected LTE Self-Optimization algorithms in a distributed mode). However, in the future, standardized or proprietary mechanisms to support distributed Multi-Technology Self-Optimization strategies may become gradually available.
- For relatively sophisticated algorithms, a centralized architecture is the only way to ensure that fully consistent optimization approaches are applied across interleaved network equipments supplied by different infrastructure manufacturers.
- Different Self-Optimization policies with conflicting targets are more naturally coordinated with a centralized architecture (see Section 3.2.3 in Chapter 3).

---

[2]A more comprehensive analysis, discussing pros and cons, is presented in Chapter 3.

On the other hand, distributed architectures facilitate the execution of Self-Optimization routines in real time. This is more challenging with a centralized architecture, in which the system's response time is limited by the speed at which the performance indicators are updated in the OSS.

5.10.1. Automatic Neighbor Relation

5.10.1.1. Intrafrequency LTE ANR

The Automatic Neighbor Relation (ANR) function is in charge of relieving the operator of manual management of Neighbor Relations (NRs) [19]. NRs are captured in the Neighbor Relation Table (NRT), which contains an entry for each NR, including the Target Cell Identifier (TCI) and some attributes (No Remove, No HO and No X2). The TCI identifies a cell by means of its E-UTRAN Cell Global Identifier (ECGI) and PCI.

The standardized ANR function works in a distributed way and resides in the eNodeB, which keeps a NRT for each cell. The ANR function can carry out the following tasks: (i) add and remove NRs; (ii) provide management mechanism to the Operation And Maintenance (O&M) system to add or update NRs; and (iii) inform the O&M system about changes in the NRT. The process to detect and add a new intrafrequency LTE neighbor is outlined below:

1. The eNodeB sends each connected UE a list of neighbor PCIs with their cell individual offsets (*Ocn*) and configures the conditions that will trigger the events associated to the corresponding measurements.
2. When the UE detects that the received signal of a given cell becomes stronger than that of the serving cell by more than a certain offset, the PCI of that cell is reported to the eNodeB, together with the associated measurement report. UEs carry out this procedure independently of whether the reported PCIs are part of the NRT.
3. If a reported PCI is not in the NRT, the eNodeB orders the UE to decode the ECGI of the newly discovered PCI, as well as the Tracking Area Code (TAC) and all available Public Land Mobile Network (PLMN) IDs. For this to happen, the eNodeB may schedule idle periods to allow the UE to read the ECGI that is broadcasted by the new neighbor associated with the detected PCI.
4. After this process has been completed, the UE reports the ECGI of the new neighbor to the eNodeB.
5. The eNodeB processes this information and may decide to update its NRT. Eventually, it may setup (if needed) a new X2 connection towards the new neighboring eNodeB. This new NR has its default attributes configured in such a way that HO, X2 connection setup and ANR actions to remove this NR are allowed [20].

This process is summarized in Figure 5.15.

5.10.1.2. ANR for iRAT and Interfrequency LTE Handovers

ANR allows LTE cells to find out the Cell Global Identifier (CGI) of 2G/3G/LTE neighbors cells autonomously, through UE measurements. This functionality is available for LTE eNodeBs to detect LTE neighbors since Release 8 [21, 22], and it has not changed in Release 9 [19, 20].

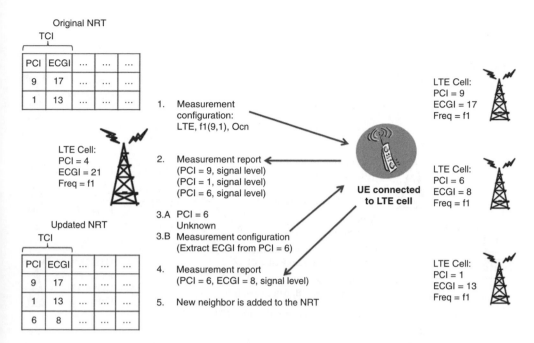

Figure 5.15   Intrafrequency LTE ANR procedure.

However, 2G/3G cells do not currently have any standardized mechanism to autonomously identify the CGI of 2G/3G/LTE neighbors by means of UE measurements.

The multi-technology support in ANR, which also works in a distributed way, allows an eNodeB to uniquely identify new iRAT and interfrequency LTE neighbor cells. Previously, eNodeBs must be configured with the iRAT (and interfrequency) carriers in which the UE will be mandated to search for new neighbors, and only neighbors in the preconfigured carriers will be identified [19]. The process to detect and add a new iRAT or interfrequency LTE neighbor is described below:

1. The eNodeB sends each connected UE a list of frequencies to measure, and configures the conditions that will trigger the commencement of iRAT and interfrequency measurements. Together with each one of the frequencies to measure, a list of PCIs[3], corresponding to already-defined neighbors, can be sent as well.
2. At some point, in one of the monitored frequencies/technologies, the UE detects a PCI and this is reported to the eNodeB, together with the associated measurement report. UEs can detect PCIs independently of whether they are part of the NRT for the corresponding technology and frequency.

---

[3] The information contained in the PCI depends on the cell's technology. UTRAN Frequency Division Duplex (FDD): carrier frequency and Primary Scrambling Code (PSC). UTRAN Time Division Duplex (TDD): carrier frequency and cell parameter ID. GSM Enhanced Data rates for GSM Evolution (EDGE) Radio Access Network (GERAN): Band Indicator + Base Station Identity Code (BSIC) + Broadcast Control Channel (BCCH) Absolute Radio Frequency Channel Number (ARFCN).

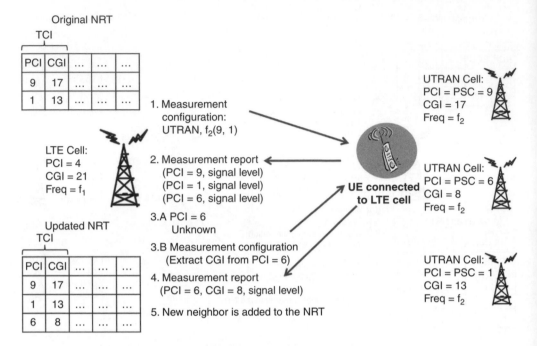

Figure 5.16 iRAT ANR procedure.

3. If a reported PCI is not part of the NRT table, the eNodeB orders the UE to decode the CGI and area code[4] of the cell associated with the newly discovered PCI. For this to happen, the eNodeB may schedule idle periods to allow the UE to read the CGI that is broadcasted by the cell associated with the detected PCI.
4. Once this process has been completed, the UE reports the CGI of the new neighbor to the eNodeB[5].
5. The eNodeB processes this information and may decide to update its NRT.

This process is summarized in Figure 5.16.

In a multi-technology environment, the operator is currently responsible for defining the neighbors of the non-Evolved Universal Terrestrial Radio Access Network (E-UTRAN) cells. To extend the ANR solution to UTRAN cells, in order to automatically detect 2G/3G/LTE neighbors of 3G cells and thereby reduce the operator's workload, a new Work Item (WI) has been created, which is expected to be resolved for Release 10 [23].

5.10.2. Mobility Load Balance

As already discussed in Section 5.8.1, in real world mobile communication systems, network load is determined by the spatial distribution of users. To balance network load with the available capacity in order to get suitable network performance across the network is one of

---

[4] Routing Area Code (RAC) for GERAN, Location Area Code (LAC) and RAC for UTRAN and TAC for E-UTRAN.
[5] If the detected cell is a Closed Subscriber Group (CSG) or hybrid cell, according to 3GPP TS 32.300, the UE also reports the CSG ID to the serving cell.

the targets of detailed network planning by, for example, introducing higher density of cells in areas where higher traffic is expected. However, even with detailed network planning, dynamic changes of the load over time cannot be taken into account. For example, higher traffic in office locations during working hours versus higher traffic in residential areas during the evening; or sudden affluence of traffic in certain transportation facilities immediately after a special event in the surroundings. This dynamic adjustment is considered to be within the scope of network operation, and dynamic load balance aims at optimizing network parameters, so that load can be balanced between adjacent cells, which means, in essence, that load from highly loaded cells is shifted to low loaded cells.

Whereas in 2G and 3G networks only some nonstandard solutions can be implemented for load balancing purposes, in LTE mechanisms for distributed MLB have been standardized by 3GPP within the context of SON. Therefore, load balancing can be done autonomously, based on measurements and by using standardized interfaces.

In the LTE System Architecture Evolution (SAE), an eNodeB is directly connected to the core network through the S1 interface, and with its neighboring eNodeBs via the X2 interface. This distributed architecture enables direct and fast communication between neighboring eNodeBs, also for load balancing purposes. Leveraging these mechanisms, an overloaded eNodeB can obtain direct information from its neighbors about their current radio resource utilization. The overall description of intra LTE load balancing procedures and general requirements can be found in [13] and [19].

5.10.2.1. 3GPP Support for LTE Mobility Load Balance

In 3GPP, MLB has been envisaged as a distributed functionality, in which the algorithms are executed in the eNodeBs. In order to support this Use Case, the following mechanisms have been standardized to facilitate the construction of distributed MLB functionalities, which will be vendor-specific [19]:

* Load reports: by using this mechanism, eNodeBs are capable to exchange load information with other eNodeBs through the X2 interface (for the intra-LTE scenario [13]) or with base stations with different radio access technologies through the S1 interface (for the iRAT scenario [24]). One base station can request others to send load reports (only once or with configurable periodicity) or to stop sending such reports.
* Load balance actions based on HOs: an excessively loaded cell can initiate the HO process for a certain set of users, indicating that the HO reason is 'Reduce Load in Serving Cell'.
* HO and cell reselection parameters adjustment: in order to facilitate a coordinated adjustment, a negotiation mechanism between pairs of eNodeBs has been specified through the X2 interface. According to this negotiation mechanism, the source cell proposes a new parameter value for the target cell, and this parameter change is expressed in relative terms (either as a positive or negative delta, taking the current setting as an implicit reference). Upon this proposal, the target cell can either accept or reject the suggested change. If the modification is accepted, the parameter value needs to be adjusted. Otherwise, the target cell must inform the source cell about the valid parameter range. Note that, in Release 9, there is no specification for such mechanism between an LTE eNodeB and a 2G/3G base station.

Additionally, management mechanisms have been standardized in order to control the behavior of this functionality through the Northbound Interface (Itf-N). With this mechanism, it is possible to activate/deactivate this functionality, configure the optimization policy and collect performance metrics. To specify an optimization policy, operators can define a set of objectives with different priorities, such as allocating the highest priority to maintain the ratio of dropped calls due to excessive load (for example) below 2%, while trying to keep another KPI below (for example) 3% with lower priority. The complete list of potential objectives, their units and their allowed values are available in [25], together with the different performance metrics of MLB.

5.10.2.2.   General Approach for LTE Mobility Load Balance[6]

Exhibit a in Figure 5.17 shows an example of clear traffic imbalance between two cells – belonging to Source eNodeB (SeNB) and Target eNodeB (TeNB), which leads to an overload situation in the cell belonging to SeNB.

To attain the load balancing goal, some of the users have to be handed over (from the source to the target cell) against the natural decision dictated by radio conditions (e.g. being served by the cell with the maximum Reference Signal Received Power, denoted by RSRP), which can be achieved by means of two mechanisms:

1. An additional HO offset between the cells (see Exhibit b and c in Figure 5.17, with dashed line), effectively displacing the border between both cells.
2. The command of load based HOs to a specific set of users. In general, when this is done, the fine tuning of HO offsets is required in order to effectively move the cell border, thereby ensuring that UEs that have been forced to carry out a HO remain connected to the target cell and are not handed over back to the source cell, which is their natural best server, i.e. the cell from which the highest RSRP is received.

5.10.2.3.   Mobility Load Balance Procedure

An eNodeB triggers the load balance procedure when the load in one of its cells exceeds a certain threshold. Via the X2 interface, the eNodeB initiates a load reporting process with the eNodeBs controlling neighbor cells to obtain information about available capacity for load balancing purposes and to create a list of potential target cells. As part of this process, the SeNB sends a RESOURCE STATUS REQUEST message to the eNodeBs controlling neighbor cells and, upon request acceptance, these send periodic reports informing about their load status to the SeNB [13]. Then the SeNB sends requests to the UE to collect HO related measurements concerning potential TeNBs.

In addition to these pieces of information, the MLB algorithm needs to take into account that transferring traffic to neighbor cells has an impact on their loading. Indeed, if the load to be transferred to a certain cell is estimated to exceed the reported available capacity, admission and

---

[6]The work presented in Sections 5.10.2.2, 5.10.2.3, 5.10.2.4 and 5.10.2.5 was carried out within the FP7 SOCRATES project [26], which is partially funded by the Commission of the European Union.

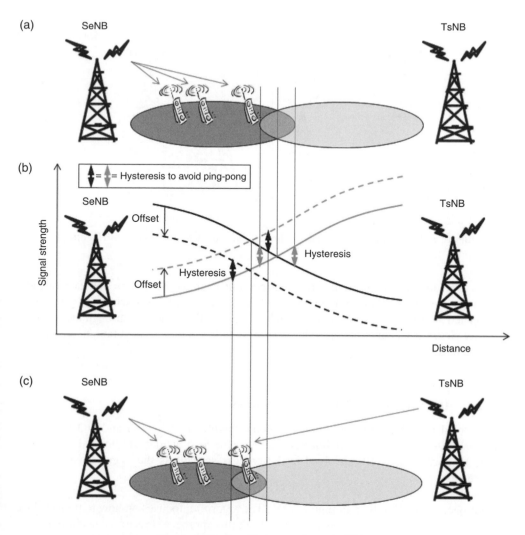

Figure 5.17   Load balance scheme in LTE.

congestion control mechanisms will reject the HO requests, which will increase the required time to reach a satisfactory load distribution and cause unnecessary signaling overhead. When estimating the impact on the load in the target cell, it is necessary to consider that the amount of required resources to serve each UE is strongly related to the Signal to Interference-plus-Noise Ratio (SINR) it is experiencing, which may change after executing the HO.

Using the aforementioned load increase estimation process, proprietary algorithms are used at the SeNB in order to identify the most convenient set of HO offsets, as well as the appropriate set of UEs that needs to be handed over to each TeNB under consideration. Next, the corresponding HO commands are issued to the selected UEs. Since these HOs are triggered by load reasons rather than pure radio conditions, this cause is detailed in the HO request.

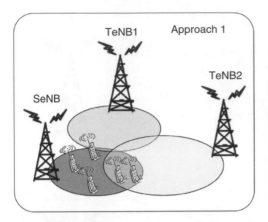 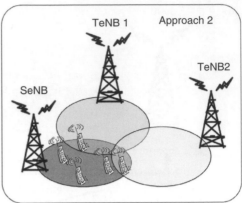

Figure 5.18   Overlapping areas depending on the approach to set the MLB HO offsets.

Moreover, in order to prevent UEs from coming back to the source cell due to radio conditions, the new artificial cell border must be maintained, which is achieved by means of coordinated adjustment of HO parameters at both sides of the adjacency through the specified Mobility Settings Change X2 procedure. More details about this negotiation process are provided in Section 5.10.2.1.

5.10.2.4.   Handover Offset Allocation Algorithms

The detailed algorithms to derive the most convenient set of TeNBs and HO offsets are vendor-specific. For the sake of illustration, two generic strategies are defined, depending on the number of eNodeBs involved in the process as well as the aggressiveness of the adjustments.

1. Approach 1 (see Exhibit a in Figure 5.18). The algorithm focuses on how to transfer load to the lowest number of neighboring cells. With this approach, less of an increase in signaling overhead is expected due to the smaller number of required target cells. For more heavily overloaded cells, the algorithm rapidly moves the operating point closer to the maximum HO offset value. However, high HO offset values involve more drastic changes and may lead to less accurate estimations of how the load in the target cells will evolve after the traffic balance, resulting in rejections in the HO requests and longer convergence times.
2. Approach 2 (see Exhibit b in Figure 5.18). This algorithm redistributes load more equally among neighbors than the first strategy. With this approach, higher additional signaling overhead is expected over the X2 interface because the SeNB needs to communicate with more TeNBs. However, this algorithm operates on lower HO offsets, which involve more accurate estimation of how the load in the neighbors will increase after the load balance. Simulations in Section 5.10.2.5 will assume that this is the selected strategy.

5.10.2.5.   Simulations

The performance of the described MLB functionality was assessed within the scope of the EU FP7 SOCRATES project [27]. The load balancing algorithm that was developed within this project as well as a mathematical framework for load increase estimation can be found in [28]. The general scope and purpose of the simulation campaigns, which utilize the number of unsatisfied users as a key performance assessment metric, is to show the algorithm self-adaptation abilities under various network conditions and evaluate its gain in terms of performance improvement.

Number of Unsatisfied Users due to Resource Limitations
This metric characterizes the number of users in the simulated area that cannot achieve, due to resource limitations, the required link performance that would be necessary to meet the specified QoS targets. Load balancing aims at decreasing this metric by shifting load from overloaded cells to other intrafrequency neighbor cells. The total number of unsatisfied users in the whole network is the sum of unsatisfied users per cell [28]:

$$z_{\text{load}} = \sum_{\forall c} max\left[0, M_c \cdot \left(1 - \frac{1}{\rho_c}\right)\right] \tag{5.1}$$

where M_c and ρ_c represent the number of users and the virtual load in cell c, respectively. The virtual load metric reflects the load situation in cell c and, since it is expressed as the sum of the required resources by all users in that cell, this metric may exceed 100% ($\rho_c > 1$), which would indicate an overload situation. A mathematical framework to derive the average number of satisfied users using the concept of virtual load is explained in detail in [29].

Number of Unsatisfied Users due to Transmission Power Limitation
In case of UL transmission, a UE may not be able to meet its QoS requirements for a given traffic demand even if a number of Physical Resource Blocks (PRBs) is available since the UEs cannot transmit with higher power than the maximum allowed one (*Pmax*). The scheduler should not grant more resources than the amount that the UE can handle considering *Pmax* and the UL target received power per PRB. The total number of unsatisfied users due to power limitation reads as follows:

$$z_{power} = \sum_{\forall c} \sum_{(u|X(u)=c)} \left| M_{max,u} < \frac{D_u}{R(SINR_u)} \right| \tag{5.2}$$

As can be seen, this metric is expressed as the sum of power limited users in every single cell, where Du is a given average data rate requirement per user u, the throughput mapping function $R(SINR)$ yields the data rate per PRB for a given SINR, $SINR_u$ is the SINR of user u, and $M_{max,u}$ denotes the maximum number of PRBs that can be granted to user u. The serving cell c of each user u is determined by the connection function $X(u)$.

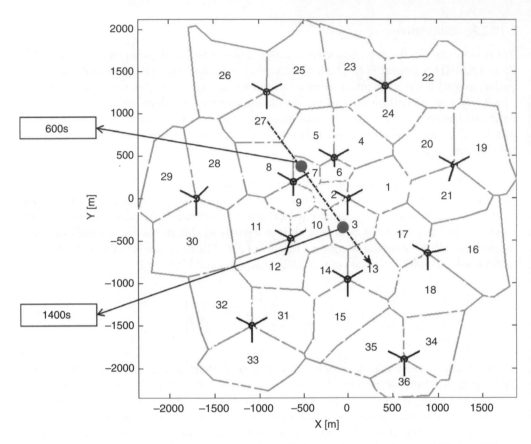

Figure 5.19 In a nonregular network layout, the traffic hotspot is moving from cell 27 to cell 13.

Simulation Assumptions

The simulations were conducted in scenarios where a concentrated group of users (hotspot) moves through the network (Figure 5.19). The network layout includes 12 base stations with different intersite distances between them. Intuitively, such a scenario nicely corresponds with a quite frequent deployment in a town surrounded by area that is more rural. Approach 2 for MLB is selected and the system bandwidth is assumed to be 5 MHz.

In the simulations, the hotspot of 400 users moves from cell 27 to cell 13 in 30 min, which is the observation time for each simulation. On top of this, in each cell, forty dynamic background users move in random direction with a speed of 3 km/h (the total number of users in simulated system is 1840). Each user requires a throughput of 30 kbps.

Simulation Results

Figure 5.20 shows DL simulation results, which illustrate the way in which the MLB functionality in this scenario improves network performance by reducing the number of unsatisfied users.

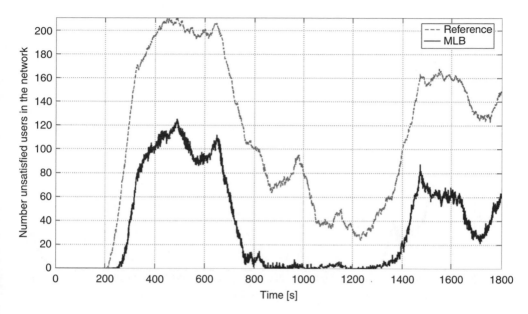

Figure 5.20   DL MLB simulation results with a moving hotspot through a nonregular network.

At the beginning of the simulation timeline the hotspot is created and starts its route in the middle size cell 27. When the hotspot moves close to the cell border the MLB algorithm redistributes hotspot users to cells 8 and 5, and background users to other neighboring cells, achieving the highest gain (~100 fewer unsatisfied users). After 600 s (first landmark in Figure 5.19), the user concentration has fully entered the network part where the cell coverage footprints are smaller due to higher eNodeBs deployment density. In this situation, the MLB algorithm is able to achieve and keep (until the users leave this area) the number of unsatisfied users close to zero, which means that all users are able to obtain the requested throughput of 30 kbps. When 1400 s have elapsed since the simulation start, the user concentration is again in a network area with larger cell coverage footprints, which makes it impossible to satisfy all users by means of MLB techniques, although the positive impact of the MLB functionality is still remarkable. As can be seen, simulations show that the efficiency of the MLB functionality depends on the hotspot position and the associated cell size. UL simulation results are depicted in Figure 5.21.

5.10.2.6.   3GPP Support for iRAT Mobility Load Balance

As will be discussed in Section 5.11, MLB can be implemented in 2G and 3G networks by means of vendor-specific solutions that have been proven to work for intertechnology load balancing and can be extended to scenarios including LTE and other radio access technologies. As stated in Chapter 2, one of the main SON related objectives for Technical Specification Group (TSG) RAN Working Group (WG) 3 in Release 10 is to enhance MLB in order to support multi-technology environments.

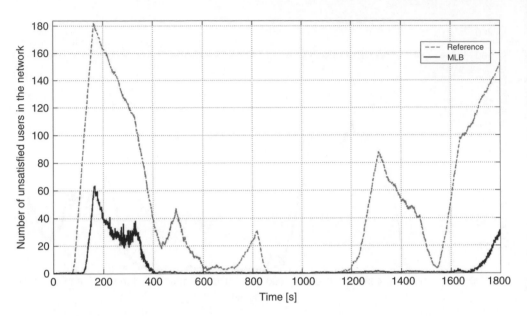

Figure 5.21 UL MLB simulation results with a moving hotspot through a nonregular network.

5.10.3. Mobility Robustness Optimization

5.10.3.1. Intra-LTE MRO

The MRO function is responsible for optimizing HO parameters in order to minimize the number of HO failures, as well as the inefficient utilization of network resources, e.g. in case of ping-pong HOs or cell border misalignment between idle and connected modes [30]. Altogether, this will contribute to improve end-user experience in terms of higher data rates, lower DCR and lower Radio Link Failure (RLF) probability, for example.

This functionality can optimize hysteresis, time to trigger, cell individual offset, frequency-specific offset and cell reselection parameters [31] in order to influence the triggering conditions of the following events: A3 (neighbor cell becomes better than serving cell plus an offset), A4 (neighbor cell becomes better than a threshold), A5 (serving cell becomes worse than a threshold and neighbor cell becomes better than another threshold), B1 (iRAT neighbor cell becomes better than a threshold) and B2 (serving cell becomes worse than a threshold and iRAT neighbor cell becomes better than another threshold). A complete list of the parameters proposed for optimization can be found in [25], and an explanation of the different events, measurement reports and associated parameters is provided in [32].

HO related failures can be classified in three categories: too early HOs, too late HOs and HOs to a wrong cell. For the distributed detection of those failures between cells belonging to different eNodeBs, the exchange of RLF INDICATION and HANDOVER REPORT messages through the X2 interface has been specified. Detailed information about these messages can be found in [13], and a brief description will be provided later in this section. In the following, the process to detect different types of HO failures is explained within the context of a HO from a cell in eNodeB1 to a cell in eNodeB2 [19] (see Figure 5.22 for a graphical illustration):

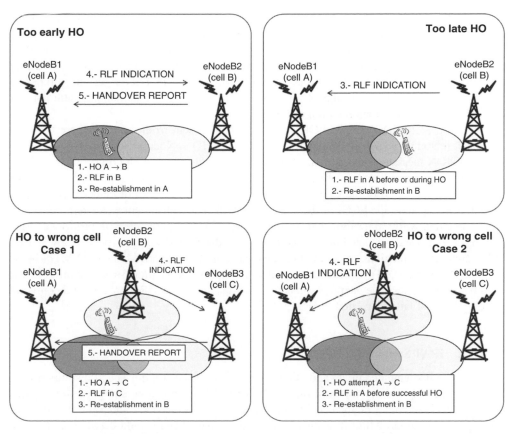

Figure 5.22 Too early HO, too late HO and HO to a wrong cell.

- Too early HOs: a RLF occurs during the HO procedure or shortly after the completion of the HO procedure and the UE attempts to reestablish the connection in the source cell (belonging to eNodeB1). Next, eNodeB1 sends an RLF INDICATION message to eNodeB2, which, upon the reception of this message, sends a HANDOVER REPORT message to eNodeB1 if the HO was not completed or was completed less than $Tstore\_UE\_cntxt$ seconds ago, where $Tstore\_UE\_cntxt$ is a configurable parameter.
- Too late HOs: a RLF occurs before or during the HO procedure and the UE attempts to reestablish the connection in the target cell (belonging to eNodeB2), which sends an RLF INDICATION message to eNodeB1.
- HO to a wrong cell: a RLF occurs shortly after or during a HO towards a third cell (belonging to eNodeB3) and the UE reestablishes the connection in eNodeB2. If the HO was successful (case 1 in Figure 5.22), eNodeB3 may send a HANDOVER REPORT message to eNodeB1 upon the reception of an RLF INDICATION message from eNodeB2. If the HO was unsuccessful and the UE reestablishes the connection in eNodeB2 (case 2 in Figure 5.22), then eNodeB2 may send an RLF INDICATION message to eNodeB1.

When the involved cells belong to the same eNodeB, the detection of the above failures does not require the exchange of messages through the X2 interface.

As already introduced, a RLF INDICATION message may be sent from eNodeB_B to eNodeB_A when a UE attempts to reestablish the connection in eNodeB_B after an RLF in eNodeB_A. The content of the message is the following: (i) the PCI of the serving cell before the occurrence of the RLF; (ii) the ECGI of the cell in which the reestablishment attempt was carried out; and (iii) an identification of the UE.

Moreover, a HANDOVER REPORT message may be sent from the cell where the RLF occurred (eNodeB_RLF) to the source cell of the HO (eNodeB_Source) when a RLF INDICATION message is received at eNodeB_RLF shortly after the HO procedure has finalized. The message will contain: the kind of detected HO problem ('too early HO' or 'HO to a wrong cell'), the ECGI of the source and target cells and, if the kind of detected problem is 'HO to a wrong cell', the ECGI of the cell in which the reestablishment was attempted.

The detailed MRO algorithm that decides how to change parameters is proprietary (i.e. vendor-specific) and is located in the eNodeB (in the distributed solution specified by 3GPP). Other concurrent functions are assumed to be located in the Network Manager (NM) or in the Domain Manager (DM) in order to monitor the optimization process and configure the optimization policies. The operator can activate or deactivate the MRO process and configure the target rate of failures related to HO [25].

5.10.3.2. iRAT Support for MRO

MRO can be implemented in 2G and 3G networks by means of vendor-specific solutions that have been proven to work for intra-/intertechnology HOs and can be extended to scenarios including LTE and other radio access technologies (see Sections 5.8.2 and 5.9.2.9). For LTE, procedures to support MRO in a distributed manner have been standardized in 3GPP (see Section 5.10.3.1).

Although in Release 9 of the 3GPP specifications there is no standardized support for MRO in a multi-technology environment, it is an objective for Release 10 to enhance this functionality to address iRAT scenarios [33].

5.10.4. Coverage and Capacity Optimization

One straightforward manner to optimize and balance capacity and coverage in LTE is by means of adjustments in the antenna settings. This can be done by means of open-loop Self-Optimization techniques or, if RET is available, by applying autonomous closed-loop schemes. Focusing on the electrical tilts, the trade-offs are similar to those that hold for UMTS:

- When decreasing the electrical tilt (i.e. when uptilting the antenna), the coverage area of the cell is expanded, but problems can be created or aggravated such as overshooting, cell umbrella, excessive intercell interference or capacity shortage.
- When increasing the electrical tilt (i.e. when downtilting the antenna), the coverage footprint of the cell is reduced, which can be useful to correct overshooting problems, to correct cell umbrella problems, to prevent the cell from interfering its neighbors or to offload traffic to other sectors due to capacity problems.

However, the detailed performance implications are different, since UMTS can benefit from a moderate degree of overlapping via SHO, which does not exist in LTE. Moreover, the KPIs to assess the coverage and quality of the reference signals are different: RSRP (analog to CPICH RSCP in UMTS) and Reference Signal Receive Quality (RSRQ, analog to CPICH Ec/Io in UMTS) [34].

5.10.5. RACH Optimization

The selected configuration for the RACH is deemed to have a critical impact on network performance. On one hand, the probability of collision when accessing this channel depends directly on the selected configuration (i.e. on the periodicity of the opportunities to carry out this process), which has a strong influence on the delays associated with call setup and HO processes. In this respect, a larger amount of resources allocated to random access involves lower collision probability and therefore lower delays. However, this also involves lower UL system capacity and therefore the appropriate trade-off needs to be found for every particular environment. In addition, it is also important to ensure minimal UL interference, i.e. to make sure that preambles are not sent with more power than required. The optimal configuration in each cell depends on multiple factors such as the amount of served UEs, the HO rate, the call setup rate, etc. These factors are likely to be different for each cell and also time variant, which makes it unfeasible to find the optimal settings manually for each cell.

The parameters to be optimized are the following: preamble type, periodicity of the opportunities to send preambles, the number of preambles that are reserved for the different groups (dedicated, group A and group B), power settings (initial power and power ramp step) and backoff time. The scope of this Self-Optimization Use Case is distinctly local and can be addressed by means of a distributed or centralized architecture.

Random access performance can be expressed in terms of access probability AP(m) or access delay probability, ADP(δ), where AP(m) is the probability of accessing the system before attempt number m (with m = 1, 2, 3...), and ADP(δ) is the probability of accessing the system before δ seconds. In order to estimate these KPIs, some information needs to be provided by the UE. For this purpose, a process called 'UE information' has been defined in the 3GPP specifications [32], which allows the eNodeB to ask connected UEs to send the number of preambles that were sent before accessing the system successfully and whether there was a collision in some of them.

5.10.6. Inter-Cell Interference Coordination

LTE multiple access schemes are based on Orthogonal Frequency Division Multiplexing (OFDM). The implicit orthogonality of OFDM avoids intracell interference and the near-far problem typical of CDMA-based systems like UMTS/High Speed Packet Access (HSPA) [35]. However, the network is still sensitive to interference between cells. Therefore, Inter-Cell Interference Coordination (ICIC) becomes one of the most powerful mechanisms, together with RF optimization, to minimize the overall interference and achieve better network performance.

The level of interference experienced by a given user depends on the amount of received power from its neighbor cells impinging on the set of resources assigned to the user for data transmission and reception. Hence, ICIC can be conceptualized as a holistic strategy related to power control algorithms, smart spectrum allocation schemes or a combination of both. LTE supports coordinated dynamic adjustments [36] of power and frequency resources via standardized signaling between eNodeBs over the X2 interface. This adjustment is done slowly, i.e. within hundreds of milliseconds, in order to keep the network around the desired operating point and provide stability. On the contrary, each eNodeB can take autonomous decisions on a per Transmission Time Interval (TTI) basis to adapt to fast channel variations through packet scheduling and link adaptation policies. The different ICIC schemes can be classified as:

- Reactive schemes: they are based on measurements that monitor the experienced interference. When the measured interference exceeds certain limits some measures can be taken including transmit power adjustments or packet scheduling actions to reduce it to an acceptable level.
- Proactive schemes: they are based on the distribution of future scheduling plans among neighbor cells so that coordinated actions can be taken beforehand to control the interference.

5.10.6.1. 3GPP Mechanisms to Support ICIC

3GPP has standardized several X2 messages [13] to facilitate the implementation of both reactive and proactive ICIC schemes in a distributed way. They are defined below:

- Relative Narrowband Transmit Power (RNTP): this is an indicator for proactive ICIC schemes in DL that points out the maximum anticipated DL transmitted power per Resource Block (RB)[7]. With this indicator, neighbor cells can know which RBs are being used with the highest transmission power, so that different power patterns can be set to configure different frequency reuse models.
- High Interference Indicator (HII): this is a message for proactive ICIC schemes in UL that tells neighbor cells about the RBs in which the serving cell intends to schedule cell-edge users. Similarly to DL RNTP indicator, the HII can be used for coordinating the spectrum allocation in UL.
- Overload Indicator (OI): this is an indicator for reactive ICIC mechanisms in UL that measures the Interference-over-Thermal (IoT) ratio and reports it to neighbor cells in quantized values (low, medium and high). A potential application is to adjust UL power control parameters to maintain a certain maximum IoT level.

5.10.6.2. LTE Constraints for Spectrum and Power Allocation

In order to have a clear understanding of what can be expected from a realistic implementation of ICIC, it is worth providing an overview of power and spectrum constraints in LTE. LTE enables spectrum flexibility since various frequency bands are allowed and the transmission

---

[7] The basic time-frequency resource available for data transmission. Note that one RB corresponds to 180 kHz and 1 ms.

Table 5.5 Number of resource blocks for different LTE channel bandwidths

Channel bandwidth [MHz]	1.4	3	5	10	15	20
Resource blocks [-]	6	15	25	50	75	100

bandwidth can be selected between 1.4-20 MHz, depending on the available spectrum [36]. According to 3GPP specifications, the multiple access schemes in LTE are Orthogonal Frequency Division Multiple Access (OFDMA) for DL and Single Carrier Frequency Division Multiple Access (SC-FDMA) for UL. The UL user-specific allocation needs to be contiguous in frequency to enable single carrier transmission while, in the DL, it is possible to freely allocate RBs from different parts of the spectrum. Table 5.5 shows the number of available RBs for each standard channel bandwidth [36].

There is no power control specified for DL so that there is full freedom for the operator to assign different power density levels to different RBs with the only restriction that all the RBs allocated to a user must be transmitted at the same power [36]. However, power allocation in UL is based on a Fractional Power Control (FPC) algorithm [37]. The idea of classic power control schemes is that all users are received with the same SINR so that the transmit power adapts to fully compensate the path-loss. As an alternative, the FPC approach compensates the path-loss only partially, which means that users with higher path-loss (i.e. cell-edge users) operate at lower effective SINR as compared to cell-center users. This strategy assumes that cell-edge users will produce a higher level of interference, and thus a decrease in their transmit power will improve the overall system performance without causing excessive outage. The FPC algorithm is mainly controlled by the power output (P_0) which sets the average power level, and the path-loss compensation factor (α) that balances the power decrease between cell-edge and cell-center users. The meaning of these parameters (P_0 and α) is illustrated below by means of the expression describing the UL transmission power (P_{TX}) for each UE:

$$P_{TX} = min\{P_{max}, P_0 \cdot N_{RB} \cdot L^{\alpha}\} \tag{5.3}$$

where P_{max} is the maximum UE transmit power, N_{RB} is the number of allocated RBs and L is the DL path-loss.

5.10.6.3. Generic LTE Spectrum Allocation Strategies

Multiple spectrum allocation strategies have been proposed but the main techniques can be grouped as follows [36].

The simplest approach is the well-known Full Frequency Reuse (FFR) scheme, in which the full system bandwidth is available for all sectors and all users. FFR provides high spectrum flexibility, and high peak data rates may be achieved by allocating the whole bandwidth to a single user. However, there is no explicit control over the interference, which might lead to outage, especially in case of high load. Under these conditions ICIC is mainly based on scheduling and power control policies.

An alternative is the Hard Frequency Reuse (HFR) scheme, which aims at mitigating the interference by allocating disjoint spectrum in sectors within a site, so that the radio distance between interfering RBs increases, as shown in Figure 5.23. Although the outage may be

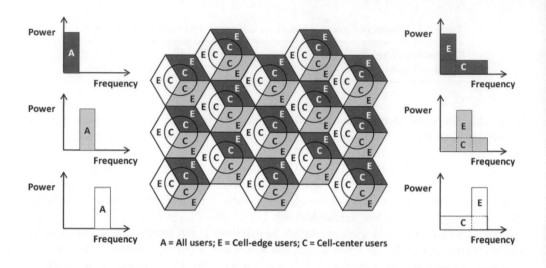

A = All users; E = Cell-edge users; C = Cell-center users

Hard Frequency Reuse   **Soft Frequency Reuse**

Figure 5.23   Hard versus soft frequency reuse schemes.

significantly improved, this technique sets an upper bound to the maximum throughput available per user since it is not possible to allocate the whole bandwidth to a single user. Therefore, the use of HFR may reduce the possibility to offer high data rate services and may also cause misuse of resources under unbalanced load conditions, due to its implicit lack of flexibility.

As a trade-off, LTE supports the so-called Soft Frequency Reuse (SFR) scheme, which pursues the maximization of the network capacity by enabling each sector to utilize the full bandwidth, but at the same time adjusts the power allocated in certain RBs to mitigate interference and, hence, improve user throughput at the cell-edge. Several SFR schemes have been studied [38] but the basic idea is illustrated in Figure 5.23. In a certain part of the spectrum, which is different for each sector within a site, higher power levels are used to transmit, while in the rest a lower transmission power level is utilized. The aim is to increase the transmission power for cell-edge users without penalizing the performance of cell center users. The latter may use full bandwidth and hence, achieve maximum data rates. Different spectrum patterns can be followed in DL and UL, but the recommendation is, for simplicity reasons, to follow a single criterion to classify users as cell-edge or cell-center. Such criterion is paramount for the SFR configuration. Multiple factors, such as the spatial distribution of users, the amount of assigned resources and the service requirements, must be carefully considered in order to configure a border between cell-edge and cell-center areas. The most straightforward criterion is the use of RSRP, which is fully linked to path-loss. However, the traditional sectorized network layout and the presence of shadow fading implies that users with higher path-loss are not necessarily the ones with worse SINR conditions. Another possibility is the use of RSRQ, which can be seen as a wideband SINR and may have better correlation with the experienced SINR. According to 3GPP specifications [32], event A1 (or A2)[8] can be configured to be triggered if either RSRP or RSRQ becomes higher (or lower) than a certain threshold.

---

[8] A1: sever becomes better than threshold. A2: server becomes worse than threshold [32].

Table 5.6   Different settings involved in ICIC and their suitability for
Self-Planning (P) and/or Self-Optimization (O); NA means 'Not Applicable'

	Spectrum allocation strategy		
Parameter	FFR	HFR	SFR
Spectrum allocation areas (UL and DL)	NA	P/O	P/O
Cell-edge threshold	NA	NA	O
DL power offset	NA	NA	O
UL power control (P_0 and α)	O	O	O

5.10.6.4.   ICIC Settings and their Suitability for Self-Planning and/or Self-Optimization

When analyzing ICIC within the scope of SON, it is necessary to differentiate between Self-Planning and Self-Optimization. Determining the generic spectrum allocation strategy to be used can be considered within the Self-Planning category since it is a high-level decision that can be taken in the long-/mid-term based on market policies, network flexibility, etc. However, when applicable in accordance with the selected allocation strategy, the set of RBs that are made available for scheduling in each sector is suitable for Self-Planning by means of the traditional frequency planning techniques discussed in Chapter 4 (before the eNodeBs are commercially activated) and further Self-Optimization. Once an eNodeB is being commercially utilized, measurements are available to fine-tune the allocation and dynamically react to changes in traffic patterns and propagation conditions, as well as modifications in the RF settings. Regarding power settings (DL power offsets and UL FPC parameters) and cell-edge thresholds, Self-Planning is not considered, except to simply apply different default settings for each generic type of cell and morphology (microcells versus macrocells, dense urban versus rural environments, different intersite distances, etc.); however, measurement-based Self-Optimization techniques are fully applicable in order to adapt the settings to each scenario on a per sector basis. A summary of this discussion can be found in Table 5.6.

Optimizing the Spectrum Allocation
The set of available resources for scheduling in a cell, which requires configuration when using HFR or SFR schemes, determines its performance to a large extent. A high number of available RBs increases the number of parallel data streams. However, it also increases the intercell interference and, as a consequence, reduces the achievable throughput per RB. From the viewpoint of Self-Optimization, it is necessary to follow a coordinated policy that determines the suitable scheduling areas for each sector according to the traffic distribution, propagation conditions and service requirements in order to maximize the network performance in the strategic direction mandated by the operator (e.g. desired trade-off between cell and individual user performance). One option is to configure the set of available resources per cell from a centralized solution for Self-Planning or Self-Optimization (depending on the stage in the lifecycle of each cell) although, on top of this, 3GPP also provides mechanisms

(e.g. exchange of RNTP indicator and HII between cells) enabling the implementation of intelligent (and proprietary) RRM (or distributed Self-Optimization) algorithms that help the system tailor its behavior, in this respect, to each scenario.

When adjusting the RB allocation in HFR, intercell interference needs to be minimized. Moreover, at the same time, implicit load balancing considerations need to be taken into account since cells with higher traffic will required more RBs to ensure the desired QoS levels. Similarly, the same applies to cells in which the underlying quality of the RF environment is particularly poor, since the reduction in the throughput per RB calls for the allocation of more RBs in order to maintain the targeted QoS.

The same general principles apply in the case of SFR, although additional requirements need to be considered since the share of resources that are available to cell-edge users for transmission with higher power need to be aligned with the cell-edge threshold that implicitly determines the distribution of users between the two areas. If the threshold is established in such a way that the cell-edge area is remarkably small, the set of RBs in which transmission with higher power is allowed needs to be dimensioned accordingly. Therefore, it is a joint optimization problem (concerning the set of RBs in the cell-edge area of every sector, as well as the threshold dividing the cell into cell-edge and cell-center areas).

Optimizing Power Settings

A well adjusted power control configuration is essential to achieve optimal UL performance. However, detailed Self-Planning of these parameters is not a trivial task because, among other factors, the interference depends on the propagation scenario and the traffic conditions. On the one hand, the power output (P_0) can be Self-Optimized based on a statistical analysis of the OI reports to get the desired level of IoT and guarantee acceptable performance for different link budgets. A cell receiving multiple OI messages is supposed to be causing high interference, and hence, a power decrease is required, while a cell receiving no OI messages might increase its power to achieve better performance without penalizing its neighbors [39]. On the other hand, the selection of the path-loss compensation factor (α) is a trade-off between capacity and outage: higher α yields better cell-edge performance at the expense of penalizing cell capacity.

As previously explained, there is no explicit power control for DL but, if a SFR scheme is adopted, it is possible to balance the performance between users with better and worse SINR conditions by adjusting the power offset, i.e. the additional power that is allowed in certain RBs to serve users in the cell-edge area to improve the overall network performance and fulfill the operator's QoS requirements.

5.10.7. Admission Control Optimization

As presented for 3G in Section 5.9.2.7, Admission Control is paramount for controlling the load of the system so that the available capacity can be exploited without compromising the system stability. At a conceptual level, the principles of Admission Control are similar for UMTS and LTE, although the exact way to measure cell load differs in both systems, since this indicator is highly technology-specific.

In LTE, one way to measure cell load is by computing the average share of time-frequency resources that are actually being used to serve traffic. In general, the translation of traffic volumes into actual cell load is not straightforward since it depends on many factors, such as the radio environment, the spatial distribution of users within the cell, the interference coming from neighboring cells and the traffic profile of each user. That is why it is more advisable to base AC decisions on loading factors than adopting an approach purely driven by traffic volumes.

The objective of a Self-Optimization algorithm for AC is to fine-tune the different AC thresholds dynamically in order to obtain the desired trade-off between capacity, coverage and quality for all the cells in the network, each one of them being operated under different circumstances in terms of propagation, interference, traffic profile and spatial users distribution.

In general, higher AC thresholds will lead to higher capacity in terms of lower blocking and HO failure rate. However, this will be obtained at the expense of degraded QoS metrics since it will become more probable that the system does not have enough resources to satisfy all users if interference degrades, users move towards less favorable locations (e.g. towards the cell-edge or indoor locations) or certain data services experience a sudden need for additional bandwidth (e.g. when using VBR services).

On the other hand, although low AC thresholds are more benign for the QoS of existing users, they are too conservative in terms of capacity and cause higher blocking and larger HO failure rate (due to rejection of incoming HOs).

The high level trade-off between capacity and quality, as well as the relative prioritization between different services, is to be decided by each operator, but then translating this directive into concrete parameters at each cell is not straightforward since each cell will experience a different situation. In general, more static conditions allow more aggressive AC thresholds and, in turn, higher system utilization. Therefore, the only way to optimize AC thresholds in order to make sure that the system is being operated optimally is to apply adaptive Self-Optimization algorithms that, in general terms, base their decisions on careful evaluation of detailed capacity and quality KPIs for the different services and make a qualified decision, taking into account the operator's policies and the technology-specific trade-offs. In principle, such Self-Optimization can be distributed or centralized. However, the complexity of the problem and the soft nature of the involved parameters call for the utilization of closed-loop architectures.

5.11. Autonomous Load Balancing for Multi-Technology Networks

A powerful technique to make mobile networks more efficient without adding extra infrastructure is to treat all technology and frequency layers as a combined global resource in which different parts have different characteristics. Typically, areas with significant coverage overlapping between technologies present remarkable equipment redundancy which should be exploited in an intelligent and flexible manner in order to increase the overall trunking efficiency of the system or improve other metrics such as coverage and quality. The aforementioned flexible utilization of the deployed infrastructure should be dynamic enough so that the network can be fully adapted to the operator's requirements and strategy at each stage of the network lifecycle.

It is a fact that not all the radio access technologies can deliver the same services and/or the same quality at the same cost. These aspects must be taken into consideration when defining a network-wide load balancing strategy in which different services (voice, streaming data calls, interactive data calls, etc.) will prioritize the different technology and frequency layers in a different manner. For instance, one possible strategy would be to prioritize, for idle mode,

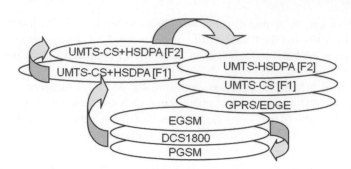

Figure 5.24   Multi-technology, multi-layer load balancing with 2 UMTS carriers and different strategies.

the most advanced technologies supported by the UE so that it can access the most advanced services. After the call is established, it can be transferred to the most suitable layer to maximize capacity and/or quality and/or coverage depending on the network operator needs.

A preliminary, high-level example of multi-technology load balancing is depicted in Figure 5.24, in which UMTS and GSM layers are further divided into different frequency sublayers. As can be seen, when two carriers are available for UMTS, each carrier can be dedicated to serve voice traffic, data traffic or a combination of both, depending on the operator strategy and the frequency band of each carrier (for instance, F1:850/900 MHz band, F2:1900/2100 MHz band).

5.11.1.   Load Balancing Driven by Capacity Reasons

Capacity problems are traditionally solved by adding more HW to the already installed equipment base but, alternatively, most infrastructure manufacturers also offer some RRM features to relieve punctual congestion problems. The main characteristic of these mechanisms is that they can react quickly to resource shortage spikes, usually at the expense of the call quality. Those RRM solutions treat the problem locally and are based on a combination of call resource reduction mechanisms (causing call quality degradation or blocking), blind offloading (facilitating or forcing outgoing HOs without considering the impact on the performance of the target cell) and call preemption. These congestion relief mechanisms are essential for punctual or short-term capacity problems (from seconds to tens of minutes), but not for persistent problems because they do not guarantee sustainability for mid-/long-term traffic growth.

One possible complementary mechanism is the utilization of an autonomous external optimization entity with a macroscopic, global vision of the network (i.e. with a centralized, closed-loop architecture as defined in Section 5.6.2). This entity would interface with all RANs to get both configuration information and performance statistics, apply a set of optimization policies that are in line with the operator's traffic management strategy and, in return, propose a new configuration to be loaded onto each RAN in order to adjust the network to the current traffic pattern. From this point forward, the discussion is focused on UTRAN, but most expla-nations can be easily extrapolated to other radio access technologies. In general, the parameters that control traffic balance mechanisms can be categorized in two different levels:

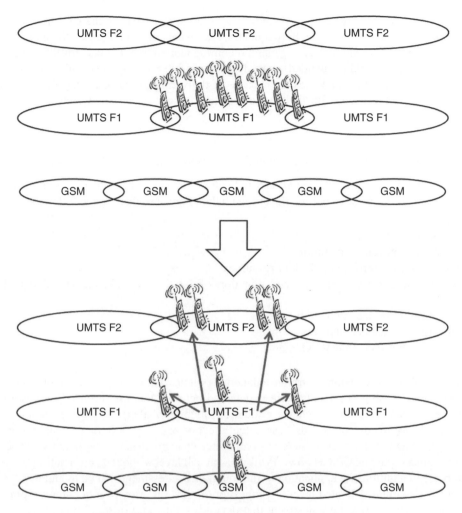

Figure 5.25 Example of load balance via cell reshaping.

- Cell level parameters, which affect the entire cell under consideration.
- Neighbor/adjacency level parameters, which only affect a certain neighbor relation. This category can be divided into the four following types:
  i. Intrafrequency neighbor related (for neighbors in the same UMTS carrier): IAF.
  ii. Interfrequency neighbor related (for neighbors in different UMTS carriers): IEF.
  iii. Intersystem (to GSM) neighbor related (for GSM neighbors): ISG.
  iv. Intersystem (to LTE) neighbor related (for LTE neighbors): ISL.

This multi-level neighboring concept is depicted in Figure 5.25 in which an example of a load balancing via cell reshaping is also illustrated.

In general, LTE should have the highest priority, followed by UMTS and GSM in this order (subject to terminal capabilities). This means that cell reselection parameters should be set to favor camping on the most advanced technology that is supported by the UE, provided that appropriate coverage is available. Subsequently, load balancing techniques will be applied to readjust the traffic distribution. As a general recommendation, moving data traffic towards an older technology should be avoided, since degradation in service quality will be noticeable. However, a priori, all technologies are capable of delivering the same quality for voice services, and therefore users with this traffic profile can be moved across technologies with higher flexibility.

5.11.1.1.   KPIs Driving the Load Balancing Process from the UTRAN side

A sample list of the most relevant UMTS indicators that can trigger the decision to modify parameters in order to carry out interlayer load balancing is provided in the following:

- DL transmitted power.
- Channelization codes utilization.
- Blocking statistics related to lack of resources.
- Statistics describing congestion and rejection of bit rate upgrade requests due to lack of resources.
- Utilization of the Iub bandwidth.
- Utilization of channel elements.
- Call processor load at the RNC.

In general, the utilization of traffic balancing policies between two neighbor sectors is triggered when a significant imbalance is detected between them in terms of congestion or resource utilization. Even though the evaluation of this imbalance is carried out on a per adjacency basis, the outcome of the assessment must be analyzed both for each neighbor and in an aggregated manner since there are parameter changes that will be neighbor-specific, whereas others (e.g. the CPICH power) will work at cell level without discriminating between neighbors. In addition, the application of the imbalance assessment will be modulated by the operator's maximum tolerated level of congestion/utilization. Moreover, a common practice is to implement a parameter adjustment that depends on the size of the stimulus, i.e. within the allowed ranges, more aggressive changes will be applied when the detected imbalance is higher.

5.11.1.2.   Mechanisms for Load Balancing from the UTRAN side

Two main mechanisms for load balancing can be used in parallel.

Cell Coverage Adjustment
This can be done via CPICH power and RET adjustments. In general, this mechanism will reduce the coverage area of highly loaded cells, which will automatically offload traffic to less loaded neighbors. On the contrary, low loaded cells will expand their coverage footprint if their neighbors are highly loaded in order to capture more traffic and also compensate for potential coverage losses caused by footprint reductions in highly loaded cells. More details about the different implications of coverage adjustments were provided in Sections 5.9.2.3 and 5.9.2.4. Since these adjustments will directly impact coverage and quality, the effects

described in those sections must be carefully considered when designing the detailed rules for capacity-based load balancing.

Handover and Cell Reselection Adjustment

This mechanism fine-tunes HO and cell reselection thresholds (at cell and adjacency level for IAF, IEF, ISG and ISL neighbors). By adjusting these parameters, cells are shrunk or expanded by triggering HOs nearer or farther away from the initial (equilibrium) point. Note that, although at high level the principle seems equivalent to the cell coverage adjustment, there is a fundamental difference, since, in this case, pilot coverage is not impacted, and therefore the optimization rules do not need to consider the potential creation of problems such as pilot coverage holes, overshooting, pilot pollution, cell umbrella, etc.

Neighbor-specific HO parameters are typically defined as an offset to be added to the radio measurement of the neighbor cell that is evaluated for the HO decision. In this way, a neighbor can be biased using a positive or negative value which incentivizes or discourages the HO decision, depending on the direction in which the cell border is to be displaced along that specific adjacency. The cell border is redefined unequally depending on the overlapping areas among neighbors, which provides full flexibility to adjust the cell size and shape to the traffic distribution and the system layout. In this respect, it must be noted that HO and cell reselection parameters should be changed consistently in order to minimize the volume of early iRAT HOs (which happen, for example, when the cell border for iRAT HO is shrunk while the idle mode border is not adjusted accordingly, since calls in the discrepancy area are started in 3G and then handed over to 2G immediately after the establishment, causing an unnecessary waste of 3G resources) and reduce the number of unnecessary location area updates (which happen, for example, when the cell border for iRAT HO is expanded while the idle mode border is not adjusted accordingly, since UEs in the discrepancy area will be quickly moved to 2G right after finishing the 3G call). Furthermore, it must be noted that this mechanism is less effective and more limited for UMTS IAF neighbors since border shifts have an important impact in call quality due to interference. For the rest of neighbors, the changes could be remarkably more aggressive due to frequency orthogonality.

IAF, IEF, ISG and ISL neighbors might be treated differently by using different threshold ranges (lower ranges will imply fewer additional HOs and/or reselections). Moreover, some neighbors can be prioritized by configuring more appealing HO conditions in order to transfer more traffic towards them. In general, cell level parameters are less flexible than adjacency-level parameters, and therefore should only be used if neighbor level parameters are not available or are insufficient to solve a capacity problem.

5.11.2. Load Balancing Driven by Coverage Reasons

When new radio access technologies are introduced, there is a stage in the network lifecycle in which the coverage of the new technology is non-contiguous, only focusing on certain coverage islands that are typically associated with traffic hot spots.

When those deployment islands are not being operated at their capacity limit, interesting results can be achieved if their coverage area is extended in order to offer more advanced and higher quality services to as many users as possible, especially for data applications. This process implies a controlled conversion of the available capacity surplus into additional coverage, mainly using cell coverage adjustment mechanisms, but also trying to extract additional capacity for future traffic increases or additional coverage expansions.

In UMTS, for example, one of the ways to convert capacity into coverage is by means of controlled antenna uptilts and pilot power increases. This needs to be done in a careful, adaptive way, aiming at capturing as much additional traffic as possible, while at the same time ensuring that quality is not degraded beyond unacceptable levels, and taking into consideration that some spare capacity needs to be maintained to cope with load and traffic peaks. In parallel, load balancing (mainly for voice calls) should be used to move calls to different layers in order to release resources for extra traffic or further coverage increases.

More details about the different implications of these coverage adjustments were provided in Sections 5.9.2.3 and 5.9.2.4. Since these adjustments will directly impact coverage and quality, the effects discussed in those sections must be carefully considered when designing the detailed rules for coverage-based load balancing.

5.11.3.   Load Balancing Driven by Quality Reasons

Most quality problems can be resolved after maximizing capacity and/or coverage. However, there are additional mechanisms that need to be considered, mainly for those cases in which quality problems are caused by interference. Typical quality problems are: dropped calls, high BLER and/or low data throughput.

In this respect, load balancing helps in reducing the interference level in a certain layer by moving traffic from bad frequencies to cleaner ones. Ideally, for voice calls, the recommendation is to count on all the technologies as potential candidates. However, for data traffic, it is not a recommended practice to move user towards a less advanced radio access technology. Within UMTS networks, a good insight about the relative quality provided by different carriers can be obtained by looking at CQI distributions and Ec/Io statistics.

After detecting quality problems and determining the direction in which traffic needs to be moved, the already proposed methodologies (cell coverage adjustment and HO/reselection adjustment) can be applied, although in the case of cell coverage adjustment the same considerations that were presented in Sections 5.9.2.3 and 5.9.2.4 need to be taken into account.

5.11.4.   Field Results

Using the general approach outlined in Section 5.6.2, the three main multi-layer load balancing strategies (capacity, coverage and quality-based) were applied in a UMTS RNC with 2 carriers (850 MHz and 1900 MHz). There was no service differentiation between both carriers (both of them carrying voice and HSDPA data traffic), and the traffic balancing algorithm was freely applied between both layers with no restrictions, although GSM was prohibited as target layer for voice load balancing as per the operator's policy. Table 5.7 shows the results after 10 incremental, adaptive optimization steps, and some of the KPIs shown in Table 5.7 are further illustrated below:

- High Resource Utilization: takes into account AC rejections, directed retries, load sharing events or any other event related to capacity shortage.
- HSDPA Throughput: is the average throughput per TTI and cell. It represents the user throughput that would be experienced if no code multiplexing was used and users were transmitting data all the time (with no silence intervals) during the measurement period.

Table 5.7 Self-Optimization KPI improvements when applying multi-layer load balancing

	Original KPI	Final KPI	KPI improvement
High Resource Utilization (%)	5.92	4.19	29.22%
Voice DCR (%)	1.91	1.46	23.56%
Voice Call Setup Failure Rate (%)	2.05	1.16	43.41%
HSDPA Throughput (kbps /cell)	1095	1200	9.59%
HSDPA Call Setup Failure Rate (%)	2.09	1.19	43.06%
Voice (Erlangs)	28944	28140	2.78%
Voice Traffic Share 1900 MHz (%)	38.51	44.93	16.67%
HSDPA Traffic Volume (Gbits)	4710	4781	1.51%
Voice ISG HO (%)	0.571	0.423	25.92%

- Voice Traffic Share 1900 MHz: represents the percentage of voice traffic carried by the 1900 MHz layer over the total voice traffic. Since the number of cells in each band was equal (and taking into consideration the propagation difference between both bands), the 1900 MHz carrier was much cleaner than the 850 MHz one. Therefore, the quality related rules in the optimization engine tried to push more traffic from 850 MHz to 1900 MHz.
- Voice ISG HO: is the percentage of UMTS voice calls that were handed over to GSM due to coverage problems. The coverage-driven policies of the optimization engine were mainly in charge of improving this KPI.

5.12. Multi-Technology Energy Saving for Green IT

Apart from metrics purely related to service performance, today's telecommunications businesses are facing increasingly and unprecedented pressure from all involved stakeholders to adopt products and operation paradigms that are environmentally friendly, and all aspects related to green IT are receiving growing attention when designing the strategy of wireless operators, handset manufactures and network infrastructure vendors. Different multi-partner research projects focused on energy efficiency for mobile networks have been conducted [40, 41], and the topic has received the attention of both NGMN and 3GPP (see Chapter 2).

In general, the society demands an environmentally respectful attitude on two main fronts. On one hand, the materials in the different hardware elements of the network should be obtained, when possible, by recycling disposed artifacts or at least the products should facilitate feasible and sustainable recycling of their different components when the lifecycle of these equipments is exhausted. On the other hand, huge attention is paid to their energy consumption, which should be minimized as much as possible. This trend is not exclusive for those equipments related to wireless communications, but more general across many industries and markets [42]. In addition, its attractiveness for cellular operators is not only related to the environmentally friendly character of these initiatives, since the business case for Energy Saving (ES) and green IT is remarkable appealing by itself, just looking at the Return On Investment (ROI). In this section, the discussion is focused on the minimization of energy consumption in wireless cellular networks, leaving manufacturing discussions aside.

5.12.1. Approaching Energy Saving through Different Angles

Different approaches can be taken in order to minimize the energy consumption associated to the operation of cellular networks, and all of them can be adopted in parallel since they are not mutually exclusive. Among others, the most relevant approaches are the following (see Figure 5.26):

- Efficient design of handsets from an energy consumption point of view. This path is fully related to hardware design and can be addressed for all current radio access technologies, since the lifecycle of terminals is much shorter than the time span during which legacy networks will be still in operation.
- Application of radio planning and optimization techniques that are at least partially oriented to maximizing the battery duration at the handsets. For example, in UMTS, such techniques include the optimization of soft parameters such as the Discontinuous Reception (DRX) Cycle Length Coefficient for paging in idle mode, the parameters controlling the execution of measurements on intrafrequency, interfrequency and iRAT neighbors in idle mode, the hysteresis for cell reselection in idle mode and the parameters related to UL power control.
- Application of radio planning and optimization techniques oriented to reduce the utilized power in radio transmission, both in terms of pilot power to provide basic signaling coverage and power allocated to user data transmission [43]. Such a scheme complements traditional power control schemes which have been widely applied to reduce/control interference.
- Making sure that the hardware design of the new base stations to be deployed is optimized from the energy consumption perspective. This is feasible for new network equipment, i.e. mainly for LTE across the world, for UMTS in some emerging markets, for UMTS900 in mature markets (deploying 3G in this band to deliver sustainable mobile broadband) and for network modernization campaigns [44].
- Adopt site construction strategies that take Air Conditioning (A/C) systems into account as an important contributor to the overall energy consumption of the site. In this respect, two main aspects need to be considered: (i) site type and location, since outdoor cabinets may only require ventilation fans[9] and well isolated rooms for indoor cabinets may help reduce A/C energy consumption; and (ii) deployment of tight control systems for the A/C activity, which should be kept to a minimum in periods with lower network activity and lower temperature.
- And last, fully aligned with the main topic in this chapter, optimizing the number of hardware units that are in operation in every moment, so that their number is minimized while still ensuring the desired QoS. In other words, switching off base stations (or modules within base stations) when they are not needed (e.g. capacity sites in office areas during the night). This strategy will constitute the main focus of the discussion throughout the rest of this section [44, 45].

---

[9] Depending on the specific weather conditions.

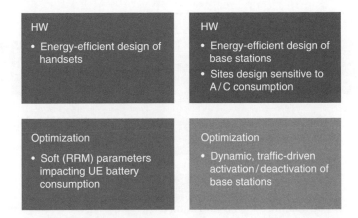

Figure 5.26   Strategies for Energy Saving in wireless cellular networks.

5.12.2.   Static Energy Saving

The simplest scheme is the one in which the times for switching on/off each one of the base stations (or modules within the base stations) are prescheduled by analyzing historical statistics that contain information about the traffic that is served by each base station at each time of the day. However, this approach cannot be applied blindly and on a fully continuous basis mainly due to two reasons.

First, the fact that a certain base station has not been serving traffic at a certain time during a long period does not mean that it can be just blindly switched off. If there are areas in which the sectors within that base station are the only feasible servers, switching off that base station means jeopardizing the network coverage, even though it is extremely unlikely that any user makes use of that coverage at the time of the day in which the base station is scheduled to be inactive. Operators need to be aware of the areas in which a base station that is candidate to be switched off behaves as an exclusive coverage provider, i.e. with no other base station that provides suitable backup coverage (in the same or other technology/band). Upon evaluation of this information, a discretionary decision needs to be made about its participation in the ES program, weighting in Quality of Service and regulatory considerations. For example, in a GSM multi-layer network, a capacity site in GSM1800 is likely to have full backup service coverage provided by one or more GSM900 sites. However, note that a high likelihood does not necessarily involve 100% probability. Moreover, it is important to highlight that such assessment needs to be made across several dimensions since the conclusion may be different for different service types. For example, it may be acceptable to deactivate a certain UMTS base station because there is suitable backup coverage for voice services. However, if the assessment is made for a certain type of data service with a concrete set of requirements, the conclusion can be completely different.

And second, even though the aforementioned lack of appropriate backup service coverage is not a problem, operators also need to take into consideration that traffic profiles are dynamic. Thus, today's valid conclusions may become invalid in a short period. Therefore, there is a clear need to revisit the assessments in this respect with certain periodicity. For example, the

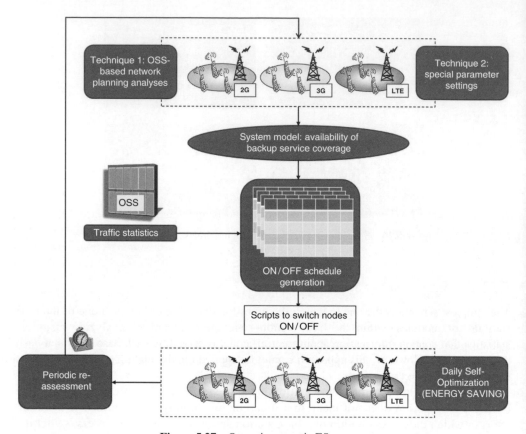

Figure 5.27 Operating a static ES system.

opening of a restaurant in a business area can completely change the traffic profile and make the deactivation of certain base stations completely unadvisable, whereas this used to be a sound decision before the new restaurant started to concentrate traffic around that area outside business hours.

Therefore, even with a static or prescheduled approach, there is the need for several steps: (i) a system modeling phase in which the existence of suitable backup coverage is evaluated for each site (or groups of sites) to be potentially deactivated during some times of the day; (ii) a historical analysis of the network statistics that, together with the previous system model, allows the generation of a calendar for activations/deactivations; and (iii) the existence of periodic reconfiguration periods in which the system modeling and the schedule generation are completely recalculated to adapt the system behavior to the real network and traffic environment. Such centralized, closed-loop scheme is summarized in Figure 5.27.

The system modeling phase can be structured in two ways: (i) by means of offline evaluations based on network topology and propagation predictions, preferably calibrated by means of OSS statistics, as proposed in Section 5.7.1 (this option is labeled as Technique 1 in Figure 5.27); or (ii) by means of online techniques that allow indirect gathering of information about the availability of suitable backup service coverage by means of special parameter settings that allow the operator to virtually remove one sector from the network unless it is absolutely necessary, i.e. unless it is the only one providing suitable (exclusive) coverage in a certain area (this option is labeled as Technique 2 in Figure 5.27). Of course, this technique does not generate any energy savings during the system modeling phase since the sector under evaluation remains active during the process and is only (eventually) excluded at logical level (unless no other sector can provide suitable coverage in a location in which this sector is the best server).

5.12.3. Dynamic Energy Saving

This approach is an extension of the static one, and is aimed at solving some of the issues associated with the reliance on a prescheduled list of actions to be taken in order to activate/ deactivate base stations. Since it takes some time to reactivate a sector, it is a technically sound idea to deactivate only those ones that are deemed suitable for deactivation after the analysis is carried out in the static approach. However, the static method does not allow the system to react to abnormal traffic patterns that lead to the reactivation of a certain base station before the scheduled time.

Such pseudoreal-time switching-on capabilities (the process will take some time in any case) can be carried out by closely monitoring updates in the OSS statistics as fast as they happen. In this way, an abnormal traffic volume can be detected during the night time in a business area (e.g. due to a special event) and this will trigger the activation of the capacity sites that would be otherwise deactivated during the night.

However, the utilization of OSS statistics for this purpose poses the two following limitations:

1. The system's reaction speed is limited by the periodicity at which the OSS statistics are updated.
2. OSS statistics lack some information. The use of terminal capabilities, for example, can be useful to decide on the activation of the LTE layer (in the previous example of the nightly event in the business area) depending on the availability of a critical mass of active terminals with LTE capabilities, even though there is a UMTS base station serving them with no congestion problems. Such activation would be decided under the assumption that these users will receive a more satisfying service if LTE is available.

The possibility of having full real-time support and taking more detailed information, such as terminal capabilities, into consideration can be enabled by deploying probes (either based on additional HW elements or SW functionality running at current network nodes) that sniff the relevant interfaces and provide access to all these information elements in real time. Another alternative is for NEs to collaborate in this process with performance related alarms (e.g. alarms associated with high traffic or alarms triggered when an LTE-capable mobile requests data services while being served by a 2G/3G cell). Such centralized, closed-loop scheme (which is an extension of the one depicted in Figure 5.27) is illustrated in Figure 5.28.

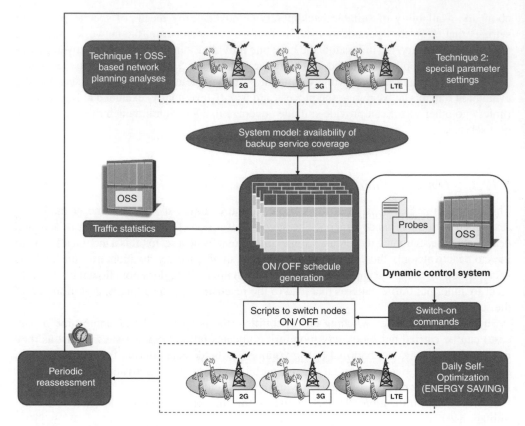

Figure 5.28   Operating a dynamic ES system.

5.12.4.   Operational Challenges

Related to the practical deployment of ES solutions in live networks, the following key operational challenges have been identified:

- A closed-loop, autonomous implementation system is required in order to ensure that the switch-on commands are properly executed without errors. For this purpose, it is critical to implement automated functionalities that check whether all required NEs are up-and-running and, if not, carry out the necessary retry actions until the problem is fully resolved.
- The ES system needs to be integrated with the alarm monitoring function at the OSS in order to filter out the alarms that are known to be caused by ES actions (e.g. sites that have been intentionally switched off and are therefore unavailable, etc.).
- Coordination is required with other OSS processes, such as maintenance or planning activities, since a NE in ES mode may not accept reconfiguration or test commands.

5.12.5.   Field Results

As part of a pilot project with a Tier 1 operator, the potential benefits of a Static ES system were assessed by analyzing real network statistics. The assessment was carried out in a cluster with 4600 sectors (2100 GSM and 2500 UMTS), and sector deactivation was only considered for 3G. Out of the 2500 UMTS sectors, potential deactivation was only deemed feasible in 1650 sectors, since there were 650 sectors labeled as business critical and an undisclosed number of sectors that lacked underlying 2G coverage. In this heterogeneous environment, comprising both urban and rural areas, the main findings were the following:

- Approximately 82% of the 3G sectors that were eligible to be selectively switched off can be deactivated during a certain period of the day.
- On average, deactivation is recommended by the system for 37% of the 3G sector-hours and 16% of the 3G site-hours, which means that remarkable energy savings are achievable by using this approach.
- The accuracy of the results (evaluated considering that an erroneous decision is made when recommending switching off a sector and finding out later that it was needed to serve certain traffic) is around 93%, which is deemed satisfactory for a static system that will be further improved when introducing dynamic mechanisms.

5.13.   Coexistence with Network Management Systems

The optimization functions covered earlier in this chapter must interwork seamlessly with other management systems for the mobile network to be operated in a truly autonomous manner. This applies also to Self-Planning and Self-Healing functions discussed in Chapter 4 and Chapter 6, respectively. First, the management systems will be presented, and then, the needs for interwork with SON functions will be discussed.

5.13.1.   Network Management System Concept and Functions

3GPP specifications define a management reference model in which the role and interfaces of Network Management Systems (NMSs), and in general of O&M systems, with respect to other entities are clarified (see [46], and Figure 5.29 below); although the practical realizations may vary depending on the technology, the concept applies to 2G, 3G and LTE. The model is integrated by:

- NE: one telecommunications entity, offering some interface for management. For instance, an RNC or a NodeB.
- EM: an EM provides one or more end-user interfaces allowing management of some NEs related to each other. One of such interfaces is the Itf-N[10], a standardized interface on which the NMS (see Figure 5.29) works. The interface between NE and EM depends on the type

---

[10]The Itf-N is a generic term in systems architecture, conceptualizing an interface that hides lower-level details. The name stems from the fact that a layer above another layer in the architectural hierarchy of a system, from which resources are used, is typically depicted on top (north) of the lower-level layer. In this section, Itf-N always refers to the interface between EM and NMS.

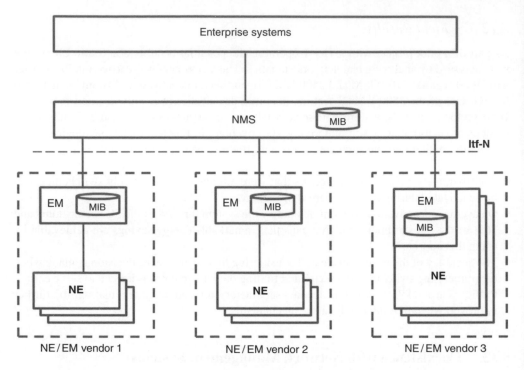

Figure 5.29 Network management reference model elements and interfaces in [46].

of node and its vendor, and is not therefore standardized. Also, the EM stores a data representation of the NEs it manages; such representation is not standardized either. The EM function may reside in the NE itself.

- NMS: provides a set of end-user functions with the responsibility for the management of a network. Such functions are built on services provided by EMs through Itf-N or through other nonstandard interfaces. Since the Itf-N is standardized, from an architectural point of view, NMSs built on Itf-Ns can connect to EMs from different vendors and technologies, providing multi-vendor and multi-technology seamless network management functions. Furthermore, the fact that the Itf-N is open allows multiple vendors to use EM services. It must be noted though that mobile infrastructure vendors provide O&M solutions that include functions referred to here as EM and NMS in one system, although typically they do not offer multi-vendor capabilities (i.e. they support only management of a mobile network of the same vendor). The functions of NMSs are detailed further below in this section.
- O&M system: the terms O&M, Operations System (OS) and OSS will be considered equivalent in this text. The O&M system is integrated by the set of EM and NMS entities available, plus other functions that are out of the scope of this text. Mobile infrastructure vendors frequently include management applications of various types in their offer for O&M systems such as planning repositories, optimization and visualization applications, etc.
- Enterprise Systems: information systems used by operators, but that are not directly or essentially related to the telecommunications aspects. Examples of enterprise systems are: Call Centers, Fraud Detection and Prevention Systems, Billing, etc.

- Management Information Base (MIB): the MIB is a database that stores the information related to management of physical and logical resources of all NEs in the mobile network, together with their hierarchical relationships. The MIB models the network by means of a topological tree made up of nodes. The nodes in the MIB are called Managed Objects (MOs) which model full NEs or logical manageable parts of an NE. For instance, an MO may group the parameters of a cell that are related to HO. The model of the MIB (i.e. the conceptual representation of the network) may be different in the NMS and in EMs (for instance, parameter labels may be different because of database performance tuning). However, the model is standardized across the Itf-N (names of parameters, managed objects, hierarchy, etc.). See, for example, [47], where the UTRAN model is defined in the form of Extensible Markup Language (XML) files, so that EMs can effectively offer management services to the NMS by exchanging properly formatted files.

Once the high-level definition of the NMS and its relationship with other entities are established, its responsibilities must be understood in order to frame how it must interwork with SON functions. NMS functions regarding PM are covered lightly in 3GPP specifications, which focus on the requirements and functions of NE, EM and Itf-N; the NMS is mandated to support Itf-N, and is expected to provide complex functions for managing and presenting PM data. Such functions are covered in Section 5.14 and (from an architectural point of view) in Chapter 3.

NMS functions related to CM are described in 3GPP specifications (see [48]) as falling into two categories: system monitoring and system modification functions.

System monitoring consists of retrieval of information and reporting. This naturally implies the existence of logs of events in the NMS, regarding object creation/deletion, attribute value change and state change. The event logs logically allow the NMS to provide historical information about the network status, which means the possibility of retrieving the network configuration at some defined moment in the past. This is useful in order to provide fallback functionality which is a high-level requirement for some standardized SON optimization functions (see [30], where a reference Use Case for Self-Optimization Monitoring and Management is provided) and is a useful feature in itself for obvious operational objectives.

The second high-level NMS function is system modification, which entails the creation (e.g. creating a new node), deletion (e.g. deleting a node) and conditioning (e.g. changing some configuration parameter of a node) of NEs. In order to properly provide this functionality, the NMS must ensure that minimum disturbance occurs in the network when these operations are realized; this may imply locking elements when the action affects the service (such operations are commonly referred to as 'offline' or 'service affecting') or even recreating NEs (in some cases, changing a configuration parameter requires recreating the MO in the MIB). Also, it is desirable that the NMS offers feedback from the EM on the success or failure of the modifications to be implemented in the network. This offers the potential to systems using NMS functions to take the necessary actions as needed if a modification fails. For instance, faulty critical modifications may require retrying or initiating alternative corrective actions.

As seen in the hierarchical diagram in Figure 5.29, system modifications from the NMS are done through the Itf-N. However, it is possible that NEs offer other interfaces to support modification. Such means are dependent on the EM/NE provider and, therefore, are closely related to the data model of the MIB in the EM. One typical example of such vendor-specific mechanism for managing the configuration is Man-Machine Language (MML) interfaces directly offered by the NE.

An important aspect related to system modification is consistency. This topic is discussed in 3GPP specifications [48] as consistency between MIBs in NMS and EMs. Assuming that the NMS stores its own MIB, ensuring consistency implies that data from the EMs must be retrieved from the NMS periodically and/or that changes in the EM are reported to the NMS and compared with the local NMS data. The discrepancies between the two views may refer to the existence of MOs (e.g. a cell may exist in the MIB of the NMS but not in the MIB of the EM) or to the state of the MOs (e.g. the value of a parameter for a given cell is different in the EM compared to the value in the NMS). Then, depending on which system, NMS or EM, is considered to hold the master configuration, discrepancies will be solved by propagating the configuration from NMS to EM (if NMS is master) or from EM to NMS (if EM is master).

Consistency may also refer to inter-EM consistency. In multi-vendor scenarios, certain changes in the configuration of a certain network may require updates in the elements of another network managed by a different EM. This is the case for certain parameters related to HO; if one of such parameters is changed in a cell that has a neighbor managed by an EM different from its own, the new parameter value needs to be updated in the other EM. The representation of a neighbor cell in a different EM is usually called 'external cell'.

Yet another valid meaning for consistency within the domain of NMS functions is ensuring that the state of the NMS MIB is correct. The correctness of the MIB depends on rules that are specific to the vendor or to the 3GPP specifications, such as:

- Parameters must have values included in the allowed set of values. For instance, the Cell Identifier of a cell must be included in the [0, 65 535] range (a 3GPP limitation, see [49] for example).
- Certain value combinations of parameter values are not allowed. Similarly, the values of certain parameters must comply with consistency equations (e.g. parameter A must be lower than parameter B).
- The cardinality of certain MOs is limited (e.g. the maximum number of cells in an RNC).

Other types of limitations may exist. Naturally, the EM may offer services through the Itf-N to verify the consistency of the NMS MIB.

Nevertheless, frequently, operators may demand additional conditions to be fulfilled by the configuration of the network, which are not known by the EM; the logical location for them (hierarchically) is the NMS, due to its being multi-vendor. Such operator-dependent rules are varied by nature; some examples are:

- Nomenclature or rules for MO identifier generation such as enforcing uniqueness of Cell Identifiers in the whole PLMN.
- Topological constraints (particularly in terms of neighbor planning).
- Narrower sets of allowed parameter values. Operators may determine which values certain sensitive parameters best serve their needs and those of the final customers, and may want to enforce using such values (parameter defaults or allowed ranges in parameter standards) in their networks.

5.13.2.   Other Management Systems

Operators have used in the past, and are likely to use in the future, a wide range of systems to meet their needs in terms of operating their networks. First, the NMS functions may be realized by several distinct systems from different providers (especially, but not only, in the case of a multi-vendor, multi-technology network). On top of this, the operational and business processes used by each operator may be supported or enforced by customized solutions. Such solutions may logically belong to the NMS or to the enterprise systems layer, referring to Figure 5.29. For instance, several repositories may be used to store critical information and to facilitate its exchange among different teams in the operator: site acquisition, hardware provisioning, transmission planning, etc. Moreover, as SON is gradually introduced, SON functions with varied degree of autonomy may and will coexist with standalone planning and optimization solutions. Again, in a multi-vendor and multi-technology environment, SON functions may not be deployed equally for the different networks, e.g. more SON functions will be deployed in 4G networks than in 2G. On the other hand, the high-level requirements of SON [50] impose that the SON architecture and implementation supports network sharing between network operators, and that the SON functions provide an easy transition from operator controlled to autonomous operation. This implicitly frames the working condition of real SON implementations in an ecosystem composed of various, inhomogeneous entities. Flexibility to achieve seamless interworking with such systems is of paramount importance to get the most out of SON.

5.13.3.   Interworking between SON Optimization Functions and NMS

Once the role of the NMS is outlined, and the existence of a cluster of external systems with which to interwork has been presented, the potential needs for SON optimization functions to interwork with them can be addressed. Naturally the focus will be placed on NMS integration, since the systems discussed in Section 5.13.2 are not known a priori as they are operator-dependent.

As explained earlier in this section, one of the NMS functions is to resolve discrepancies between MIBs at NMS and EM level. Changes in the configuration of the NEs (i.e. the live network configuration) may come from the NMS itself, which does not cause discrepancies unless the change procedure fails, or from other entities. In particular, SON optimization functions may produce a significant amount of configuration changes in the network. If the SON functions utilize NMS external interfaces to modify the system configuration, no discrepancy will happen and consistency checks, both vendor and operator-dependent, can be run. If EMs or NE interfaces are used, though, discrepancies could be caused between the EM MIB (assuming EM and NE are synchronized) and the NMS MIB, requiring the NMS users to resolve the conflict. Therefore, in order to facilitate the resolution of discrepancies such changes should be automatically notified to the NMS. Furthermore, the operator-dependent consistency rules described at the end of section 5.13.1 may not be understood by SON functions; this implies that the logical way of enforcing such rules is to have SON functions utilize the NMS to carry out CM changes.

Yet another implication of carrying out configuration changes through the NMS, is that consistency checks related to default values or allowed ranges may be communicated to the originator of the parameter change, so that these restrictions are taken into account. For instance, SON optimization functions may vary some cell parameter value, causing deviations from the parameter default. Since SON functions are allowed to do this, consistency checks may offer separate reports for parameter changes done by SON and for those carried out from other agents. Some specific examples of Self-Optimization functions requiring interwork with NMS follow:

- ANR management [20]. The inputs for ANR, as well as for certain Self-Optimization functions, may be provided not only by the current configuration and statistics/events, but also from NMS/planning, like in the case of reference neighbor plans or lists of 'no-remove' adjacencies (white list).
- Another example is closed-loop, periodical (e.g. weekly) RET optimization. The changes in RET need to be propagated to planning tools and databases so that other processes using RET as inputs (e.g. open-loop RF optimization, SON functions involving RF data, etc.) utilize updated information. The NMS in this case (in fact in most SON cases) acts as a bridge that ensures up-to-date, consistent data is used by all processes, in the SON system or outside it.
- Finally, as mentioned in [30], a general SON optimization logical flow may involve the evaluation of the results of the executed corrective actions. If the system status is not satisfactory after the execution of the corrective actions, fallback may be needed to revert the system to the previous configuration. The fallback configuration may be retrieved from a preselected previous configuration or from historical CM information in the NMS.

5.14.   Multi-Vendor Self-Optimization

As discussed earlier in this book, SON has been largely discussed in standardization forums in relation to LTE. Nevertheless, most operators will continue to carry and support two or three mobile networks. For the optimization domain, as for other SON areas, it is therefore desirable that SON extends its benefits to 2G and 3G.

However, multi-vendor optimization systems must overcome a number of challenges to become a homogeneous, autonomous Self-Optimization system entity. In general, optimization Use Cases and principles have been standardized mostly in relation to LTE. Although there is substantial research on automatic optimization algorithms for 2G and 3G networks, these initiatives stem from individual company efforts. Optimization standardization affects three distinct areas, which are discussed in the following paragraphs.

First, optimization algorithms require KPIs to determine which problems affect the network and what the potential causes can be. KPIs are made up of counters, which vary substantially across vendors, not to mention technologies. In numerous occasions, counters offered by a particular infrastructure vendor are unavailable from others. This is especially true when the counter refers to a network functionality that only exists for some vendor, or that differs significantly across vendors. In other cases, counters are similar, but vary significantly in one or more of the following aspects:

- Trigger point or event: the exact trigger point definition (i.e. the point in the message flow between network entities at which the counter is stepped) may change, resulting in numeric differences between KPIs meant to measure the same phenomenon (call setup success rate, DCR, etc.) in different vendors.

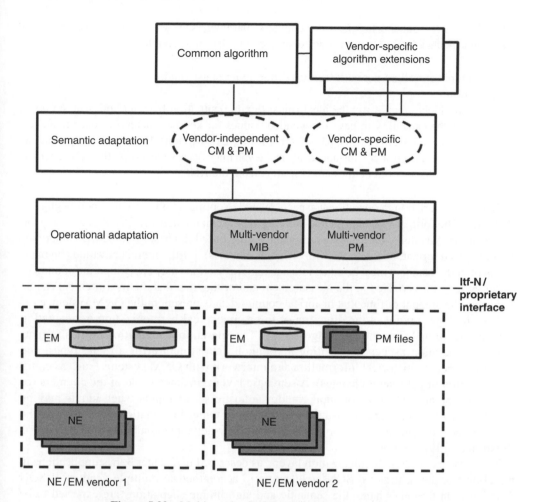

Figure 5.30   Multi-vendor Self-Optimization functional blocks.

- Managed object associated with the counter: counters are typically associated with some managed objects in the vendors' managed objects structure. From the point of view of the optimization algorithm, the lower in hierarchy the managed object (that the counter applies to), the finer granularity can be potentially achieved for optimization.
- Message cause: the trigger points are associated with some pattern of messages and/or timer expiry, although many messages convey information items that distinguish between causes that triggered the message. For instance, service type, HO causes, drop causes, etc. Depending on the implementation, each cause may be associated with a counter, or several causes may be lumped into one single counter.
- Resolution: referred to lowest time granularity measured by the counter (e.g. in histogram counters).

Secondly, a similar discussion can be applied to optimized parameters. Optimization requires not only information to diagnose problems, but also control over key aspects in the network to carry out the tuning. Like counters, parameters may vary across vendors; the variation aspects are similar to the ones discussed in relation to counters:

- Parameters controlling network features differing amongst vendors.
- Resolution.
- Managed object granularity: the ideal (most flexible) situation from an optimization point of view is the capability to have independent values for key parameters for each managed object. Nevertheless, in practice parameters may have restrictions. For instance, a parameter controlling some protocol instanced at cell level may be associated to the RNC, which implies all values would be the same for all cells in the RNC.

Finally, the variation in operational aspects related to optimization presents a non-negligible challenge. The formats in which data is exchanged between the network O&M system and the optimization functions may vary significantly (file-based interfaces in XML, binary or column-based format, or databases with different schemas). Furthermore, activation and management of the counters must be taken into account; restrictions may apply in terms of number of network entities covered in the measurement and/or the time that the measurement can be activated, the amount of time that historical counter data is present in the O&M system, etc.

The challenges point in the direction of three conceptual harmonization or adaptation layers: operational, semantic and algorithmic. The operational adaptation unifies formatting and operation aspects, presenting a common data structure for all data types, and aligning access to historical information and interwork with O&M systems (such as counter activation or parameter changes). A semantic layer translates some of the counters (or KPIs) and parameters to a common, vendor-agnostic set of inputs; such inputs may be processed by a common, vendor-agnostic algorithm. Other information items may not be harmonized, and need to be processed by vendor-dependent subalgorithms. This is illustrated in Figure 5.30.

The presence of these harmonization layers, as described in Chapter 3, has implications in the ability of the centralized SON architecture to accommodate multiple vendors of SON functions. On the other hand, the semantic and algorithmic adaptations are expected to be dependent on the vendor of the SON functions since they are highly coupled with the internal logic of its algorithms.

5.15. References

[1] Halonen, T., Romero, J. and Melero, J. (2002) *GSM, GPRS and EDGE Performance. Evolution Towards 3G/UMTS*, John Wiley & Sons, Ltd, Chichester.

[2] Robinson, J.B. (1949) *On the Hamiltonian Game (a Travelling-Salesman Problem)*, RAND Research Memorandum RM-303.

[3] Talbi, E.-G. (2009) *Metaheuristics: from Design to Implementation*, John Wiley & Sons, Ltd, Chichester.

[4] Glover, F.W. and Kochenberger, G.A. (2003) *Handbook of Metaheuristics*, International Series in Operations Research and Management Science, Springer, Heidelberg.

[5] Gen, M. and Cheng, R. (1997) *Genetic Algorithms and Engineering Design*, John Wiley & Sons, Ltd, Chichester.

[6] Braspenning, P.J., Thuijsman, F. and Weijters, A.J.M.M. (1995) *Artificial Neural Networks: an Introduction to ANN Theory and Practice*, Springer, Heidelberg.

[7] Jensen, F.V. and Nielsen, T.D. (2001) *Bayesian Networks and Decision Graphs*, Springer-Verlag, Heidelberg.

[8] Gerla, G. (2001) *Fuzzy Logic: Mathematical Tools for Approximate Reasoning*, Kluwer Academic Publishers, Dordrecht.

[9] Hata, M. and Nagatsu, T. (1980) Mobile Location Using Signal Strength Measurements in Cellular Systems, *IEEE Transactions on Vehicular Technology*, **29**, pp. 245–352.

[10] COST 231 TD(91)109 (1991) *1800 MHz Mobile Net Planning Based on 900 MHz Measurements*, http://vbn. aau.dk/en/publications/1800-mhz-mobile-netplanning-based-on-900-mhz-measurements(09388710-a55e-11db-b8eb-000ea68e967b).html (accessed 3 June 2011).

[11] Lee, W.C.Y. (1986, 1993) *Mobile Communications Design Fundamentals*, John Wiley & Sons, Inc., Indianapolis.

[12] Kostanic, I., Guerra, I., Faour, N., Zec, J. and Susanj, M. (2003) Optimization and Application of W.C.Y Lee Micro-cell Propagation Model in 850MHz Frequency Band, *Proceedings of Wireless Networking Symposium*, Austin, TX.

[13] 3GPP, Technical Specification, Technical Specification Group Service and System Aspects (2010), *X2 Application Protocol (X2AP)*, 3GPP TS 36.423 Version 9.3.0, Release 9, 15 June 2010, http://www.3gpp.org/ ftp/Specs/archive/36_series/36.423/36423-930.zip (accessed 3 June 2011).

[14] 3GPP, Technical Specification, Technical Specification Group Radio Access Network (2010) *Requirements for Support of Radio Resource Management*, 3GPP TS 25.133, Version 9.5.0, Release 9, 1 October 2010, http:// www.3gpp.org/ftp/Specs/archive/25_series/25.133/25133-950.zip (accessed 3 June 2011).

[15] Kuri, J. and Mermelstein, P. (1999) Call admission control on the uplink of a CDMA system based on total received power, *IEEE Proc. International Conference on Communications*, pp. 1431–1436, June 1999.

[16] Kumar, S. and Nanda, S. (1999) High data-rate packet communications for cellular networks using CDMA: algorithms and performance, *IEEE Journal on Selected Areas in Communications*, **17**(3), pp. 472–492.

[17] Dziong, Z., Ming, L. and Mermelstein, P. (1996) Adaptive traffic admission for integrated services in CDMA wireless-access networks, *IEEE Journal on Selected Areas in Communications*, **14** (9), pp. 1737–1747.

[18] 3GPP, Technical Specification, Technical Specification Group Radio Access Network (2008), *Radio Resource Control (RRC); Protocol Specification*, 3GPP TS 25.331 Version 8.1.0, Release 8, 28 January 2008, http:// www.3gpp.org/ftp/Specs/archive/25_series/25.331/25331-810.zip (accessed 3 June 2011).

[19] 3GPP, Technical Specification, Technical Specification Group Radio Access Network (2010), *Overall Description; Stage 2*, 3GPP TS 36.300 Version 9.5.0, Release 9, 4 October 2010, http://www.3gpp.org/ftp/ Specs/archive/36_series/36.300/36300-950.zip (accessed 3 June 2011).

[20] 3GPP, Technical Specification, Technical Specification Group Service and System Aspects (2009) *Automatic Neighbour Relation (ANR) Management; Concepts and Requirements*, 3GPP TS 32.511, Version 9.0.0, Release 9, 31 December 2009, http://www.3gpp.org/ftp/Specs/archive/32_series/32.511/32511-900.zip (accessed 3 June 2011).

[21] 3GPP, Technical Specification, Technical Specification Group Service and System Aspects (2009) *Automatic Neighbour Relation (ANR) Management; Concepts and Requirements*, 3GPP TS 32.511, Version 8.2.0, Release 8, 12 June 2009, http://www.3gpp.org/ftp/Specs/archive/32_series/32.511/32511-820.zip (accessed 3 June 2011).

[22] 3GPP, Technical Specification, Technical Specification Group Radio Access Network (2010) *Overall Description; Stage 2*, 3GPP TS 36.300 Version 8.12.0, Release 8, 21 April 2010, http://www.3gpp.org/ftp/Specs/ archive/36_series/36.300/36300-8c0.zip (accessed 3 June 2011).

[23] 3GPP, RAN Plenary Meeting #48, RP-100688 (2010) *ANRF for UTRAN WI*, www.3gpp.org (accessed 3 June 2011).

[24] 3GPP, Technical Specification, Technical Specification Group Service and System Aspects (2010) *S1 Application Protocol (S1AP)*, 3GPP TS 36.413 Version 9.3.0, Release 9, 15 June 2010, http://www.3gpp.org/ftp/Specs/ archive/36_series/36.413/36413-930.zip (accessed 3 June 2011).

[25] 3GPP, Technical Specification, Technical Specification Group Service and System Aspects (2010) *Self-Organizing Networks (SON) Policy Network Resource Model (NRM) Integration Reference Point (IRP): Information Service (IS)*, 3GPP TS 32.522 Version 9.1.0, Release 9, 8 October 2010, http://www.3gpp.org/ftp/ Specs/archive/32_series/32.522/32522-910.zip (accessed 3 June 2011).

[26] SOCRATES Project (2010) www.fp7-socrates.eu (accessed 3 June 2011).

[27] SOCRATES (2010) *Self-Optimisation and Self-Configuration in Wireless Networks*, European Research Project, http://www.fp7-socrates.eu (accessed 3 June 2011).

[28] Lobinger, A., Stefanski, S., Jansen, T. and Balan, I. (2010) Load Balancing in Downlink LTE Self-Optimizing Networks, *IEEE Proc. 71st Vehicular Technology Conference*, pp. 1–5, May 2010.

[29] Viering, I., Döttling, M. and Lobinger, A. (2009) A mathematical perspective of self-optimizing wireless networks, *IEEE International Conference on Communications 2009 (ICC)*, Dresden, Germany, June 2009.

[30] 3GPP, Technical Specification, Technical Specification Group Service and System Aspects (2010) *Self-Organizing Networks (SON) Policy Network Resource Model (NRM) Integration Reference Point (IRP);*

Requirements, 3GPP TS 32.521 Version 9.0.0, Release 9, 6 April 2010, http://www.3gpp.org/ftp/Specs/archive/32_series/32.521/32521-900.zip (accessed 3 June 2011).

[31] 3GPP, Technical Specification, Technical Specification Group Radio Access Network (2010) *Self-Configuring and Self-Optimizing Network (SON) Use Cases and Solutions*, 3GPP TS 36.902 Version 9.2.0, Release 9, 15 June 2010, http://www.3gpp.org/ftp/Specs/archive/36_series/36.902/36902-920.zip (accessed 3 June 2011).

[32] 3GPP, Technical Specification, Technical Specification Group Radio Access Network (2010) *Radio Resource Control (RRC); Protocol Specification*, 3GPP TS 36.331 Version 9.3.0, Release 9, 18 June 2010, http://www.3gpp.org/ftp/Specs/archive/36_series/36.331/36331-930.zip (accessed 3 June 2011).

[33] 3GPP (2010) *Overview of 3GPP Release 10*, Version 0.0.7, June 2010www.3gpp.org (accessed 3 June 2011).

[34] 3GPP, Technical Specification, Technical Specification Group Radio Access Network (2010) *Evolved Universal Terrestrial Radio Access (E-UTRA); Physical Layer; Measurements*, 3GPP TS 36.214 Version 9.2.0, Release 9, 14 June 2010, http://www.3gpp.org/ftp/Specs/archive/36_series/36.214/36214-920.zip (accessed 3 June 2011).

[35] Holma, H. and Toskala, A. (eds.) (2004) *WCDMA for UMTS: Radio Access for Third Generation Mobile Communications*, John Wiley & Sons, Ltd, Chichester.

[36] Holma, H. and Toskala, A. (eds.) (2009) *LTE for UMTS. OFDMA and SC-FDMA Based Radio Access*, John Wiley & Sons, Ltd, Chichester.

[37] Úbeda Castellanos, C., López Villa, D. Rosa, C., Pedersen, K.I., Calabrese, F.D., Michaelsen, P. and Michel, J. (2008) Performance of Uplink Fractional Power Control in UTRAN LTE, *IEEE Proc. 67th Vehicular Technology Conference*, pp. 2517–2521, May 2008.

[38] Xiangning, F., Si, C. and Xiaodong, Z. (2007) An Inter-Cell Interference Coordination Technique Based On Users' Ratio and Multi-Level Frequency Allocations, *Wireless Communications, Networking and Mobile Computing, 2007. WiCom 2007. International Conference on*, pp.799–802, September 2007.

[39] Úbeda Castellanos, C., Calabrese, F.D., Pedersen, K.I. and Rosa, C. (2008) Uplink Interference Control in UTRAN LTE Based on the Overload Indicator, *IEEE Proc. 68th Vehicular Technology Conference*, September 2008.

[40] Optimising Power Efficiency in Mobile RAdio Networks (Opera-Net) Project (2008) http://opera-net.org (accessed 3 June 2011).

[41] Energy Aware Radio and Network Technologies (EARTH) Project (2010) https://www.ict-earth.eu (accessed 3 June 2011).

[42] European Commission's Action Plan on Energy Efficiency (2006–2020) http://ec.europa.eu/energy/efficiency/action_plan/action_plan_en.htm (accessed 3 June 2011).

[43] Viering, I., Peltomaki, M., Tirkkonen, O., Alava, M. and Waldhauser, R. (2009) A Distributed Power Saving Algorithm for Cellular Networks, *IWSOS '09 Proceedings of the 4th IFIP TC 6 International Workshop on Self-Organizing Systems*, Springer, Heidelberg.

[44] Opera-Net Project (2010) Presentation at the Celtic Event Valencia 2010, http://opera-net.org (accessed 3 June 2011).

[45] 3GPP, Technical Report, Technical Specification Group Services and System Aspects (2010) *Study on Energy Savings Management (ESM)*, 3GPP TR 32.826 Version 10.0.0, Release 10, 6 April 2010, http://www.3gpp.org/ftp/Specs/archive/32_series/32.826/32826-a00.zip (accessed 3 June 2011).

[46] 3GPP, Technical Specification, Technical Specification Group Services and System Aspects (2010) *Principles and High Level Requirements*, 3GPP TS 32.101 Version 9.1.0, Release 9, 6 April 2010, http://www.3gpp.org/ftp/Specs/archive/32_series/32.101/32101-910.zip (accessed 3 June 2011).

[47] 3GPP, Technical Specification, Technical Specification Group Services and System Aspects (2009) *Configuration Management (CM); UTRAN Network Resources Integration Reference Point (IRP): Network Resource Model (NRM)*, 3GPP TS 32.642 Version 9.0.0, Release 9, 18 December 2009, http://www.3gpp.org/ftp/Specs/archive/32_series/32.642/32642-900.zip (accessed 3 June 2011).

[48] 3GPP, Technical Specification, Technical Specification Group Services and System Aspects (2010) *Configuration Management (CM); Concept and High-Level Requirements*, 3GPP TS 32.600 Version 10.0.0, Release 10, 18 June 2010, http://www.3gpp.org/ftp/Specs/archive/32_series/32.600/32600-a00.zip (accessed 3 June 2011).

[49] 3GPP, Technical Specification, Technical Specification Group Radio Access Network (2000) *UTRAN Iub Interface NBAP Signalling*, 3GPP TS 25.433 Version 3.3.0, Release 99, 5 October 2000, http://www.3gpp.org/ftp/Specs/archive/25_series/25.433/25433-330.zip (accessed 3 June 2011).

[50] 3GPP, Technical Specification, Technical Specification Group Services and System Aspects (2009) *Self-Organizing Networks (SON); Concepts and Requirements*, 3GPP TS 32.500 Version 9.0.0, Release 9, 31 December 2009, http://www.3gpp.org/ftp/Specs/archive/32_series/32.500/32500-900.zip (accessed 3 June 2011).

6

Multi-Technology Self-Healing

Juan Ramiro, Raúl Moya, Juan Carlos del Río, Mehdi Amirijoo,
Remco Litjens, Khalid Hamied, Nizar Faour, Carlos Úbeda,
Gustavo Hylander and Javier Muñoz

6.1. Self-Healing Requirements for 2G, 3G and LTE

As stated in Chapter 2, Self-Healing can be defined as the execution of the routine actions that keep the network operational and/or prevent disruptive problems. In this context, the Self-Healing terminology has been adopted to cover the categories in [1] that are related to Fault Management (FM), Fault Correction and Operation And Maintenance (O&M). In other words, to comprehend the detection and correction/mitigation of problems, as well as the availability of functionalities facilitating smooth system maintenance with minimal outage. This term, which has been adopted by the Third Generation Partnership Project (3GPP) along the same lines [2], may look overly ambitious. The reason is that there are occasions in which the Self-Healing functionalities can only point out the existence (and possibly the root cause) of a problem, but the automatic generation of an immediate remedy that does not require user intervention is not feasible. In other cases, however, it is feasible to automatically derive and implement a temporary solution that alleviates certain problems.

The need for automation in this field is aggravated by the deployment of new, complex technologies and the launch of new multimedia and broadband services for an ever-demanding customer base, which put significant strain on network engineers, who need to ensure the delivery of competitive Quality of Service (QoS) with shrinking Operational Expenditure (OPEX) budgets. In order to streamline performance engineering and troubleshooting processes, it is necessary to automate the continuous execution of advanced analyses that: (i) carry out intelligent cross-correlation between performance and configuration management information to detect current and upcoming problems in real-time; (ii) diagnose the root

Self-Organizing Networks: Self-Planning, Self-Optimization and Self-Healing for GSM, UMTS and LTE,
First Edition. Edited by Juan Ramiro and Khalid Hamied.
© 2012 John Wiley & Sons, Ltd. Published 2012 by John Wiley & Sons, Ltd.

causes for these problems; and (iii) automatically suggest a cure for them. Ideally, Self-Healing solutions should portray the following characteristics:

- Multi-source. Capable of leveraging all pieces of information that can be collected autonomously: performance counters from the Operations Support System (OSS), configuration management information, alarms, call traces, Charging Data Records (CDRs), etc.
- Multi-vendor. Capable of: (i) accessing and interpreting the different sources of information that are provided by each different vendor; (ii) tailoring the diagnosis to the intrinsic, implementation-specific particularities of each infrastructure supplier; and (iii) generating suggestions that are not only generic recipes, but also consider and exploit the mechanisms and restrictions that condition the operation and configuration of each equipment make.
- Multi-technology. Able to deal effectively with all relevant technologies, both separately and in a combined way. This has two implications: (i) it must be possible to analyze problems in the inter-Radio Access Technology (iRAT) handover mechanisms by analyzing comprehensively both sides of the adjacency, which have different technologies; and (ii) it must be possible to leverage multi-technology analysis capabilities to detect, for example, common problems affecting different technologies (e.g. power failure in a site facility, making all collocated technologies simultaneously unavailable).
- Flexible. Although it is expected that commercial solutions are shipped with a comprehensive set of embedded algorithms, it must be possible to incorporate the local knowledge already available at the operator by means of, for example, decision rule editors.
- Configurable. They must provide operators with mechanisms to configure high-level objectives that govern the problem detection process.
- Capable of triggering actions in other Self-Organizing Networks (SON) modules. As a result of the conducted analyses, the recommendation might be, for example, to carry out an urgent adjustment of the Radio Frequency (RF) environment in certain network elements in order to temporarily fix a coverage hole. In principle, the functionality capable to carry out such adjustment might be a Self-Optimization component that, in general, might be configured to run with certain periodicity. Since in this case an urgent, *ad hoc* Self-Optimization is required, the process needs to be initiated by the Self-Healing component, which might as well provide guidelines, policies and objectives for the *ad hoc* execution of the Self-Optimization action.

6.2. The Self-Healing Process

Due to the growing complexity and the increasing number of layers in mobile networks (see Figure 6.1), engineers are facing a situation in which they have to manage more Key Performance Indicators (KPIs), more network settings, and more functionality without additional human resources. In order to reduce the necessary operating costs (OPEX) under these conditions, operators need to find more efficient methods to identify the root cause of problems in order to rapidly solve them and prevent future network issues.

A technical study about Self-Healing was produced in Release 9 [3], in which a general workflow for Self-Healing is provided and different recovery actions are identified for different types of system problems. Additionally, three concrete Self-Healing Use Cases are defined: Self-Recovery of Network Element (NE) software, Self-Healing of board faults and Self-Healing of cell outage.

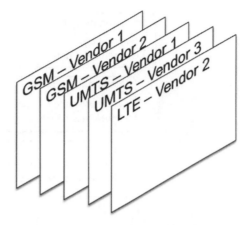

Figure 6.1   Increasing number of network layers and growing complexity.

The nature of the problem opens the door to virtually any of the existing self-detection techniques that have been researched and applied to mobile networks. This section describes how the classical FM processes, also known as TroubleShooting (TS), can be applied to Self-Healing within the context of SON for cellular networks. The Self-Healing process involves the following steps (see Figure 6.2):

- Detection: realization that a problem is occurring or about to occur.
- Diagnosis: finding the root cause of the detected problem.
- Cure: identification and application of the appropriate corrective actions to (fully or partially, definitely or temporarily) restore the service.

6.2.1.   Detection

Cell underperformance and outage are understood as the presence of problems impacting QoS beyond acceptable levels, which, among other factors, has a strong influence on churn. Outages in cellular networks may occur with different degrees of significance and may be due to different causes. The failure of a site, a sector or a physical channel/signal can be caused by, for example, hardware or software failures (radio board failure, channel processing implementation error, etc.), external problems such as lack of power supply or network connectivity issues, or even erroneous configuration. A Self-Healing system shall be capable to detect existing explicit problems (reactive detection), as well as to anticipate the occurrence of future, upcoming problems based on intelligent analysis and cross-correlation of data trends (proactive detection).

Reactive detection is the simplest option and can be implemented by, for example, benchmarking the main KPIs against a set of reference thresholds, triggering an outage indication when one or more KPIs do not meet the specified requirements. Alternatively, a unified Health Indicator (HI) can be generated by means of a weighted combination of a set of relevant KPIs. Such HI needs to be carefully designed in order not to mask network issues, for example, in

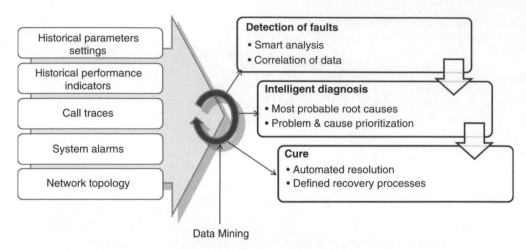

Figure 6.2   Self-Healing process.

the case in which the HI is built by three KPIs (X, Y and Z), and severe problems in Z are masked by outstanding performance for X and Y.

Proactive detection requires more elaborated analysis of the available inputs, since the target is to identify problems before they become explicit. This analysis is based on probabilistic estimations, typically relying on the utilization of data trends (indicating performance degradation) and lessons learned from previous experiences to predict future behavior of network performance, as well as the moment in which a cell may start experiencing explicit outage.

6.2.2.   Diagnosis

In this context, diagnosis is understood as the identification of the root cause of the detected problems, which may range from the introduction of human errors (that lead to wrong network settings) to the occurrence of equipment failures driven by software or hardware issues, in addition to the ever changing particularities of the radio propagation environment. In order to cope with the complex nature of the problem, the application of Artificial Intelligence techniques in Self-Healing becomes evident. Such techniques cover different degrees of complexity and sophistication, from the application of simple rule-based systems to more elaborated schemes based on, for example, neural networks.

6.2.3.   Cure

From the three phases of the Self-Healing process, providing a cure is perhaps the most challenging one. Corrective actions are usually applied manually, and introducing automated techniques will require remarkable changes in the operator's internal processes.

In addition to organizational issues, the automated derivation and implementation of corrective actions have also many technical challenges, like the required adaptation to a wide range of element and system interfaces, as well as the need to validate the successful execution of the cure. Moreover, sometimes the suggested cure cannot be implemented remotely (for example, hardware replacement, change of mechanical tilt, etc.). For this reason, even in the best possible scenario, it may not be feasible to achieve a completely autonomous resolution for all kinds of problems.

6.3. Inputs for Self-Healing

Due to the dynamic nature of mobile networks and the number of elements involved, many possible reasons for cell outage exist: from hardware problems to configuration issues and even temporary degradation due to external factors, such as weather conditions, which do not require any corrective action. Relatively often, the reason for a problem is not an isolated cause but a combination of multiple ones that are simultaneously active and cannot be decoupled. Therefore, analyzing the cause of a problem in a cell is a complex, cumbersome task, which requires examining a wide range of potential problems and thus requires having access to a wide, heterogeneous range of input information sources. As shown in Figure 6.2, such information can be classified in several categories, which will be presented in the following:

- Configuration Management (CM) data, including historical records.
- Performance Management (PM) data, including historical records.
- FM data or system alarms.
- Call Traces (CT).
- Network topology information.

6.4. Self-Healing for Multi-Layer 2G Networks

6.4.1. Detecting Problems

The general process to detect problems in the network is similar for all radio access technologies. See Section 6.2.1.

6.4.2. Diagnosis

Upon the detection of a problem, the different elements in 2nd Generation (2G) networks need to be analyzed at different levels, from Transceiver (TRX) to Base Station Controller (BSC). Guided by the nature of the detected symptoms, the analyses will yield a set of probable causes. These analyses cover three main differentiated areas:

- System alarms.
- Analysis of configuration parameters.
- Correlation of performance indicators.

6.4.2.1. Analyzing 2G System Alarms

System alarms are normally created by different elements and report a wide range of problems, e.g. Pulse Code Modulation (PCM) link failures, high temperature, faulty TRXs, etc. They usually contain very valuable information about the event that raised the alarm, which represents a crucial input for the analysis. Thus, the Self-Healing function shall correlate any detected performance degradation with reported alarms.

6.4.2.2. Analyzing 2G Configuration Parameters

It is well-known that most causes of performance degradation can be attributed to configuration changes (i.e. changes in parameter settings). Therefore, one of the first analyses to be conducted is the correlation between performance degradation and recent configuration changes in the same area.

Moreover, it is necessary to: (i) identify parameters that are out of their allowed (or recommended) ranges; and (ii) run comprehensive consistency checks. Some recommended configuration consistency checks are listed below.

MSC/BSC Level Parameters
- The Self-Healing function should check the consistency of the duration of the timers defined in the Mobile Switching Center (MSC). For example, the timer for implicit detach of a mobile subscriber must be greater or equal to the longest timer for periodic location area update in the interworking BSCs.
- Neighbor relationships in which source and target cells belong to different MSCs/BSCs should be periodically audited in order to ensure that the parameters defining the target cell, e.g. location area and cell ID, are properly specified in the source MSC/BSC. Otherwise handovers may not occur.
- Depending on the vendor, there are some features that must be enabled at BSC level before they can be activated at cell level. A typical example is enabling Half Rate feature for GSM sectors with the goal of increasing capacity, and failing to enable it at BSC level. As a result, the feature will not be active in cells belonging to the corresponding BSC.

Frequency Planning Parameters
The frequency assignment in the GSM network has a significant impact on end-user experience, since co- and adjacent-channel interference affect voice quality, performance of signaling channels, etc. Several rules can be applied in the Self-Healing function in order to trigger the corresponding Self-Optimization functions in case any of these rules are broken. Examples are:

- Sectors of same site should not reuse same or adjacent channels due to potential signal coupling between transmission and reception modules.
- At sector level, channels assigned to different TRXs must fulfill a minimum spectral separation due to requirements imposed by the combiners to which they are connected.
- Certain basic rules should be followed in the frequency assignment, since the number of channel reuses within neighboring cells must be minimized as much as possible depending on the type of reuses, such as Broadcast Control Channel (BCCH) cochannels are strongly discouraged while Traffic Channel (TCH) adjacent channels can be allowed.

Neighbor Planning Parameters
- The Self-Healing function should verify that the cell is not being declared as a neighbor to itself in order to prevent triggering a useless handover that would cause a dropped call.
- For a given sector, other sectors belonging to the same site should be included in the sector's neighbor list as the received signal levels from cosited sectors are usually the highest.

- It is recommended that neighbor relationships are declared symmetrically, i.e. if sector A is in the neighbor list of sector B, then sector B should also be in the neighbor list of sector A.
- The number of neighbors should not be too small in order to prevent situations where calls are not handed over to the best possible cell. On the other hand, an excessive number of neighbors will produce an overloaded BCCH Allocation List (BAL) that may affect BCCH-BSIC decoding.
- It is recommended to avoid too many inter-band 2G neighbors unless it is strictly necessary. In network deployments with wide dual-band coverage, it is recommended to hand over to a cell in the same band for mobility purposes and to a cell in different band for traffic sharing.
- There should be a minimum number of 3rd Generation (3G) neighbors to allow traffic sharing to relief congestion in the 2G cell.
- Neighbors that are far away from the source cell are unlikely to be good handover candidates. Thus, it is recommended to delete those neighbors from the BAL.
- The following needs to be verified for handover control parameters for multi-band networks: (i) escape quality and level handovers are defined; (ii) hierarchical priorities are consistently defined; and (iii) the inter-band handover feature is activated.

6.4.2.3.   Correlating 2G Performance Indicators

Advanced analysis of the main KPIs can be performed by the Self-Healing function as part of the process to estimate the root cause of the experienced degradation, in case that degradation cannot be directly attributed to an evident parameter configuration problem. It is recommended that cells under investigation are analyzed together with their neighbors.

Correlation of pairs of KPIs provides valuable information that must be exploited for accurate diagnosis. Some nonexhaustive examples are provided:

- Received signal Quality (RxQual) versus Received signal Level (RxLev). Correlation of these two indicators allows the Self-Healing module to identify whether bad quality is due to poor coverage or excessive interference (if the reported received signal level is high).
- Dropped Call Rate (DCR) versus RxQual. One of the most common causes for dropped calls is the lack of a dominant best server. Thus, if DCR increases after experiencing degradation in RxQual, then a recovery action is needed to improve the radio frequency conditions. Depending on the aforementioned correlation between RxQual and RxLev, different Self-Optimization functions will be recommended, e.g. optimization of the RF settings such as antenna tilts or fine-tuning of the frequency plan.
- DCR versus Handover Failure Rate. An increment in the number of dropped calls is usually correlated with an increase in the handover failure rate, often due to problems in the most suitable target cell. Further analysis on the neighbor cells may detect sleeping cells or faulty TRXs, which are likely to lead to congestion and handover blocking.
- DCR versus Inter-Band Traffic Sharing. In dual-band sectors, a variation in the traffic distribution between bands may lead to degradation in the DCR if load has been shifted towards the layer with worse radio quality. Upon the detection of this problem, two recovery actions can be recommended to reestablish the original situation: (i) to fine-tune the inter-band handover thresholds; or (ii) to optimize electrical tilts.

- Number of measurement reports per TRX versus blocking in the cell. In most infrastructure vendors, RxQual and RxLev reports are collected at TRX level. Therefore, when a drastic reduction of the number of collected measurement reports is accompanied with higher blocking in the cell, it is likely that this is caused by a faulty TRX.
- (Enhanced) General Packet Radio Service, denoted by (E)GPRS, codec usage versus RxQual. A reduction in the usage of high-speed codecs for a cell can be attributed to low quality conditions, which require higher coding protection. If degradation in RxQual is observed in this scenario, then it is necessary to optimize the frequency plan and/or the antenna settings.
- Handovers due to UL RxQual versus Timing Advance (TA). If the number of handovers due to UL quality increases and TA statistics indicate that the cell is serving users that are too far, the most likely cause of the problem is an abnormal RF configuration (i.e. overshooting). If TA statistics do not indicate overshooting, the frequency plan needs to be optimized.

6.4.3. Cure

Once the root cause of the detected performance degradation has been identified, the Self-Healing function needs to orchestrate the generation of a cure for the diagnosed problem. The degree of automation of the proposed solution varies and will depend on its nature and specifics, since not all corrective actions can be carried out without human intervention.

The following are potential actions to fix performance degradation:

- Reversion of parameter settings back to previous configurations.
- Assigning a default value for out-of-range parameter.
- Activating certain features that were disabled.
- Perform complex actions that involve multiple cells such as the generation of a new frequency plan or the optimization of the antenna settings. That can be taken care of by Self-Optimization functions.

Self-Healing functions must interwork collaboratively and should utilize existing Self-Planning and Self-Optimization components in order to guarantee a powerful, cross-functional action with the widest possible scope.

6.5. Self-Healing for Multi-Layer 3G Networks

6.5.1. Detecting Problems

The general process to detect problems in the network is similar for all radio access technologies (see Section 6.2.1).

6.5.2. Diagnosis

Although the technology-specific details are different, 3G analyses also cover three main differentiated areas:

- System alarms.
- Analysis of configuration parameters.
- Correlation of performance indicators.

6.5.2.1. Analyzing 3G System Alarms

As stated in Section 6.4.2.1, the Self-Healing function must also consider alarms generated by the fault management system to collect all events raised by a malfunctioning element (e.g. power amplifier, base band processor, Iub card, etc.) and use that information to correlate any potential performance degradation with the reported alarms.

6.5.2.2. Analyzing 3G Configuration Parameters

As in the 2G case, there are key consistency rules that should be applied to the network parameters to identify potential problems or risks. These consistency checks are summarized in the following categories:

CN/RNC Level Parameters
- Timers' consistency should be checked among different NEs, e.g. Core Network (CN), Radio Network Controller (RNC) and User Equipment (UE), to make sure that the waiting time is not too short to finalize other nested procedures along all involved network elements. These timers are normally stored in different databases (normally UE timers are defined in the CN and the RNC, and sent to the UE during the access signaling). They comprise functionalities such as: Connection Establishment, Channel Switching, Idle Mode and Common Channel procedures, Handovers and Synchronization.
- If the infrastructure vendor does not provide automatic synchronization of neighbor relationships with source and target cells belonging to different RNCs, neighbors should be periodically audited in order to assure that the parameters defining the target cell, e.g. location area, cell ID, scrambling code, etc., are properly specified in the source RNC. Otherwise handovers may not occur. This is also valid in case of GSM neighbor relations between any RNC and surrounding BSCs, with similar parameters defining the target cell: BCCH, BSIC, Location Area Code (LAC), cell ID, etc.
- There are also some features and licenses that should be consistently activated at different levels: e.g. RNC, NodeB and cell.

Scrambling Code Parameters
A tight reuse of Primary Scrambling Codes (PSCs) may cause some conflicts that need to be addressed. One of these conflicts is related to the composite neighbor list sent to the UE by the serving RNC. A composite neighbor list is the one that combines all neighbor lists of all cells that are in the Active Set when the UE is in Soft Handover (SHO). Duplicate scrambling codes in the composite neighbor list may cause drops due to handoff with wrong neighbors, since the RNC cannot uniquely resolve duplicate PSCs. This check can be complex, depending on the maximum number of allowed cells in SHO, i.e. the size of the Active Set. If a neighbor is added to the Active Set, then the neighbors of this neighbor should be checked for PSC collision. This type of collision is referred to as 2nd order collisions. In general, N-order PSC collisions should be verified, where N equals the maximum Active Set size. However, practically the Self-Healing function should check for up to 2nd order collisions as shown in Figure 6.3. See Section 4.11 for more details.

Another potential conflict is code interference, which happens when PSC reuse distance is small in terms of radio coupling. Code interference is especially harmful for the UE since it makes it hard to decode the signal from any of the cells that use same PSC. This circumstance

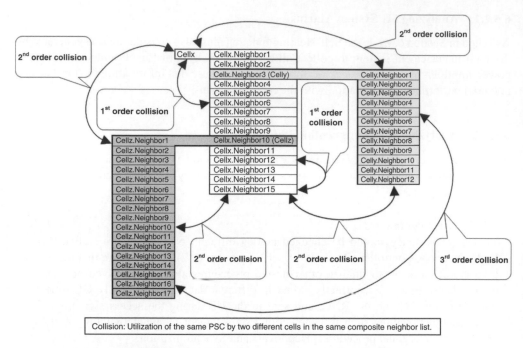

Collision: Utilization of the same PSC by two different cells in the same composite neighbor list.

Figure 6.3   Example of 1st, 2nd, and 3rd order collisions.

shall be detected and pointed out by the Self-Healing function, which may trigger PSC optimization to solve the problem.

Neighbor Planning Parameters

- Three different types of neighbors can be defined: intrafrequency, interfrequency and iRAT. Since the neighbor lists can be combined in a composite neighbor list from the cells in SHO, the Self-Healing function should check all high order collisions in any of the physical parameters sent to the UE in order to avoid them. These physical parameters are:
  - PSC for intrafrequency neighbors.
  - PSC-UARFNC pair for interfrequency neighbors.
  - BCCH-BSIC pair for GSM neighbors.
- The Self-Healing function should check that colocated interfrequency and/or iRAT neighbors are defined.
- The Self-Healing function should verify that intrafrequency co-sited sectors are defined as neighbors.
- The Self-Healing function should check for neighbor reciprocity. Although this is not so critical for handovers (since neighbor lists are combined), it is recommended that neighbor relationships are declared symmetrically due to the fact that, in idle mode and for cell Forward Access Channel (FACH), asymmetries may create areas with 3G service outage due to strong interference.
- Depending on the network configuration, there could be limited bandwidth to transmit all neighbor lists (intrafrequency, interfrequency and iRAT) and their associated parameters for cell reselection if Broadcast Channel (BCH) System Information Block (SIB) 11[1] only is

---

[1] System Information Block type 11, which contains measurement control information for UEs in idle mode.

used (SIB11bis extension is available from 3GPP Release 6 specifications [4]). This means that, in practice, it may not be possible to use the maximum size of each neighbor type (i.e. 31 for intrafrequency, 32 for interfrequency and 32 for iRAT) due to this bandwidth limitation. Therefore, it is necessary to ensure that SIB11 is not overloaded to prevent load increases when changing neighbor lists or neighbor parameters.

- There should be a minimum number of 2G neighbors to allow voice traffic balancing that reliefs the congestion in a 3G cell, and for coverage continuity. The same rule should be applied in multicarrier networks with different propagation bands (850/900 MHz versus 1900/2100 MHz) for interfrequency neighbors.

Handover and Cell Reselection Parameters

- In idle mode it is essential to ensure a minimum hysteresis to prevent unnecessary ping-pongs causing excessive battery consumption, network signaling or paging losses while the UE is busy with mobility procedures. These consistency checks comprise all neighbor types. However, the most critical ones are those causing excessive signaling due to LAC/Routing Area Code (RAC) changes. In some cases, not only reselection parameters should be checked on both sides but also a cross-checking between handover and reselection parameters needs to be carried out, since Packet Switched (PS) calls in UMTS use reselection parameters when transferred to GPRS.
- In the same way, consistency rules shall be applied to detect potential handover ping-pong problems for each neighbor type. Unlike reselection, in this case different Radio Access Bearers (RABs) or physical channels may have different settings and even different parameters are applied (for instance, HSDPA may use some specific parameters for handovers).

6.5.2.3.   Correlating 3G Performance Indicators

As in 2G, the correlation of pairs of 3G KPIs provides very insightful information to estimate the root cause for the detected performance problem. Some nonexhaustive examples are provided:

- DCR/Average Channel Quality Indicator (CQI) versus iRAT Traffic Sharing. Areas with poor 3G coverage are those in which the traffic sharing ratio, measured as the traffic carried on 3G over the total traffic, is lower than the network average. If quality indicators, such as DCR or average CQI (obtained from HSDPA call statistics), are also bad, this is most likely caused by the detected coverage problem. In this case, coverage optimization via Common Pilot Channel (CPICH) powers and antenna tilts must be considered in combination with iRAT reselection and handover parameter adjustments.
- DCR/Average CQI versus Propagation Delay. Another typical reason for poor performance is cell overshooting. To detect this, a good indicator is the propagation delay distribution obtained during Random Access Channel (RACH) access. If the percentage of samples that indicate excessive cell radius is significant and quality is poor (in terms of DCR or average CQI), most likely that a particular cell is overshooting some areas, causing poor quality. That can only be solved through coverage optimization.
- DCR/Average CQI versus Inter-Band Traffic Sharing. In multi-carrier networks with dual-band sectors (e.g. 850/900 MHz plus 1900/2100 MHz), a variation of the Inter-Band Traffic

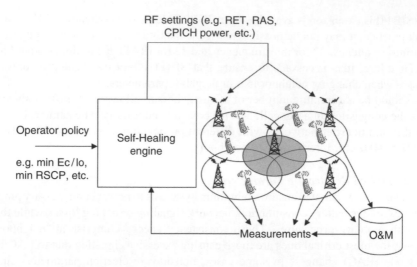

Figure 6.4   3G Self-healing system based on RF adaptations.

Sharing can easily cause quality degradation (in terms of average CQI and/or DCR) if traffic has been shifted towards the worst band. This can be solved by either tuning the inter-band handover thresholds or optimizing electrical tilts to reestablish the original situation.

6.5.3.   Cure

This process is similar to the one described for 2G (see Section 6.4.3). In the following, further details are provided in order to illustrate an example in which the Self-Healing function detects that a 3G NodeB has ceased working, causing a coverage hole. Assuming that ensuring pilot coverage is critical, the impact of an automated modification of the antenna tilts in the surrounding NodeBs is illustrated. The high level philosophy of the utilized system is depicted in Figure 6.4.

For the sake of illustration, a sample case is described for a network with 19 NodeBs (57 sectors), in which the original antenna tilt is 4°. The maximum transmit power per sector is assumed to be 46 dBm and the addition window for soft handover is 3 dB. The noise figure is assumed to be 5 dB at the NodeB and 10 dB at the UE. Moreover, for this analysis the required minimum Received Signal Code Power (RSCP) and Ec/Io to have UMTS service coverage is assumed to be −103 dBm and −15 dB, respectively. For those spatial locations in which these two conditions are not satisfied, no UMTS coverage is available and UEs will not access the UMTS layer. The RSCP coverage of the baseline scenario is illustrated in Figure 6.5.

The data traffic model is given by a requested Downlink (DL) and Uplink (UL) throughput of 512 kbps and 64 kbps, respectively. The activity factor is assumed to be 10% for data users, and 50% for voice users. Data subscribers experiencing bitrates below the requirements are dropped and considered unsatisfied. In the polygon depicted in Figure 6.5, the pilot coverage is 98.8% and, after simulating a mix of 2200 voice users and 750 data connections with 100 Monte Carlo runs, the percentage of satisfied data users is 98.8%, whereas the percentage of

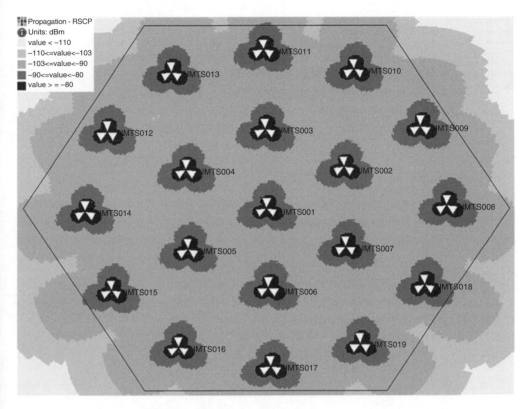

Figure 6.5   RSCP coverage of the baseline scenario.

satisfied voice users is 98.7%. However, when the central site becomes inactive, the resulting RSCP coverage map is the one depicted in Figure 6.6.

After deactivating the central site, the pilot coverage in the depicted polygon is 97.7%, and the percentage of satisfied voice and data users is 96.9% and 96.0%, respectively (meaning a relative increase in unsatisfied users rate of 158% and 208%, respectively), mainly due to the fact that some locations that were previously served by the central site are now covered with other sites that, although providing sufficient pilot coverage, serve those locations with worse radio conditions as compared to the radio conditions originally provided by the central site. However, when a Self-Healing function is applied in order to fine tune the antenna tilts of the surrounding sectors, the RSCP coverage map depicted in Figure 6.7 is experienced.

After activating the Self-Healing function, the pilot outage became 98.76%, i.e. the original coverage was restored almost completely. However, the percentage of satisfied users further decreased to 95.0% and 94.6% for voice and data, respectively (meaning a relative increase in unsatisfied users rate of 61% and 35%, respectively, as compared to the situation in which the faulty NodeB is unavailable but no measures have been taken), which means that the coverage recovery comes at the expense of clear quality degradation, as expected. Note that coverage expansions capture users with worse-than-average radio conditions.

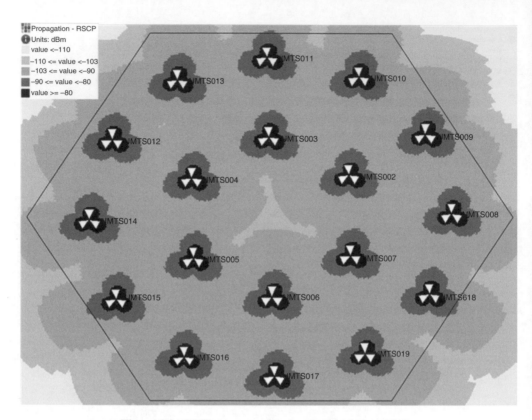

Figure 6.6   RSCP coverage after deactivating the central site.

6.6.   Self-Healing for Multi-Layer LTE Networks[2]

This section is focused on cell outage management, with emphasis on cell outage compensation for Long Term Evolution (LTE). While some cell outage cases can be detected by the O&M system through performance counters and/or alarms, others may remain undetected for hours or even days, and are finally detected through long term analyses and/or investigations triggered by subscriber complaints. Once detected, the underlying problems shall be solved, although significant coverage and capacity degradations may be experienced in the meanwhile, which is likely to have a negative impact on customer satisfaction and revenue.

Cell Outage Management (COM) comprises mechanisms for both Cell Outage Detection (COD) and Cell Outage Compensation (COC), and is an integral part of the SON concept in Evolved Universal Terrestrial Radio Access Network (E-UTRAN) [2, 6, 7], with the objective to enhance network robustness and resilience. Figure 6.8 depicts the different elements

---

[2] The work presented in Section 6.6, except for the content of Section 6.6.4, was carried out within the FP7 SOCRATES project [5], which is partially funded by the Commission of the European Union.

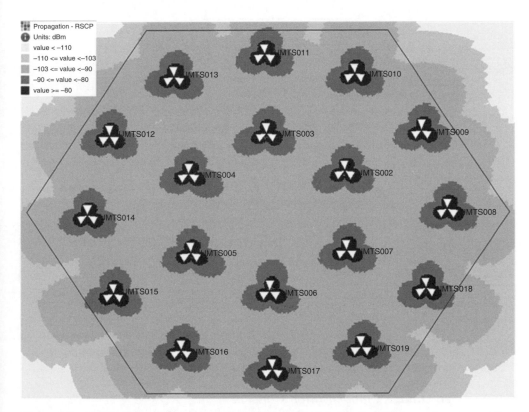

Figure 6.7 RSRP coverage after activating the Self-Healing module that fine-tunes the antenna tilts in the sectors surrounding the faulty NodeB.

and workflow of COM in future cellular networks. The depicted example is characterized by a site outage, whose preoutage service area is indicated in gray. A variety of measurements, e.g. alarms, counters or key performance indicators, are gathered by the UEs, the enhanced NodeBs (eNodeBs) and/or the O&M center and fed as input to the COM algorithms (COD and COC). Taking these measurements as inputs, the COD function then automatically identifies the occurrence and scope of an outage problem, and triggers both the COC function as well as the corresponding process in the operator's maintenance department for potential manual recovery actions. The COC function derives compensation measures in order to automatically mitigate the incurred coverage degradations by adapting one or more control parameters (e.g. antenna tilt and/or power settings) in the cells surrounding the site that has failed. Such compensation is governed by the operator-formulated policy regarding the trade-off between different local and regional performance effects, and is characterized by an iterative process of radio parameter adjustment and evaluation of the performance impact until convergence is reached. Important feedback is provided by the so-called X-map estimation function (see Appendix B), which processes measurements including location information in order to generate, for example, coverage or performance maps. Subsequent discussion will be focused on the COC function.

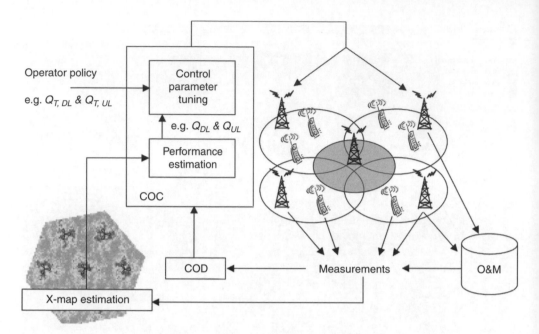

Figure 6.8   Overview of Cell Outage Management.

6.6.1. Cell Outage Compensation Concepts

Key elements of the COC mechanism are the control parameters, the measurement functionality, the operator policy and the compensation algorithm.

In principle, all radio parameters that somehow affect coverage and the spatial aspects of capacity and service quality are potentially relevant control parameters from a COC perspective. The promising control parameters include:

- The DL power split between the common channel (including the Reference Signal) and the Physical Downlink Shared Channel (carrying user data): by increasing the common channel power in cells surrounding an outage area, the service area of those cells can be potentially extended to cover all or part of the outage area. Since less power is available for actual user data transmissions, this coverage extension comes at a cost of a reduced traffic handling capacity and hence service quality experience, which stresses the involved trade-off.
- The target received power density of Physical Uplink Shared Channel (carrying user data): UL transmit powers are typically derived from a target received power density P_0, in combination with a path-loss compensation factor. A reduction of P_0 lowers UL intercell interference levels and hence increases coverage. This coverage gain comes at a cost of reduced channel rates, increased cell loads and hence lower per user throughputs.
- Antenna parameters: modern antenna design allows electrical adaptation of both the orientation of the main antenna lobe and the antenna pattern, e.g. via remote electrical tilt or beamforming techniques. Since the antenna tilt is known to be a highly responsive lever to shape the cell footprint, it is a promising candidate for use in COC.

The continuous collection of measurements and analysis of radio parameters, counters, KPIs, statistics, alarms and timers are an indispensable precondition for COM, and these pieces of information are provided by the Performance Estimator. These measurements may be obtained from various sources, including the eNodeBs (e.g. cell loads, radio link/handover failure statistics, interference levels, etc.), the UEs (e.g. reference signal received power levels, failure reports, etc.) and the O&M system providing quality-based KPIs, e.g. the tenth UL (DL) throughput percentile Q_{UL} (Q_{DL}). A key challenge is to perform real-time coverage estimation as part of the COC feedback loop. An option discussed in more detail in Appendix B is to generate coverage maps based on UE measurement reports in combination with positioning information (the usage of this approach is, however, not considered below).

Network performance comprises multiple conflicting aspects. At the highest abstraction level, a distinction may be made between factors impacting user experience, e.g. coverage, capacity and quality. A distinguishing characteristic of different network operators is how they choose to trade investment/operational costs for network performance, as well as their desired trade-off between the different performance aspects. This is captured by the operator policy. An example of an operator policy that is used in the following examples is to maximize coverage provided that certain minimum user throughput conditions are met, in terms of tenth UL and DL throughput percentiles, that is, $Q_{UL} \geq Q_{T,UL}$ and $Q_{DL} \geq Q_{T,DL}$, where $Q_{T,UL}$ and $Q_{T,UL}$ are the requirements for UL and DL, respectively. An alternative objective may be to optimize a weighted sum of multiple performance metrics. Moreover, an operator policy may depend on location, time and/or the operational state of a cell, e.g. whether it is in normal operational state or currently involved in compensation actions to solve an outage in its vicinity.

An extensive controllability study has been reported in [8], which assesses the compensation potential of several control parameters in different scenarios. The results of this study have suggested P_0 and the antenna tilt as the most promising control parameters for COC purposes.

6.6.2. Cell Outage Compensation Algorithms

One promising idea is to adjust UL power control parameters, e.g. P_0, in the vicinity of the eNodeB or sectors in outage, in order to decrease the UL intercell interference and thereby increase the Signal to Interference-plus-Noise Ratio (SINR) for the UEs in the outage area that are power-limited. By appropriately decreasing P_0, the UL coverage is thus improved, thereby decreasing the size of the outage area. On the other hand, by decreasing P_0, the UL quality will be reduced, due to the decreased SINR for UEs that are not power-limited and the increased number of served users. Therefore, each cell in the vicinity of the outage cell(s) needs to continuously measure the DL and UL quality indicators, e.g. the tenth DL and UL throughput percentiles and effectively minimize P_0, while ensuring that the UL and DL quality indicators meet the specified requirements. For an elaborate description of the algorithm refer to [9].

Alternatively, the antenna tilt of the neighboring cells can be adjusted in order to increase the coverage around the base station that has failed. By reducing the tilt (i.e. by uptilting), the coverage can be improved at the cost of lower quality in the neighboring cells. This is due to reduced spectral efficiency (higher intercell interference), as well as larger number of served users. A similar approach to P_0 tuning can be employed in which the quality is continuously measured and the antenna tilt is reduced to the largest possible extent, provided that high intercell interference is avoided and UL and DL quality meets the specified requirements.

6.6.3.   Results for P_0 Tuning

The aforementioned approach has been evaluated using a Monte Carlo-based LTE network simulator using a hexagonal layout of 19×3 cells. A capacity-driven network layout is assumed with an intersite distance of 500 m. The path-loss is computed as $128.1 + 37.6 \log_{10}(r)$, where r is expressed in km. A shadowing map with a standard deviation of 8 dB and a shadow decorrelation distance of 33 m is added to the path-loss. The 3D antenna model given in [7] is assumed, and the initial antenna tilt is set to 15°. The data traffic model is characterized by a generic elastic data service with a requested DL and UL throughput of 1 Mbps and 250 kbps, respectively. A fair rate scheduler is used to distribute resources in UL and DL. The outage quality targets set to $Q_{T,UL} =$ 64 kbps and $Q_{T,DL} = 128$ kbps, respectively.

The results in Figure 6.9 and Figure 6.10 illustrate the case in which P_0 is tuned in order to alleviate the coverage holes caused by an outage during high load. In these simulations, an outage occurs at iteration 50, resulting in a reduction of number of served UEs and the creation of coverage holes, as shown in Figure 6.10. Furthermore, the DL and UL quality is reduced, since users previously served by the failed eNodeB now perform cell reselection to neighboring cells, which results in further quality degradation (see Figure 6.9). Compensation starts at iteration 100, when P_0 starts being gradually decreased, resulting in a reduction of mainly UL quality (see Figure 6.9) and in an increase in coverage and number of served users (see Figure 6.10).

Simulation studies show that automatic compensation using P_0 is feasible in a capacity-driven layout. The compensation potential in terms of coverage improves as load decreases and can reach 85% user recovery in scenarios with low load. This comes with a cost, namely a reduction of quality in the neighboring cells, where the quality degradation is most visible for high and medium loads, and can be as low as 50% of the quality experienced when no compensation actions are taken after the outage. The approach has also been evaluated in a coverage-driven layout with 2200 m intersite distance. The degree of compensation is, however, not as large as in the capacity-driven layout.

6.6.4.   Results for Antenna Tilt Optimization

The approach in which antenna tilts of the sectors surrounding the faulty eNodeB are adjusted has proven to be also effective to compensate for the coverage holes that are created when faulty sectors cease working. For the sake of illustration, a sample case is described for a network with 19 eNodeBs (57 sectors), in which the original antenna tilt is 4°. The available spectrum in this case is 10 MHz Frequency Division Duplex (FDD) and Multiple Input Multiple Output (MIMO) 2×2 is assumed to be used in the DL, whereas MIMO 1×2 is used in the UL. The UE maximum transmit power is assumed to be 23 dBm and UL power control is configured with $P_0 = -100$ dBm and path-loss compensation factor $\alpha = 1.0$. The noise figure is assumed to be 2 dB at the eNodeB and 10 dB at the UE. Moreover, for this analysis the

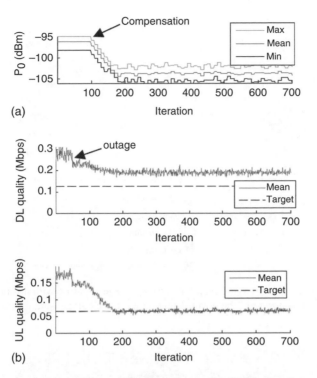

Figure 6.9 Temporal behavior of the P_0 tuning algorithm during high load, where outage occurs at iteration 50 and compensation is initiated at iteration 100: (a) P_0 of the compensating cells; (b) Mean of the DL and UL quality metrics computed over the compensating cells.

required minimum Reference Signal Received Power (RSRP) and Reference Signal Received Quality (RSRQ) to have LTE service coverage is assumed to be −121 dBm and −17 dB, respectively. For those spatial locations in which these two conditions are not satisfied, no LTE coverage is available and UEs will not access the LTE layer. The RSRP coverage of the baseline scenario is illustrated in Figure 6.11.

The data traffic model is given by a requested DL and UL throughput of 1 Mbps and 256 kbps, respectively. The minimum bitrates for the DL and UL are 256 kbps and 64 kbps, respectively. Users experiencing bitrates below the minimum are dropped and considered unsatisfied. In the polygon depicted in Figure 6.11, the pilot coverage is 98.8% and, after simulating 3000 users with 100 Monte Carlo runs, the percentage of satisfied users is 98.6%. However, when the central site becomes inactive, the resulting RSRP coverage map is the one depicted in Figure 6.12.

After deactivating the central site, the pilot coverage became 97.8%, and the percentage of satisfied users became 94.9%. (meaning a relative increase in unsatisfied users rate of 264%), mainly due to the fact that some locations that were previously served by the central site are now covered with other sites that, although providing sufficient pilot coverage, serve those locations with worse radio conditions as compared to the radio conditions originally provided by the

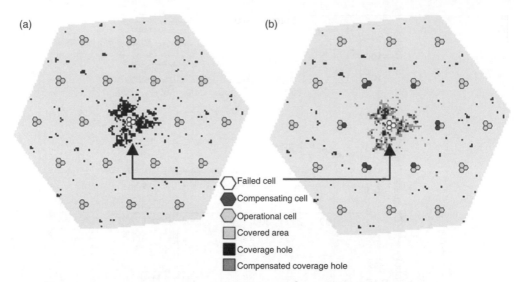

Figure 6.10 P_0 tuning algorithm during high load: (a) outage situation with no compensation (iterations 50–99); (b) snapshot of the outage situation when the compensation has converged (iteration > 200).

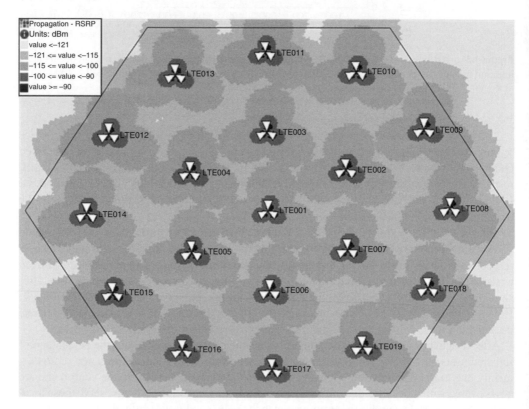

Figure 6.11 RSRP coverage of the baseline scenario.

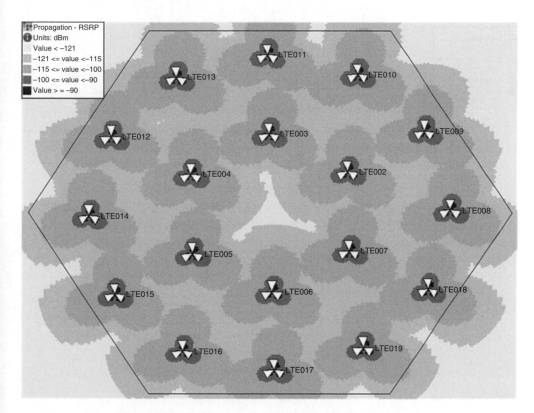

Figure 6.12 RSRP coverage after deactivating the central site.

central site. However, when a Self-Healing function is applied in order to fine tune the antenna tilts of the surrounding sectors, the RSRP coverage map depicted in Figure 6.13 is experienced.

After activating the Self-Healing function, the pilot outage becomes 98.8%, i.e. the original coverage was totally restored. However, the percentage of satisfied users further decreased to 91.7% (meaning a relative increase in unsatisfied users rate of 63%, as compared to the situation in which the faulty eNodeB is unavailable but no measures have been taken), which means that the coverage recovery comes at the expense of clear quality degradation, as expected. Note that coverage expansions capture users with worse-than-average radio conditions.

6.7. Multi-Vendor Self-Healing

The fact that network equipment is supplied by different infrastructure vendors adds extra complexity to the definition of effective Self-Healing functionalities. As already mentioned throughout the book, the standardization of SON is mainly focused on LTE, which means that in the case of GSM and UMTS proprietary interfaces need to be developed in order to handle every network element, monitor their performance, decide the corrective actions that are required and implement them. The amount of examples related to the need for explicit

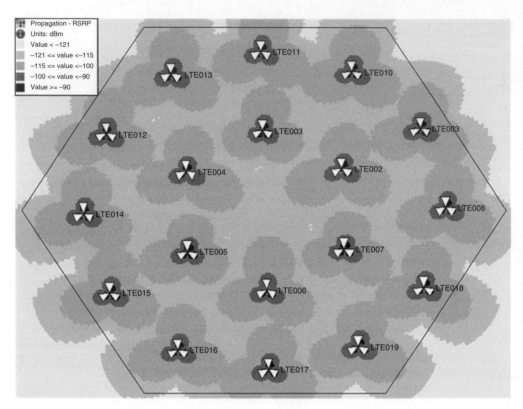

Figure 6.13 RSRP coverage after activating the Self-Healing module that fine-tunes the antenna tilts in the sectors surrounding the faulty eNodeB.

multi-vendor support in SON solutions is massive. Narrowing down the scope to Self-Healing, the following nonexhaustive examples are provided:

- Every vendor implementation provides a different granularity to set up the value of certain configuration parameters, e.g. some vendors allow adjustment of power related parameters, such as the CPICH, with steps of 0.1 dB, whereas others only allow this with steps of 0.5 dB.
- In some vendors, the existence of counters for specific network events (e.g. dropped call reasons, such as bad quality, bad signal level, etc.) facilitates the identification of the most probable cause of the problem, and therefore the effectiveness of the cure generation process. Unfortunately, other vendors do not provide similar counters.
- When identifying UMTS cells with abnormally large coverage area, the utilization of Physical Random Access Channel (PRACH) propagation delay measurements is paramount. However, in some vendors these measurements are not available in the set of OSS performance counters and must be obtained by means of call traces, with the subsequent need for additional collection and preprocessing mechanisms.
- When considering certain corrective recovery actions, the fact that some vendor implementations force the restart of the reconfigured network element needs to be considered. In

these cases, the Self-Healing function shall schedule the changes (if possible at all and depending on the severity of the detected problem, i.e. if the operator can afford waiting) during time windows with minimal service impact (i.e. typically during the night).

These examples make it clear that multi-vendor support and flexibility is a key requirement for a Self-Healing module, which will need to read input information in different formats, carry out analyses that consider vendor-specific implementations and generate corrective actions that are in line with the mechanisms and limitations of the equipments supplied by different infrastructure vendors.

6.8. References

[1] Next Generation Mobile Networks (NGMN) Alliance, Requirement Specification (2008) *NGMN Recommendation on SON and O&M Requirements*, Version 1.23, December 2008, www.ngmn.org (accessed 3 June 2011).

[2] 3GPP, Technical Specification, Technical Specification Group Services and System Aspects (2010) *Self-healing Concepts and Requirements*, 3GPP TS 32.541 Version 1.4.0, Release 10, 6 August 2010, http://www.3gpp.org/ftp/Specs/archive/32_series/32.541/32541-140.zip (accessed 3 June 2011).

[3] 3GPP, Technical Report, Technical Specification Group Service and System Aspects (2009) *Study on Self-Healing*, 3GPP TR 32.823 Version 9.0.0, Release 9, 1 October 2009, http://www.3gpp.org/ftp/Specs/archive/32_series/32.823/32823-900.zip (accessed 3 June 2011).

[4] 3GPP, Technical Specification, Technical Specification Group Radio Access Network (2008) *Radio Resource Control (RRC); Protocol Specification*, 3GPP TS 25.331 Version 8.1.0, Release 8, 28 January 2008, http://www.3gpp.org/ftp/Specs/archive/25_series/25.331/25331-810.zip (accessed 3 June 2011).

[5] SOCRATES Project (2010) www.fp7-socrates.eu (accessed 3 June 2011).

[6] 3GPP, Technical Specification, Technical Specification Group Radio Access Network (2010), *Self-Configuring and Self-Optimizing Network (SON) Use Cases and Solutions*, 3GPP TS 36.902 Version 9.2.0, Release 9, 15 June 2010, http://www.3gpp.org/ftp/Specs/archive/36_series/36.902/36902-920.zip (accessed 3 June 2011).

[7] Amirijoo, M., Jorguseski, L., Kürner, T., Litjens, R., Neuland, M., Schmelz, L.C. and Türke, U. (2009) Cell Outage Management in LTE Networks, *Proceedings of ISWCS '09*, Siena, Italy.

[8] Amirijoo, M., Jorguseski, L., Litjens, R. and Nascimento, R. (2011) Effectiveness of Cell Outage Compensation in LTE Networks, *Proceedings of CCNC 2011*, Las Vegas, USA.

[9] Amirijoo, M., Jorguseski, L., Litjens, R., and Schmelz, L.C. (2011) Cell Outage Compensation in LTE Networks: Algorithms and Performance Assessment, *International Workshop on Self-Organizing Networks (IWSON)*, Budapest, Hungary.

7

Return on Investment (ROI) for Multi-Technology SON

Juan Ramiro, Mark Austin and Khalid Hamied

7.1. Overview of SON Benefits

Explosive market penetration of smartphones and laptop computers with mobile broadband connections is creating a critical scenario in which traffic volumes carried by networks grow at a significant pace whereas Average Revenue Per User (ARPU) is reaching saturation. In such context, the Self-Organizing Networks (SON) management paradigm is proposed mainly as a mechanism to facilitate the delivery of mobile broadband services that are sustainable for wireless carriers from a financial perspective. The main benefits of SON are summarized in Figure 7.1.

As can be seen in Figure 7.1, Operational Expenditure (OPEX) savings heavily rely on automation, which is the main keyword associated with SON. However, as will be discussed later, not all automation initiatives result, by nature, in a financial benefit. This will only be the case if the task was originally being executed manually in the absence of SON. In other words, there are no implicit OPEX savings associated with the introduction of a new automated task. Other sources of OPEX savings come from energy efficiency and decreased need for leased resources (e.g. transmission lines), typically achieved by optimizing the utilization of the available capacity.

Another key benefit from SON is Capital Expenditure (CAPEX) savings, which can be achieved by means of a more accurate capacity planning process and by increasing the effective capacity of the existing network infrastructure. This capacity increase allows operators to delay and/or reduce the deployment of costly capacity expansions, and can be obtained by means of automated planning and optimization of the different settings of each network element: physical location, Radio Frequency (RF) configuration, soft parameters, etc. Wireless

Self-Organizing Networks: Self-Planning, Self-Optimization and Self-Healing for GSM, UMTS and LTE,
First Edition. Edited by Juan Ramiro and Khalid Hamied.
© 2012 John Wiley & Sons, Ltd. Published 2012 by John Wiley & Sons, Ltd.

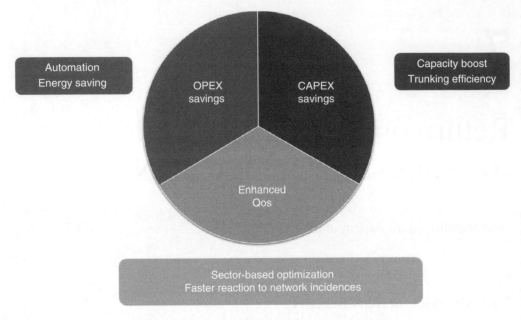

Figure 7.1 Benefits from multi-technology SON.

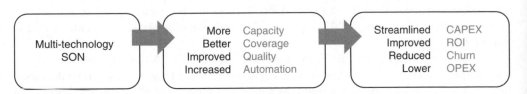

Figure 7.2 From technical to financial benefits.

networks are extremely complex systems, and it is well known that capacity, coverage and quality are many times conflicting targets in wireless network design and operation. Therefore, the capacity enhancing techniques under discussion will be only beneficial from an operational viewpoint if they can provide capacity enhancements while at the same time maintaining satisfactory coverage and quality.

Moreover, another remarkable benefit of the application of SON techniques comes from improved Quality of Service (QoS), which is mainly due to two reasons. First, the availability of Self-Planning and Self-Optimization techniques ensures that the network is always delivering the desired QoS while maintaining reasonable operational costs, which would otherwise be extremely challenging to achieve by means of manual network maintenance activities. Secondly, the deployment of Self-Healing routines minimizes the time to solve, or at least compensate for, the disturbing consequences of network incidences. Ultimately, quality improvements result in increased customer loyalty and satisfaction, thereby reducing churn and directly impacting the operators' Earnings Before Interest, Taxes, Depreciation and Amortization (EBITDA) positively (see Figure 7.2).

7.2. General Model for ROI Calculation

In order to estimate the financial benefits for an operator deploying a set of SON functions in its different network layers, a faithful method is to look at their impact on the company's cash flow, ignoring accounting concepts such as profit/loss and solely focusing on cash calculations. Additionally, there are taxation implications involved in this discussion. However, since these are country-specific, they are not considered in this generic analysis.

When introducing a set of SON functions, it is likely that the impact on the operator's cash flow is observed over several years. In some cases, e.g. when a process is simply automated and there are obvious and recurrent OPEX savings associated with that process, the description of the benefits can be simplified by just reporting annual savings. However, in other cases, such simplification is not straightforward and a more comprehensive methodology needs to be adopted. For example, let us assume that a SON function is activated and, as a consequence, the need for new CAPEX expansions is reduced by 40% during the first year, 30% during the second year and 20% during the third and fourth years. In this case, it becomes clear that a mere annual calculation of the savings is no longer an illustrative tool. In these more complex cases, a more suitable methodology is the computation of the Net Present Value (NPV) of the Cash Flow (CF) that can be differentially attributed to SON over a meaningful time horizon. This calculation, which is often utilized to assess long term projects taking into account the time value of money, is executed as follows:

$$\text{NPV} = \sum_{n=1}^{\infty} CF_n \frac{1}{(1+r)^n} \qquad (7.1)$$

where CF_n is the differential cash flow (i.e. incoming cash flow minus outgoing cash flow, only for those contributions that are directly attributable to SON) associated with year n (taking the entire year as evaluation period and measuring magnitudes at the end of the year) and r is the discount rate, which is the rate of return that investors would typically expect from a project with a risk profile similar to that of the process under evaluation. In principle, the summation should be carried out for an infinite time horizon. However, in practice, this is neither feasible nor important since the contribution of CF_n for large values of n (i.e. many years into the future) does not have a strong influence on the final result due to the negative exponential nature of the factor that it is multiplied with (see Equation 7.1).

For each year, the differential cash flow is calculated as the net amount of extra cash (positive or negative) that the operator is going to have at the end of the year due to the deployment of a certain set of SON functions. Typical components of this magnitude are: (i) additional revenue due to extra quality and therefore reduced churn; (ii) OPEX reduction through automation of tasks that used to be executed manually before the introduction of SON; (iii) OPEX

	Year 1	Year 2	Year 3	Year 4	Year 5
CAPEX Forecast without SON (€ million)	100	100	0	0	0
CAPEX Forecast with SON (€ million)	0	100	100	0	0
Differential Cash Flow due to SON (€ million)	100	0	−100	0	0
NPV (€ million)	25				
Discount rate (%)	20				

Figure 7.3 Differential cash flow and NPV associated with the delay of the need for CAPEX expansions.

reduction through better resource utilization and the subsequent lower need for leased lines for transmission; (iv) fewer CAPEX expansions due to capacity increases through careful adaptive optimization; and (v) more rational and accurate decision processes for CAPEX expansions.

As can be seen, due to the exponential nature of the equation that describes the computation of the NPV, the mere delay of the need for a CAPEX expansion creates value for the operator. Let us assume a simplistic sample scenario in which the forecasted CAPEX expansions for years [1-5] are €[100, 100, 0, 0, 0] million (measured at the end of each year; Figure 7.3). If the impact of applying SON techniques causes a delay of one year in the need for capacity expansions, the CAPEX forecast will become €[0, 100, 100, 0, 0] million. In this case, the differential cash flows for years [1-5] due to the application of SON is €[100, 0, −100, 0, 0] million. However, the different annual components do not just cancel out, since each one of them is multiplied by a different factor that depends on time and therefore reduces the relative weight of contributions with a larger value of n, i.e. reduces the relative importance of capital outlays in the future (see Figure 7.3). With a discount rate of 20%, the NPV of this delay is €25 million.

The calculations described so far do not consider the cost of the SON solution. This cost needs to be deducted from the NPV of its forecasted benefits in order to compute the actual value that the decision to purchase and deploy a SON solution will bring to the operator. Note that such a simple calculation process (i.e. deducting the cost of the solution from the NPV of the forecasted benefits) implies that the solution has been purchased (and paid for) at the beginning of the assessment period and that the associated benefits are evaluated on an annual basis, starting with a first calculation at the end of the first year. If the cost of the solution is incurred in different stages, this needs to be considered in the calculations, since the differential cash flows for the operator will change. Similarly, higher accuracy could be obtained by increasing the resolution of the analysis, e.g. using monthly periods instead of annual ones.

Another illustrative indicator for ROI is the payback period, defined as the time the solution takes to pay for itself. This method is simpler and does not include risks or value depreciation factors. At each point in time, the cumulative cash flow is computed by simply adding the differential cash flows of all the already elapsed periods. When this magnitude equals the cost that was incurred to deploy the solution, the payback time has elapsed (see an example in Figure 7.4).

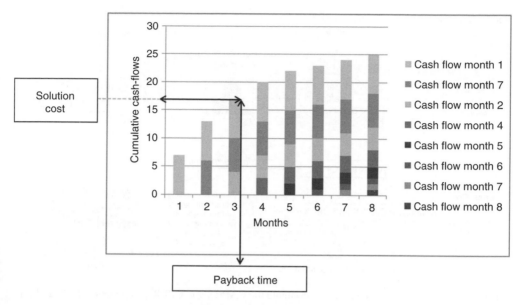

Figure 7.4   Payback time.

7.3.   Case Study: ROI for Self-Planning

The content of this section is organized as follows: Section 7.3.1 discusses two alternative scopes for Self-Planning and centers the discussion on the automated capacity planning process, which is described in Section 7.3.2 and modeled in Section 7.3.3. For a meaningful application of the presented model, a versatile mathematical representation of the per sector traffic distribution is required in order to consider heterogeneous traffic volumes across the network, which has a direct impact on the calculation of the needs for capacity expansions. Such model is presented in Section 7.3.4, and Section 7.3.5 describes its application to assess the capacity needs in a generic network, depending on its initial configuration, per sector traffic distribution and forecasted traffic growth. Based on the application of the described models, the way to compute annual CAPEX and OPEX with and without SON is discussed in Section 7.3.6 and Section 7.3.7, respectively. Finally, a concrete sample scenario is presented and the proposed models are applied in order to compute the ROI of the selected Self-Planning process, as well as its sensitivity to some key inputs and assumptions. The way in which the aforementioned models are applied in a combined way to compute the ROI is summarized in Figure 7.5.

7.3.1.   Scope of Self-Planning and ROI Components

Even in the absence of SON, cellular operators need to plan their networks. Therefore, for Self-Planning, it is fully justifiable to consider OPEX savings due to automation of the planning tasks since this expenditure used to be required and the deployment of SON functionalities will eliminate or reduce it. The same applies in the case of extra revenue due to

For every year in the considered time horizon...

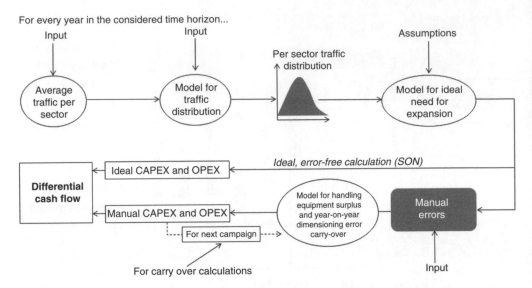

Figure 7.5 Summarizing the process to compute the CAPEX and OPEX components of the ROI for Self-Planning.

reduced churn since this real difference in cash flow is fully attributable to the introduction of SON.

The nature of the CAPEX savings that can be attributed to the application of Self-Planning techniques needs to be classified in two main areas before addressing the calculations in a systematic manner:

1. Self-Planning functions that, in essence, orchestrate qualified decisions about the amount of Hardware (HW), network elements and transmission links to be deployed (no matter whether they are purchased as CAPEX or leased as OPEX). Among these, an illustrative example is the automated capacity planning process that allows operators to make a timely and accurate decision on capacity expansions, i.e. on the extra 3G carrier transceivers to be deployed and the E1 links to be activated in the Iub interfaces in order to cope with the forecasted traffic growth.
2. Self-Planning functions that derive, for example, the antenna settings (e.g. tilt) of new base stations. In this case the benefits from Self-Planning can be materialized as a capacity increase of the resulting network, which implies that more traffic can be carried with the same HW. Obviously, this is implicitly related to the future need for HW expansions to cope with traffic growth. The higher the effective capacity of the current infrastructure, the fewer capacity additions will be required in the future, which has a strong influence on the CAPEX contributions to the different cash flows.

The analysis in this section will be focused on automated capacity planning, which is an example of the first group, since the dynamics involved in the second case are very similar to the ones that will be considered when addressing the ROI for Self-Optimization and they will be extensively covered when discussing that topic.

Figure 7.6   Self-Planning process for capacity extensions.

7.3.2.   Automated Capacity Planning

In this case study, the ROI of an accurate capacity planning process for Universal Mobile Telecommunications System (UMTS)/High Speed Packet Access (HSPA) networks is illustrated. This process is summarized in Figure 7.6, and more details are available in Chapter 4.

After calibrating the system models to be fully aligned with the information in the Operations Support System (OSS), a traffic forecast is generated, which needs to be consistent with the available historical data and include seasonal trends, as well as the impact of planned actions, such as the launch of new services or disruptive marketing initiatives. Based on this process, the forecasted bottlenecks are characterized, i.e. the SON function identifies where and when new capacity expansions are going to be required in order to cope with the forecasted traffic growth at the desired QoS levels. In order to carry out accurate evaluations, initial calibration with current network performance for today's traffic levels is paramount since otherwise calculations might be biased by unrealistic theoretical assumptions.

In this example, the capacity expansion actions under consideration are: (i) extra carrier additions in those sectors that are expected to be overloaded; and (ii) bandwidth expansions in the Iub interfaces by adding more E1 links at those sites in which the Iub is expected to start causing congestion or excessive latency. Doing this in a fully reactive manner would allow (almost) perfect accuracy, at the expense of degrading QoS due to lack of anticipation, resulting in the initiation of corrective actions when problems are already explicit. However, a proactive mode in which capacity expansions are planned beforehand is more advisable since it allows more accurate budgeting and prevents the aforementioned QoS degradation.

7.3.3.   Modeling SON for Automated Capacity Planning

The SON function under evaluation is the one that executes the calibrated traffic forecast and carries out a qualified assessment of when and where certain capacity expansions are going to be required. Apart from the obvious OPEX savings through automation, this functionality will bring significant CAPEX savings by means of a much more accurate capacity planning process. Therefore, in order to quantify these benefits, it is paramount to characterize the accuracy of automated processes compared to traditional, manual calculations. In this respect, SON processes will be assumed to be ideal, i.e. error-free, whereas manual actions contain implicit errors that are described throughout the rest of this subsection.

Field trials conducted by knowledgeable wireless operators have shown scenarios in which the manual execution of the process under analysis would yield an overestimation rate (o) of up to 42%, whereas the underestimation rate (u) would be around 2%. This means that, in this case, when an engineer needs to manually estimate the amount of additional capacity to be deployed in the network, this figure may be inflated by up to 42% due to forecast/calculation errors. At the same time, while there is idle capacity that has been unnecessarily installed in some parts of the network due to errors in the manual dimensioning process, it is also typical to have certain parts of the network where the installed capacity is clearly insufficient. In field trials, the missing capacity in these under-dimensioned parts of the network has been found to be around 2% of the overall required capacity.

In practice, this over-dimensioning process cannot continue ad infinitum, since the engineers of the operator would obviously notice and correct the situation. For simplicity, the following assumptions are adopted for this analysis of the manual dimensioning process in terms of absorption of capacity surplus and dimensioning error carry-over from year to year:

- Only one capacity planning campaign per year is conducted, which is assumed to happen towards the beginning of the year.
- All under-dimensioned items are manually detected and corrected the following year by adding the necessary capacity expansions (on top of the additional ones that are required due to the traffic growth that has been experienced). The corrections targeted at fixing the under-dimensioning errors of year *i* are done during the capacity planning process of year *i+1*. Since this particular compensation process is reactive, it is assumed to be fully accurate, even when executed manually.
- All over-dimensioned items constitute a capacity surplus that will be gradually consumed as the traffic growth in the subsequent years requires extra capacity. No additional expenditure will be carried out until that surplus is exhausted. For the sake of simplicity, full flexibility for HW relocation is assumed, which means that the capacity surplus can be consumed at no extra cost in order to address further expansion needs and manual compensation of under-dimensioned items. Note that in reality this is not the case, since there are HW relocation costs. However, such effect is considered to be out of the scope of this analysis.

7.3.4. Characterizing the Traffic Profile

7.3.4.1. Traffic Distributions

The traffic profile is an important piece of information for ROI calculation. When calculating the ROI of Self-Planning functionalities, two types of traffic (voice over UMTS Release 99, denoted by R99, and data over High Speed Downlink Packet Access, denoted by HSDPA) will be considered and, for each one of them, two main properties need to be carefully taken into consideration: (i) temporal evolution of the average traffic per sector during the busy hour; and (ii) distribution of the traffic per sector across the network during the busy hour.

The first one is a key input to the process, which is typically obtained through traffic forecasts, taking the operator's business plan into account. The second one is needed to model the fact that networks are not uniformly loaded. When analyzing the distribution of the traffic per sector in the busy hour, results from real networks have shown that it can be represented by means of a Gamma function in which the standard deviation is roughly proportional to the average (although the factor relating both magnitudes will change from network to network). The Gamma distribution reads as follows:

$$f(x) = x^{k-1} \frac{exp\left(-x/\theta\right)}{\Gamma(k)\theta^k} \tag{7.2}$$

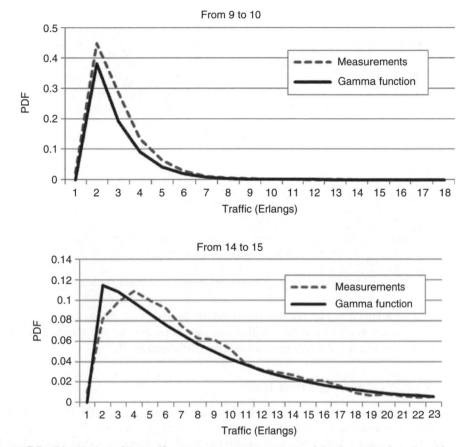

Figure 7.7   Distribution of the traffic per sector at two moments of the day: (top) from 9 to 10; (bottom) from 14 to 15.

where Γ is the Gamma function and the parameters θ and k can be derived from the desired mean and standard deviation by means of the following equations:

$$\theta = \frac{Var(x)}{E(x)} \tag{7.3}$$

$$k = \frac{E(x)}{\theta} \tag{7.4}$$

Figure 7.7 depicts the distribution of the traffic per sector, measured in a commercial network for two different periods of one hour, together with the way in which the Gamma function models such distribution, taking the average and standard deviation of the traffic per sector as an input.

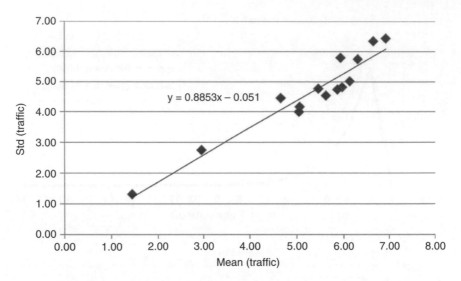

Figure 7.8 Average versus standard deviation of the traffic per sector.

Figure 7.8 correlates the average and the standard deviation of the measured traffic per sector for different periods of one hour within one day, showing that the standard deviation grows with the average in a roughly proportional way (dictated by θ).

7.3.4.2. Combining Voice and Data Traffic

Since live operational networks have a combination of voice and data traffic, the combination of both needs to be addressed in order to compute aggregated, normalized traffic metrics in the radio interface and in E1 links. When assessing the need for capacity expansions, it can be considered that the maximum amount of normalized traffic per sector that can be served through a single carrier is 100 'normalized carrier traffic units' (200 for two carriers, 300 for three carriers, etc.), although more conservative exercises can be done, for example assuming that the maximum traffic per carrier (from a radio perspective) is 80 or 90 'normalized carrier traffic units'. Similarly, the maximum amount of normalized traffic that can be served through a single E1 link is 100 'normalized E1 traffic units' (200 for two E1 links, 300 for three E1 links, etc.) and conservative (lower) thresholds can also be applied. A straightforward and simplistic way to combine and normalize voice and data traffic is described in the following equations.

$$T_{sector} = 100 \cdot \left(\frac{T_{VoicePerSector}}{T_{MaxVoicePerCarrier}} + \frac{T_{DataPerSector}}{T_{MaxDataPerCarrier}} \right) \tag{7.5}$$

$$T_{Iub} = 100 \cdot \left(\frac{T_{VoicePerSite}}{T_{MaxVoicePerE1}} + \frac{T_{DataPerSite}}{T_{MaxDataPerE1}} \right) \tag{7.6}$$

where T_{sector} is the normalized radio traffic per sector, $T_{VoicePerSector}$ is the voice traffic per sector (in Erlangs), $T_{MaxVoicePerCarrier}$ is the maximum voice traffic per sector and carrier (in Erlangs) when there is no data traffic, $T_{DataPerSector}$ is the data traffic per sector (in kbps) and $T_{MaxDataPerCarrier}$ is the maximum data traffic per sector and carrier (in kbps) when there is no voice traffic. Similarly, T_{Iub} is the normalized Iub traffic per site, $T_{VoicePerSite}$ is the voice traffic per site (in Erlangs), $T_{MaxVoicePerE1}$ is the maximum voice-only traffic per E1 link (in Erlangs), $T_{DataPerSite}$ is the data traffic per site (in kbps) and $T_{MaxDataPerE1}$ is the maximum data-only traffic per E1 link (in kbps).

7.3.5.   Modeling the Need for Capacity Expansions

Assuming an ideal (i.e. error-free) capacity planning process, the first step is to compute the amount of capacity expansions that need to be carried out in order to serve the forecasted traffic. To make this process simpler, it is assumed that the traffic volume is always growing or at some point stable, but never decreasing. Furthermore, the initial calculations are conducted assuming a perfect capacity planning process, and the impact of manual errors will be introduced at a later stage.

Let us assume that all sectors are equipped with infrastructure to manage one UMTS carrier (in this context, a carrier is a transceiver that allows one sector to manage an independent UMTS carrier). At a certain year i, the amount of additional carriers $C(i)$ as compared with the initial configuration of one carrier per sector can be computed as follows:

$$C(i) = N * \sum_{n=1}^{\infty} (n-1) \cdot P_n(i) \qquad (7.7)$$

where N is the number of sectors in the network, and $P_n(i)$ is the rate of sectors that need at least the activation of the nth carrier in order to (partially or fully) cope with the amount of traffic that they have during year i. The calculation of $P_n(i)$ is based on the traffic distributions presented in Section 7.3.4.1 (see Figure 7.9 and Figure 7.10). Note that the reason why the

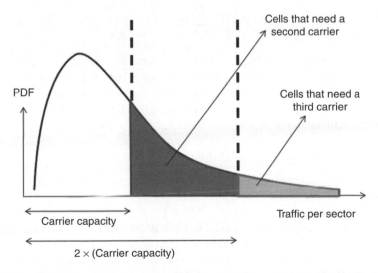

Figure 7.9   Computing the amount of extra carriers required for a certain traffic distribution (as compared with the original configuration).

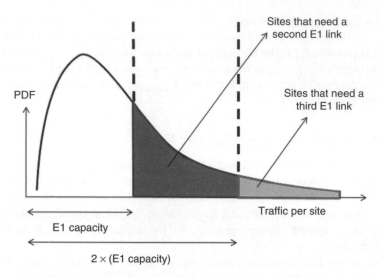

Figure 7.10 Computing the amount of extra E1 links required for a certain traffic distribution (as compared with the original configuration).

aforementioned percentage differs from year to year is that the traffic distribution changes as time goes by. In addition, note that for this simple calculation, no spectrum limitation is assumed (this will be addressed in the Self-Optimization example).

Similarly, if the starting point is that all sectors are equipped with HW to manage M carriers, the structure of the process would be similar, and the resulting equation (defining the number of extra carriers as compared with the starting point) would be the following:

$$C(i) = N * \sum_{n=M}^{\infty} (n - M) \cdot P_n(i) \qquad (7.8)$$

At this point, it is important to highlight that the calculated number of additional carriers, $C(i)$, that are required for year i represents the extra HW that is required for that year as compared with the initial configuration (i.e. with the equipment that is installed at the beginning of the time period for the cash flow discount calculations, also denoted by $C(0)$). Therefore, for a given year i, the actual number of new additional carriers $A(i)$ is given by:

$$A(i) = C(i) - C(i - 1) \qquad (7.9)$$

When evaluating capacity expansions in the Iub interfaces, the mechanism is similar, although some differences need to be considered:

- Since Iub interfaces are physically related to NodeBs, not to sectors, it is assumed that their resources are pooled among sectors in the same NodeB.
- Therefore, a certain Iub needs a capacity expansion when the aggregated traffic of its corresponding NodeB exceeds the currently installed capacity, although some safety margins to make calculations more conservative can be introduced.

- When characterizing the traffic per site, there are several options, depending on the local homogeneity of the network. In the typical case of three sectors per NodeB, it can be assumed that the average traffic per NodeB in the busy hour equals three times the average traffic per sector, which implicitly endorses the assumption that loaded sectors tend to be relatively grouped together, whereas less loaded sectors also tend to be locally grouped within areas of low loaded NodeBs. This is the assumption that has been considered in this example, although other distributions can be used if more detailed knowledge is available about the traffic distribution across the network under consideration.
- When analyzing the investment in new carriers, only extra, additional carriers were considered every year since, in this context, each investment in extra carrier equipment is a one-off payment. However, in this example, let us assume that the E1 links that provide capacity to Iub interfaces are leased lines. In this case, an annual fee needs to be paid for each and every leased line, no matter whether that line was already active during the previous years or not.

7.3.6. CAPEX Computations

In this example, CAPEX contributions to the differential cash flow are computed by direct comparison between an ideal SON functionality that executes error-free capacity planning and a manual process that involves errors as described in Section 7.3.3. Applying the methodologies and models described throughout this chapter, for every year i, the differential cash flow component due to a more rational CAPEX allocation process, $CF_{CAPEX}(i)$, is computed as:

$$CF_{CAPEX}(i) =$$
$$= Cost_{Carrier} * (A_{Manual}(i) - A_{SON}(i)) \tag{7.10}$$

where $Cost_{Carrier}$ is the monetary cost per additional carrier equipment; $A_{SON}(i)$ is the amount of additional carriers that would be installed in the network during year i for the case in which a perfect SON function is available to carry out this process without any errors; and $A_{Manual}(i)$ is the equivalent number when capacity planning is assumed to be done manually (according to the process described in Section 7.3.3).

7.3.7. OPEX Computations

In this example, the OPEX savings associated with SON can be structured around two main categories that are totally cumulative: (i) lower annual fees for leased E1 links through a more rational process to decide on their activation; and (ii) lower workload through automation.

7.3.7.1. Lower Annual Fees for Leased Transmission Lines

This calculation follows the same process as the one described for the computation of the individual CAPEX contributions to the annual cash flow. As already stated, the only difference is that leased transmission lines constitute an annual recurring cost, i.e. they are not paid for only once. Thus, every year, the calculation for each one of the methodologies (manual planning or SON) needs to consider all the installed leased lines, no matter whether they were installed that year or during previous ones.

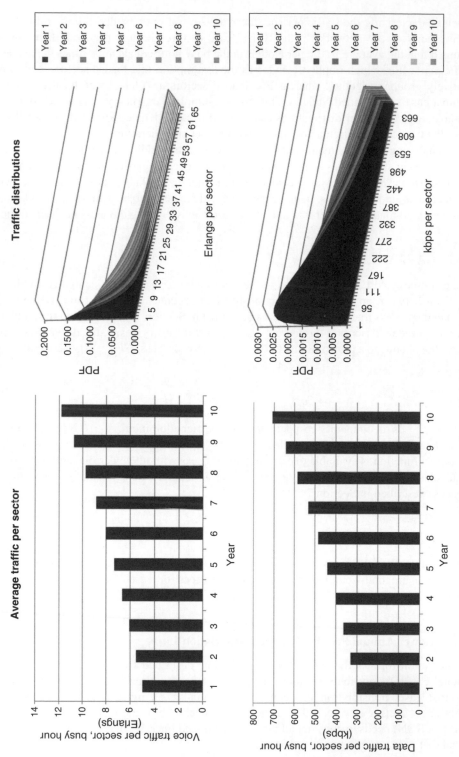

Figure 7.11 Assumed traffic patterns for voice and data.

Applying the methodologies and models described along this chapter, for every year i, the differential cash flow component due to more rational leased line installation process $CF_{OPEX,links}(i)$ is computed as:

$$CF_{OPEX,links}(i) = Cost_{Link} * (L_{\text{Manual}}(i) - L_{SON}(i))$$ (7.11)

where $Cost_{Link}$ is the annual cost per installed transmission link (per E1 in this specific case); $L_{SON}(i)$ is the total amount of links (E1 links in this case) that are being operated in the network during year i (no matter whether they were installed that year or during the previous ones, since they are assumed to be leased lines and therefore imply annual, recurrent payments) for the case in which a perfect SON function is available to derive the amount of required links without any errors; and $L_{\text{Manual}}(i)$ is the equivalent number when capacity planning is assumed to be done manually (according to the process described in Section 7.3.3).

7.3.7.2. Lower Workload through Automation

These OPEX savings are straightforward to calculate. The computation is based on the number of man-hours that are required to carry out the aforementioned capacity planning tasks, the nominal cost per man-hour and the number of man-hours (if any) that are required in order to operate the corresponding SON functionality. Note that some SON functionalities, especially in the Self-Planning domain, are automatic and not always autonomous, i.e. some human intervention may be needed in order to initiate the processes. These OPEX savings can be summarized as follows:

$$CF_{OPEX,auto}(i) =$$
$$= Cost_{Man\text{-}hour} * (Effort_{Manual}(i) - Effort_{SON}(i))$$ (7.12)

where $CF_{OPEX,auto}(i)$ is the differential cash flow component during year i due to automation; $Cost_{Man\text{-}hour}$ is the monetary cost per man-hour; $Effort_{Manual}(i)$ is that number of man-hours required to carry out the aforementioned manual capacity planning tasks during year i; and $Effort_{SON}(i)$ is the equivalent number for the case in which a perfect SON function is available to carry out the capacity planning process in an automated manner.

7.3.8. Sample Scenario and ROI

7.3.8.1. ROI Calculation

The assumptions for this sample calculation are presented in Table 7.1. The assumed traffic growth profile is illustrated in Figure 7.11 for both voice and traffic, and the overall amount of carriers that are required (at network level) to serve all the offered traffic is depicted in Figure 7.12. As can be seen, this number grows steadily when traffic increases.

Let us start considering the ideal, error-free capacity planning process. The assumptions in Table 7.1 indicate that the network is initially equipped with 60000 carriers (10000 sites × 3 sectors/site × 2 carriers/sector). However, as shown in Figure 7.12, during the first year a total amount of 60166 carriers are required and therefore 166 additional ones need to be added. The number of incremental carrier expansions that are required every year to cope with the

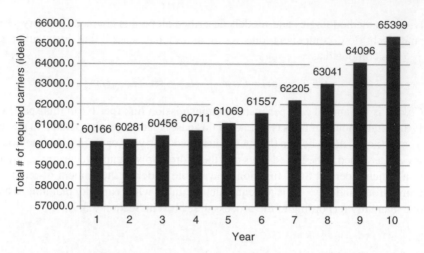

Figure 7.12 Overall number of carriers that are required to serve all the offered traffic.

Table 7.1 Assumptions for the ROI assessment of Self-Planning

Network size and initial configuration	
Number of sites	10000
Sectors per site	3
Initial number of carriers per sector (year 1)	2
Initial number of E1 links per site (year 1)	2
Manual process accuracy *(conservative error rates, in line with field observations)*	
Over-dimensioning rate (o)	20%
Under-dimensioning rate (u)	2%
Initial average traffic per sector in the busy hour (year 1)	
Voice (R99)	5 Erlangs
Data (HSDPA)	300 kbps
Traffic growth	
Relative traffic growth pattern (%; years 2-10)	[10, 10, 10, 10, 10, 10, 10, 10, 10] %
Cost structure	
Annual OPEX per E1link	€5000
Man-days to dimension 1000 sites	40
OPEX per engineer and day	€300
OPEX to operate an ideal SON solution	€0
CAPEX for extra carrier per sector	€15000
Capacity figures and other radio indicators	
Radio interface capacity for voice only (R99)	21 Erlangs
Radio interface capacity for data only (HSDPA)	1600 kbps
Nett Iub capacity for voice only (R99, per E1 link)	25 Erlangs
Nett Iub capacity for data only (HSDPA, per E1 link)	630 kbps
Soft Handover Overhead	40%

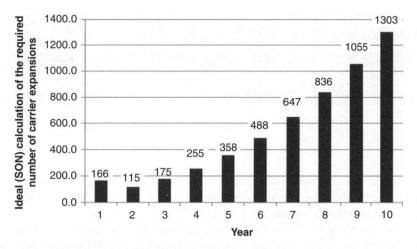

Figure 7.13 Ideal calculation (carried out by the Self-Planning function) of the required carrier expansions (per year) to cope with the experienced traffic growth.

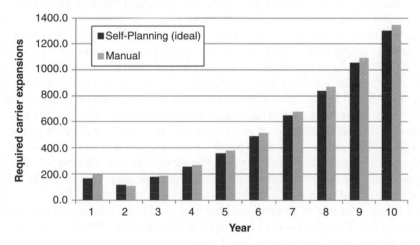

Figure 7.14 Ideal versus manual calculation of the required carrier expansions (per year) to cope with the experienced traffic growth.

offered traffic is illustrated in Figure 7.13. As can be seen, this number equals 166 for year 1, and then evolves as a function of the annual traffic growth. The need for carrier additions in the first year is larger than in the second year because of the initial configuration of the network, which is under-equipped. Since traffic is assumed to grow 10% each year, the subsequent need for new carriers grows exponentially, due to the fact that annual absolute traffic increases become higher every year.

For the manual case, the amount of extra carrier additions that imply the actual purchase of additional equipment is illustrated in Figure 7.14. This calculation is conducted according to the assumptions presented in Section 7.3.3 about the manual dimensioning process regarding the handling of capacity surplus and dimensioning errors carry-over from year to year.

	Year 1	Year 2	Year 3	Year 4	Year 5	Year 6	Year 7	Year 8	Year 9	Year 10
CAPEX reduction	0.45	−0.14	0.16	0.22	0.28	0.35	0.43	0.51	0.59	0.67
Extra carriers	0.45	−0.14	0.16	0.22	0.28	0.35	0.43	0.51	0.59	0.67
OPEX reduction	8.32	1.66	1.86	2.08	2.32	2.57	2.84	3.13	3.42	3.72
E1 expansions	8.20	1.54	1.74	1.96	2.20	2.45	2.72	3.01	3.30	3.60
Automation	0.12	0.12	0.12	0.12	0.12	0.12	0.12	0.12	0.12	0.12
Total (€ million)	8.77	1.52	2.02	2.30	2.60	2.92	3.27	3.64	4.01	4.39

Figure 7.15   Cash flow projection for Self-Planning.

Note that, to facilitate the comparison, Figure 7.14 also reproduces the number of carrier additions in the case in which the capacity planning process is error-free due to the application of advanced Self-Planning techniques.

As can be seen in Figure 7.14, the difference between ideal and manual calculations is not as big as the 20% over-estimation rate assumed in Table 7.1, due to the carry-over and compensation mechanisms that have been assumed for over-dimensioned items (see Section 7.3.3), which basically imply that the unnecessary carriers that are added during one year can be flexibly used (through cost-free HW relocation) during subsequent years, thereby decreasing the effective amount of purchases that need to be carried out. Also due to the application of these carry-over mechanisms, during the second year, the need for additional carriers with the manual methodology is lower than with Self-Planning, which is translated into a negative contribution of the CAPEX-related differential cash flow component of that year (this is illustrated in Figure 7.15). Note that this effect is present every year. However, in the second year the combination of a big need for investment during the first year (due to the original network under-dimensioning) followed by a modest traffic increase in the second year has lead to such extreme values that the value of the differential cash flow related to CAPEX becomes negative. In other words, the manual process over-invested heavily during the first year and then made use of that investment in the second year. As stated before, this means that the application of Self-Planning techniques delays the need for investment, which has a positive impact on the NPV, even though some individual cash flow components are negative (see Figure 7.3).

Once the amount of IIW to be purchased has been calculated, it is possible to compute the corresponding cash flow components by simply multiplying the amount of required carriers per year with the assumed monetary cost per carrier.

Going forward, a very similar illustration can be done for the addition of E1 links, which are considered as annual recurrent OPEX in this exercise. Since they are very similar processes, details are not provided here for the sake of brevity. Applying the described methodology, the cash flow projection presented in Figure 7.15 can be built and, assuming a discount rate of 20%, the resulting NPV (described in Equation 7.1) is €15.92 million.

7.3.8.2.   Sensitivity Analysis

In the following, a brief sensitivity analysis is provided in order to illustrate the dependence of the NPV on the different input parameters. First of all, the dependence on the traffic growth pattern is analyzed, and Table 7.2 shows three different patterns for traffic growth together with their associated NPV.

Table 7.2   NPV of Self-Planning versus traffic growth

Relative traffic growth pattern (%; years 2-10)	NPV (€ million)
[10, 10, 10, 10, 10, 10, 10, 10, 10]	15.92
[20, 20, 20, 20, 20, 20, 20, 20, 20]	32.35
[30, 30, 30, 30, 30, 30, 30, 30, 30]	51.07

Table 7.3   NPV of Self-Planning versus over-dimensioning rate

Over-dimensioning factor (o)	NPV (€ million)
20%	15.92
30%	25.74
40%	36.87

Table 7.4   NPV of Self-Planning versus initial number of carriers per sector

Initial number of carriers per sector	NPV (€ million)
1	18.64
2	15.92
3	15.13

The main conclusion from Table 7.2 is that more aggressive traffic growth patterns result in more ROI from the Self-Planning functionalities, since more aggressive traffic growth leads to more need for capacity expansions and therefore more room for benefitting from an error-free capacity planning process.

Table 7.3 shows the dependency of the NPV on the value of the over-dimensioning rate (o in Table 7.1). For this calculation, the first traffic growth pattern in Table 7.2 is used: [10, 10, 10, 10, 10, 10, 10, 10, 10]%. As expected, higher over-dimensioning rates for the manual process lead to more ROI from the ideal Self-Planning solution.

In Table 7.4, the dependency of the NPV on the initial number of carriers per sector is analyzed. For this exercise, the first traffic growth pattern in Table 7.2 is used: [10, 10, 10, 10, 10, 10, 10, 10, 10]%, and the over-dimensioning rate (o) is assumed to be 20%. As can be seen, the value of an ideal solution is higher when the network under consideration is initially equipped with fewer carriers, since this leads to more need for capacity expansions and therefore more room for benefitting from an error-free capacity planning process.

7.4.   Case Study: ROI for Self-Optimization

7.4.1.   Self-Optimization and ROI Components

The kind of Self-Optimization covered in this chapter is conceived as an online, adaptive process that runs continuously and autonomously, i.e. with no human intervention at all. With this scheme, inputs are gathered from several sources (performance counters, current parameter settings, alarms, call traces, etc.) and qualified decisions on how to modify the key parameter settings on a per network element (or even adjacency) basis are automatically made and implemented in order to adapt those settings to the varying environmental and traffic conditions, ensuring optimum performance. Manual execution of such process is typically unfeasible due to the unrealistic engineering bandwidth requirements associated with it. Therefore, even though it is a fully autonomous process, there are no OPEX savings directly linked to automation, since these tasks are not typically carried out in the absence of SON.

Moreover, the presence of active Self-Optimization processes results in measurable quality improvements, which are translated to additional revenue and cash flow through reduced churn, which will be taken into account in this analysis. Apart from quality enhancement, the discussion of the benefits associated with Self-Optimization is going to be focused around the fact that a properly optimized network experiences a noticeable capacity increase (mainly on the radio side, but also in the transmission domain through increased trunking efficiency by means of load balancing techniques), which means that more traffic can be served with the same amount of HW. This capacity increase lowers the need for future HW and transmission expansions due to capacity reasons and therefore has a strong influence on the CAPEX and OPEX contributions to the different cash flows.

Although both Self-Planning and Self-Optimization reduce the CAPEX related to HW expansions, they do it by exploiting different mechanisms. Whereas Self-Optimization increases the amount of traffic that can be served with each radio equipment unit, Self-Planning avoids unnecessary investments by accurately calculating where and when additional capacity is going to be required.

7.4.2.   Modeling SON for Self-Optimization

From a modeling perspective, the impact of Self-Optimization is going to be structured around the following components:

- Reduced dropped call rate, from DCR_{Manual} to DCR_{SON}, which will reduce churn, increase loyalty and therefore improve EBITDA.
- Higher sector capacity from a radio perspective, understood as an increase in the amount of traffic that can be served with each carrier in each sector. This gain will reduce/delay the need for new carrier additions. Applying SON techniques, the radio capacity per sector will increase by a factor of G_{SON}, which is assumed here to be equal for voice and data traffic.
- Reduced dispersion of the per sector traffic distribution through load balancing techniques oriented to alleviate Iub congestion problems. Field trials have shown that the standard deviation of the traffic distribution can be reduced by a factor (denoted by X_{SON}) of around 20% after applying inter-NodeB load balancing techniques for UMTS.

7.4.3.   Characterizing the Traffic Profile

The exact same models and considerations as in Section 7.3.4 are applied here. The only new element in the model is that, whereas in the Self-Planning analysis the traffic distributions were assumed to be the same before and after the application of SON techniques, in this case the

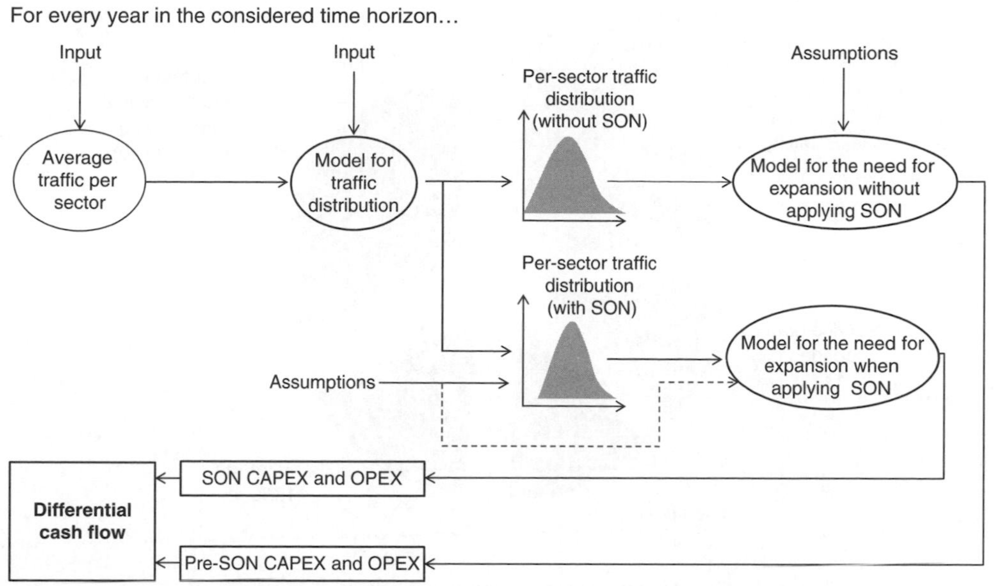

For every year in the considered time horizon…

Figure 7.16 Summarizing the process to compute the capacity-driven CAPEX and OPEX components of the ROI for Self-Optimization.

utilization of Self-Optimization will reduce the standard deviation of the per sector traffic distribution with a factor denoted by X_{SON} (see Section 7.4.2), thereby narrowing the distribution.

7.4.4. Modeling the Need for Capacity Expansions

In general, the same methodology that was described in Section 7.3.5 is applied here. In this case, evaluations with and without SON will apply different values for the sector capacity (from a radio perspective) and different traffic distributions, as described in Sections 7.4.2 and 7.4.3. Morcover, two new effects are introduced into the model:

1. Limitations in terms of available spectrum and maximum space for radio equipment in the physical site facility. Although these two effects are unrelated by nature, they are both jointly materialized into a simple limitation in terms of maximum number of carriers per sector, denoted by $MAX_{carriers}$. When a sector requires more than $MAX_{carriers}$ carriers, a new site is assumed to be deployed in its vicinity due to capacity reasons.
2. Radio Network Controller (RNC) capacity limitations, characterized by the maximum number of logical cells ($RNC_{capacity}$) that can be handled. In this context, the addition of a new carrier in a given sector implies the creation of a new logical UMTS cell. When this limit is exceeded, an additional RNC is installed.

The entire process to compute the ROI components due to lower need for capacity expansions is summarized in Figure 7.16.

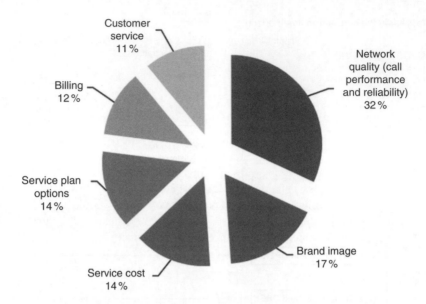

Figure 7.17   J.D. Power overall customer satisfaction breakdown [3].

7.4.5.   *Quality, Churn and Revenue*

As alluded to previously, network quality has an impact with respect to overall churn. While there have been a variety of ways to estimate those who will churn in a wireless network [1, 2], it is first important to distinguish different reasons why a customer would churn (or leave a certain operator for another), and later identify the reasons that are impacted by SON. For example, the reasons for churn can be classified into the following categories (in line with [3]):

* Service cost. Out of the total number of churners, the fraction due to this reason is in the following denoted by *cfcost*, which can take a value between 0 and 1.
* Billing (fraction = *cfbilling*, between 0 and 1).
* Brand image (fraction = *cfbrand*, between 0 and 1).
* Customer service (fraction = *cfservice*, between 0 and 1).
* Service plan options (fraction = *cfoptions*, between 0 and 1).
* Network quality (fraction = *cfnetwork* = *cfn*, between 0 and 1).

Mathematically, the following expression must hold:

$$cfcost + cfbilling + cfbrand + cfservice + cfoptions + cfn = 1 \qquad (7.13)$$

While a churning customer may actually have several reasons to churn, one reason is typically dominant. To obtain a first approximation about the relative importance of the different reasons to churn, satisfaction surveys, such as those by J.D. Power, Nielsen or others, can be reviewed. A typical finding in these reports is that network quality is normally the most important metric driving overall customer satisfaction. Indeed, J.D. Power [3] breaks down the drivers of overall satisfaction as shown in Figure 7.17, where it can be seen that network quality has a contribution of 32% to overall customer satisfaction. In this respect, it is assumed that improving network quality via SON or other approaches will reduce primarily the churn due to network quality issues.

To characterize the benefit of improving network quality from a churn perspective, the first step is to compute the overall percentage of users that are churning due to network quality, denoted by cn, which is obtained by multiplying the total churn rate (denoted by c) with the fraction of churners that are due to network quality (denoted by cfn).

$$cn = c \cdot cfn \tag{7.14}$$

There can be several ways to historically estimate cfn, such as surveying customers that churned and asking them to classify the primary reason why they churned, looking at proxies for complaints due to network quality (for example, based on calls into customer care) or relying on external third party surveys measuring network satisfaction. These approaches work well for understanding what happened after the customer has already switched carriers. However, to predict how much the network churn rate (c) can be reduced by improving network quality, a model is required as a function of Key Performance Indicators (KPIs) that can be measured using network statistics. Since the objective here is to formulate a model for ROI purposes, let us consider using a simplistic linear model to estimate the value of cfn as:

$$cfn = w_1 \times KPI_1 + w_2 \times KPI_2 + \ldots w_n \times KPI_n + b \tag{7.15}$$

where $\{w_i\}$ denotes the set of weights characterizing the relative importance of each KPI, and b is a constant. As an example, let us assume that that the following three KPIs will be considered to model churn:

- Percentage of customers which are unhappy due to coverage.
- Percentage of calls that are dropped due to network quality issues.
- Percentage of calls that cannot be started due to accessibility issues.

Specifically, how to measure these three KPIs on a given network will vary depending on the network infrastructure vendor and the wireless access technology. For instance in a UMTS system the following definitions can be applied:

$$KPI_1 = \% \, calls \, with \left\{ (RSCP < -100 \, dBm) \, or \left(\frac{Ec}{Io} < -15 \, dB \right) \right\} = Outage \tag{7.16}$$

$$KPI_2 = Drop \, Call \, Rate = DCR \tag{7.17}$$

$$KPI_3 = Call \, Setup \, Failure \, Rate = CSFR \tag{7.18}$$

If multiple measurements (K) of $cfn(i=1 \ldots K)$ are available, for example from surveys or statistics related to customer care complaints in different geographic areas, then the optimal values for the weights $\{w_i\}$ and constant b in Equation (7.15) can be easily be computed through standard multi-regression techniques[1]. In matrix form, the least squares solution for the weights can be illustrated as follows:

$$\begin{bmatrix} cfn(1) \\ \vdots \\ cfn(K) \end{bmatrix} = \begin{bmatrix} KPI_1(1) & KPI_2(1) & \cdots KPI_n(1) \\ \vdots & \vdots & \vdots \\ KPI_1(K) & KPI_2(K) & \cdots KPI_n(K) \end{bmatrix} \begin{bmatrix} w_1 \\ w_2 \\ \vdots \\ w_n \end{bmatrix} + \begin{bmatrix} b \\ \vdots \\ b \end{bmatrix} \tag{7.19}$$

---

[1] For example, Microsoft Excel has a simple multi-regression solver with the LINEST() command.

Simplifying in matrix form:

$$\%CFN = KPI \times \bar{W} + \bar{B}$$

(7.20)

As explained in [4], the least squares solution for the weights can be found by applying the following equation:

$$\bar{W} = (KPI^{T}KPI)^{-1} KPI^{T}\overline{\%CFN}$$

(7.21)

where $\bar{B}$ can be calculated as the residual error vector.

Let us then assume that the application of SON multiplies each KPI by a factor G_i (typically with $G_i \leq 1$, implying a *KPI* improvement). In this case, combining (7.14) and (7.15), the overall reduction in the churning rate due to quality (denoted by Δ_{churn}) would be as follows:

$$\Delta_{churn} = cn_{OLD} - cn_{NEW} = c \cdot (cfn_{OLD} - cfn_{NEW}) = c \cdot \sum_{i=1}^{n} w_i \cdot KPI_i \cdot (1 - G_i)$$

(7.22)

And the annual revenue loss that is avoided due to quality improvement, denoted by R, can be formulated in the following simplistic way:

$$R = \Delta_{churn} \cdot S \cdot ARPU \cdot 12$$

(7.23)

where S stands for the number of subscribers and the ARPU is measured on a monthly basis.

7.4.6.   CAPEX Computations

In this case, there are three different CAPEX components to be taken into consideration: new carriers, new capacity sites and new RNCs. For each year i, the differential cash flow component due to CAPEX savings due to capacity increase, $CF_{CAPEX}(i)$, can be written as:

$$CF_{CAPEX}(i) =$$
$$= Cost_{Carrier}{}^{*} (A_{Manual}(i) - A_{SON}(i)) +$$
$$+ Cost_{Site}{}^{*} (S_{Manual}(i) - S_{SON}(i)) +$$
$$+ Cost_{RNC}{}^{*} (RNC_{Manual}(i) - RNC_{SON}(i))$$

(7.24)

where $Cost_{Carrier}$, $Cost_{Site}$ and $Cost_{RNC}$ are the monetary costs per additional carrier equipment, capacity site and RNC, respectively; $A_{SON}(i)$, $S_{SON}(i)$ and $RNC_{SON}(i)$ stand for the amount of additional carriers, capacity sites and RNCs that would be installed in the network during year i when Self-Optimization processes are applied; and $A_{Manual}(i)$, $S_{Manual}(i)$ and $RNC_{Manual}(i)$ are the equivalent magnitudes when no Self-Optimization processes are applied. For this exercise, the costs per carrier, capacity site and RNC are assumed to be constant with time. For more details about the way to compute the need for capacity expansions due to traffic growth, see Section 7.4.4.

7.4.7.   OPEX Computations

As compared with Self-Planning, in this case, the only considered component is the lower annual fees for leased transmission lines. At high level, the way to compute this cash flow component is similar to the one described in Section 7.3.7.1. However, in that case the reason for the OPEX savings was the application of a more accurate capacity planning process that avoided over-dimensioning, whereas in this case the reason behind these savings is the application of load balancing techniques between NodeBs that increase the trunking efficiency of the system and therefore reduce the need for extra E1 links as traffic grows. As already described in Section 7.4.2, this is modeled mathematically by means of a reduction in the standard deviation of the per sector traffic distribution, i.e. by narrowing the distribution and therefore reducing the percentage of NodeBs requiring a higher number of E1 links. Thus, the differential cash flow component due to Self-Optimization, $CF_{OPEX,links}(i)$, is computed as:

$$CF_{OPEX,links}(i) = Cost_{Link} * (L_{Manual}(i) - L_{SON}(i))   \qquad (7.25)$$

where $Cost_{Link}$ is the annual cost per installed transmission link (per E1 in this specific case); $L_{SON}(i)$ is the total amount of links (E1 links in this case) that are being operated in the network during year i (no matter whether they were installed that year or during the previous ones, since they are assumed to be leased lines and therefore imply annual, recurrent payments) for the case in which Self-Optimization is applied; and $L_{Manual}(i)$ is the equivalent number when such techniques are not utilized. For this exercise, the annual cost per link is assumed to be constant in time.

7.4.8.   Sample Scenario and ROI

The assumptions for this sample calculation are listed in Table 7.5, and the assumed traffic growth profile is illustrated in Figure 7.18 (for both voice and traffic). As can be seen, the model is applied over a time window of 10 years. Figure 7.19 shows the number of additional carriers, E1 links, capacity sites and RNCs that need to be added to cope with the growing traffic, for both manual network management and Self-Optimization. The same trend is observed in all curves. As can be seen, the capacity gain provided by the application of Self-Optimization algorithms delays and reduces the need for HW expansions, since each network element is capable to carry more traffic. Moreover, the extra revenue per year due to reduced churn is depicted in Figure 7.20.

Applying the described methodology, the cash flow projection depicted in Figure 7.21 can be built and, assuming a discount rate of 20%, the resulting NPV (described in Equation 7.1) is €138.36 million, split in different categories as shown in Figure 7.22.

Table 7.5   Assumptions for the ROI assessment of Self-Optimization

Network size, limitations and initial configuration	
Number of sites	10000
Sectors per site	3
Initial number of carriers per sector (year 1)	2
Initial number of E1 links per site (year 1)	2
$RNC_{capacity}$	500
$MAX_{carriers}$	3
Subscribers, ARPU and churn	
Number of subscribers	20 million
Monthly ARPU	€10
Churn rate (c)	10%
Fraction of churners due to bad network quality (cfn)	0.1
Quality improvement due to SON	
DCR_{Manual}	0.7%
DCR_{SON}	0.5%
Capacity gain figures	
G_{SON}	50%
X_{SON}	20%
Initial average traffic per sector in the busy hour (year 1)	
Voice (R99)	5 Erlangs
Data (HSDPA)	300 kbps
Traffic growth	
Relative traffic growth pattern (%; years 2-10)	[20, 20, 20, 20, 20, 20, 20, 20, 20] %
Cost structure	
Annual OPEX per E1 link	€5000
CAPEX for extra carrier per sector	€15000
CAPEX for extra capacity site	€50000
CAPEX for extra RNC	€100000
Capacity figures and other radio indicators	
Radio interface capacity for only voice (R99)	21 Erlangs
Radio interface capacity for only data (HSDPA)	1600 kbps
Nett Iub capacity for only voice (R99, per E1 link)	25 Erlangs
Nett Iub capacity for only data (HSDPA, per E1 link)	630 kbps
Soft Handover Overhead	40%

In the following, a brief sensitivity analysis is provided in order to illustrate the dependence of the NPV on the traffic growth pattern (see Table 7.6). As can be seen, more aggressive traffic growth patterns result in higher ROI from the Self-Optimization functionalities, since a more aggressive traffic growth requires more capacity expansions and this constitutes a more adequate scenario to realize the benefits of capacity improvements.

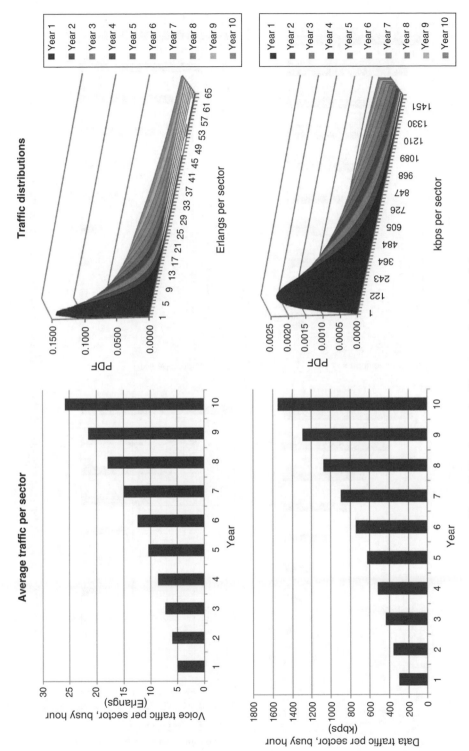

Figure 7.18 Assumed traffic patterns for voice and data.

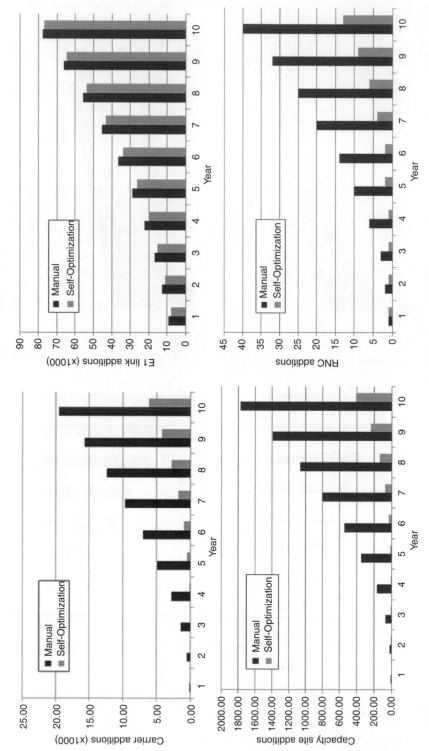

Figure 7.19 Carrier, E1 link, capacity site and RNC additions that are required to serve all the offered traffic.

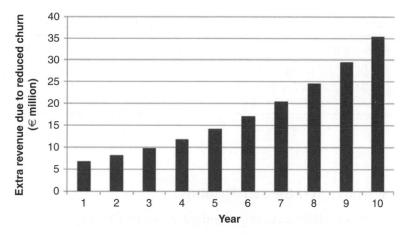

Figure 7.20   Extra revenue per year due to reduced churn.

	Year 1	Year 2	Year 3	Year 4	Year 5	Year 6	Year 7	Year 8	Year 9	Year 10
CAPEX reduction	**0.98**	**2.68**	**6.97**	**11.11**	**17.62**	**17.24**	**20.81**	**19.03**	**20.60**	**20.81**
Carriers	0.83	1.87	4.31	6.02	8.81	8.08	9.49	8.59	9.29	9.57
Site facilities	0.16	0.72	2.56	4.79	8.51	8.76	10.92	10.14	10.91	10.85
RNCs	0.00	0.10	0.10	0.30	0.30	0.40	0.40	0.30	0.40	0.40
OPEX reduction	**6.55**	**7.62**	**9.31**	**10.95**	**12.54**	**13.70**	**11.57**	**9.26**	**8.60**	**3.70**
Fewer E1	6.55	7.62	9.31	10.95	12.54	13.70	11.57	9.26	8.60	3.70
Extra revenue	**6.86**	**8.23**	**9.87**	**11.85**	**14.22**	**17.06**	**20.48**	**24.57**	**29.48**	**35.38**
Lower churn	6.86	8.23	9.87	11.85	14.22	17.06	20.48	24.57	29.48	35.38
Total (€ MEUR)	**14.39**	**18.53**	**26.15**	**33.92**	**44.37**	**48.00**	**52.85**	**52.87**	**58.68**	**59.90**

Figure 7.21   Cash flow projection for Self-Optimization.

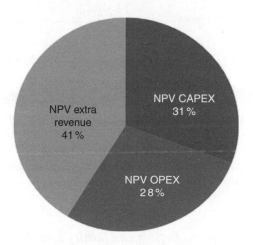

Figure 7.22   ROI breakdown.

Table 7.6   NPV of Self-Optimization versus traffic growth

Relative traffic growth pattern (%; years 2–10)	NPV (€ million)
[10, 10, 10, 10, 10, 10, 10, 10, 10]	101.77
[20, 20, 20, 20, 20, 20, 20, 20, 20]	138.36
[30, 30, 30, 30, 30, 30, 30, 30, 30]	175.17

7.5.   Case Study: ROI for Self-Healing

The benefits from Self-Healing can be categorized in two main areas:

1. OPEX reduction due to reduced workload through automation of monitoring and troubleshooting tasks that were also being carried out when the network was managed manually. In this respect, field experience has shown that, by means of smart automation, the time to find the root cause for the 50 worst-performing sectors in a 2G/3G network with 2000 sites can be reduced up to 90%.
2. Extra revenue due to improved QoS through faster resolution of network incidences, which leads to reduced churn.

7.5.1.   OPEX Reduction through Automation

The method for computing annual, recurrent OPEX savings ($OPEX_{Savings}$) due to reduced workload can be simply summarized by means of the following equation, assuming that troubleshooting activities are normally carried out on a daily basis:

$$OPEX_{Savings} = \frac{Num_{Sites}}{Num_{SitesPerEngineer}} \cdot$$
$$\cdot I \cdot OPEX_{PerEngineerAndDay} \cdot Num_{WorkingDaysPerYear} \cdot Reduction_{Workload} \qquad (7.26)$$

where Num_{Sites} is number of sites; $Num_{SitesPerEngineer}$ is the number of sites that every engineer is assigned to monitor and troubleshoot on a daily basis; I is the job intensity (100% for a full-time job, 50% for a part-time job, etc.); $OPEX_{PerEngineerAndDay}$ is the daily OPEX per engineer; $Num_{WorkingDaysPerYear}$ is the effective number of days in which noncritical maintenance and monitoring activities are carried out; and $Reduction_{Workload}$ is an efficiency factor describing the percentage of workload that can be reduced through automation.

7.5.2.   Extra Revenue due to Improved Quality and Reduced Churn

The same model that was introduced in Section 7.4.5 can be used.

Table 7.7 Assumptions for the ROI assessment of Self-Healing

Network size	
Number of sites	10000
Cost structure for OPEX	
Number of sites per engineer	100
Job intensity (100% for full time job; 50% for part-time job)	50%
OPEX per engineer and day	€300
Number of working days per year	52×5=260
Self-Healing impact	
Workload reduction factor (in line with field observations)	90%
DCR_{Manual}	0.7%
DCR_{SON}	0.5%
Subscribers, ARPU and churn	
Number of subscribers	20 million
Monthly ARPU	€10
Churn rate	10%
Percentage of churners due to bad network quality	10% (out of the 10% above, i.e. 1%)

	Year 1	Year 2	Year 3	Year 4	Year 5	Year 6	Year 7	Year 8	Year 9	Year 10
OPEX reduction	**3.51**	**3.51**	**3.51**	**3.51**	**3.51**	**3.51**	**3.51**	**3.51**	**3.51**	**3.51**
Automation	3.51	3.51	3.51	3.51	3.51	3.51	3.51	3.51	3.51	3.51
Extra revenue	6.86	6.86	6.86	6.86	6.86	6.86	6.86	6.86	6.86	6.86
Lower churn	6.86	6.86	6.86	6.86	6.86	6.86	6.86	6.86	6.86	6.86
Total (€ million)	**10.37**	**10.37**	**10.37**	**10.37**	**10.37**	**10.37**	**10.37**	**10.37**	**10.37**	**10.37**

Figure 7.23 Cash flow projection for Self-Healing.

7.5.3. Sample Scenario and ROI

The assumptions for this sample calculation are presented in Table 7.7. Based on the models presented in Section 7.5.1 and Section 7.5.2, the annual OPEX savings due to Self-Healing is €3.51 million and the annual extra revenue is €6.86 million. Adopting a time horizon of 10 years, and applying the same methodology that was used for Self-Planning and Self-Optimization, the cash flow projection in Figure 7.23 can be built. Assuming a discount rate of 20%, the resulting NPV (described in Equation 7.1) is €43.48 million.

7.6. References

[1] Ferreira, J., Vellasco, M., Pacheco, M. and Barbosa, C. (2004) Data Mining Techniques on the Evaluation of Wireless Churn, *European Symposium on Artificial Neural Networks Bruges*, Belgium, 28–30 April 2004.
[2] Pendharkar, P. C. (2009) Genetic algorithm based neural network approaches for predicting churn in cellular wireless network services, *Expert Systems with Applications: An International Journal*, **36**(3), pp. 6714–6720.
[3] J.D. Power and Associates Reports (2007) *Call Quality Plays an Increasingly Important Role in Customer Satisfaction With the Wireless Phone Experience*, 19 April 2007, http://www.prnewswire.com/news-releases/jd-power-and-associates-reports-call-quality-plays-an-increasingly-important-role-in-customer-satisfaction-with-the-wireless-phone-experience-58607222.html (accessed 3 June 2011).
[4] Lawson, C. L. and Hanson, R. J. (1974) *Solving Least Squares Problems*, Prentice-Hall, New Jersey.

Appendix A

Geo-Location Technology for UMTS

Carlos Úbeda

A.1. Introduction

The use of geo-located Measurement Reports (MRs) has been proven to be useful in applications like traffic map generation or accurate radio propagation modeling. A first approach is to use Received Signal Level (RSL) measurements, which are widely and easily available. However, distance-dependency is measured with very high intrinsic uncertainty leading to poor geo-location accuracy, and therefore current trends support the use of time-delay measurements [1, 2] or a combination of both [3]. Some studies [4, 5] also claim that tracking techniques instead of geo-location of single events provide better accuracy.

Geo-location algorithms based on time-delay measurements make use of the so-called Observed Time Differences (OTDs) [6] reported every time an MR is triggered. OTDs are subject to a certain number of constraints that will affect the accuracy when geo-locating the User Equipment (UE), such as multipath propagation, limited number of measured Base Transceiver Stations (BTSs), nonperfect BTS synchronization recovery, measurement errors and MRs not being sent in a continuous way but in an event-driven fashion.

Although there is wide literature [1-5] on geo-location and tracking algorithms for mobile networks, those studies mostly analyze theoretical scenarios using unrealistic tracking paths and too optimistic sets of MRs. This Appendix quantifies the effect of the aforementioned limitations in real networks, proposes several techniques to mitigate them, and finally analyzes in detail their impact on the geo-location accuracy.

This Appendix is organized as follows: Section A.2 briefly describes the use of OTDs applied to geo-location. Section A.3 provides a detailed description of the geo-location algorithm. Section A.4 states the assumptions used to assess the algorithm performance.

Self-Organizing Networks: Self-Planning, Self-Optimization and Self-Healing for GSM, UMTS and LTE,
First Edition. Edited by Juan Ramiro and Khalid Hamied.
© 2012 John Wiley & Sons, Ltd. Published 2012 by John Wiley & Sons, Ltd.

Section A.5 evaluates the performance of the proposed algorithm under different scenarios. Finally, A.6 summarizes the main conclusions from this Appendix.

A.2.   Observed Time Differences (OTDs)

OTDs are reported by UEs to the Radio Network Controller (RNC) for Active Set management every time an MR message is sent. According to the Third Generation Partnership Project (3GPP) [6], an OTD sent from a given UE that reports sector i can be expressed in chips[1] as:

$$T_M[i] = Q + R[i] - T_P[i] \tag{A.1}$$

where $T_M[i]$ is ranged between $[0,38399]$, Q is the internal time reference of the UE, $R[i]$ is the synchronization reference at sector i, and $T_P[i]$ is the propagation time, which is proportional to the distance $d[i]$ between the UE and sector i as follows:

$$T_P[i] = \frac{1}{\rho} \cdot d[i] = \frac{1}{\rho} \cdot \sqrt{(x_i - x_0)^2 + (y_i - y_0)^2} \tag{A.2}$$

where ρ is the chip length in meters, and $[x_i, y_i]$ and $[x_0, y_0]$ are respectively the sector and UE locations.

A.3.   Algorithm Description

OTDs are proportional to the propagation distance between the UE and the reported sector, and hence can be used for geo-location of events. Details are described in the following paragraphs.

A.3.1.   Geo-Location of Events

OTDs reporting the same sectors within a site (or BTS) do not provide any new linearly independent equation because the distance between those sectors and the UE is exactly the same. Therefore, it is possible to simplify the problem by grouping MRs in terms of sites as follows:

$$T_{NM}[i_S] = T_M[i] + 256 \cdot T_{cell}[i] \tag{A.3}$$

where $T_{NM}[i]$ is the normalized OTD, i_S is the site to which sector i belongs, and T_{cell} is a known network parameter that relates the synchronization reference of the different sectors within a site.

Considering that the internal time reference of the UE is unknown, relative values must be taken, so that the relative distance between sites i_S and j_S regarding a given UE can be written as:

$$d[i_S, j_S] = \rho \cdot (T_{NM}[j_S, i_S] + R[i_S, j_S]) \tag{A.4}$$

where the operator $X[a,b] = X[a] - X[b]$.

---

[1] A chip corresponds to 0.26 ms or 78 m.

Assuming that the synchronization reference of each site is known, the location of the UE $[x_o, y_o]$ can be derived by solving the following system of nonlinear equations:

$$\begin{pmatrix} d[1,2] = \rho \cdot (T_{NM}[2,1] + R[1,2]) \\ \vdots \\ d[1,n] = \rho \cdot (T_{NM}[n,1] + R[1,n]) \end{pmatrix} \tag{A.5}$$

where n is the number of different reported sites. Firstly a Taylor series [7] is applied for linearizing, and then the system of equations is solved using the well-known Recursive Least Squares (RLS) method [8]. It is worth highlighting that at least three different sites, that is, $n \geq 3$, must be reported in a given MR for the system of equations to be solvable. In case there are only two sites reported, the location of the UE could be calculated using Propagation Delay (PD) measurements, which report the distance to the call-setup sector, so that an extra equation could be added to Equation (A.5) as below:

$$T_{PD}[i_S] = \frac{1}{\rho} \cdot d[i_S] \tag{A.6}$$

where $T_{PD}[i_S]$ is the PD between the UE and the call-setup sector associated to site i_S. The use of PD is limited by the fact that it is reported with a granularity of three chips, and that it can be used only close enough in time to the call establishment.

A.3.2. Synchronization Recovery

As previously discussed, OTD measurements can be easily used for geo-location once the synchronization reference of the different sites is known. However, such synchronization is not often available in a real network, so that it is necessary to develop a practical methodology to estimate it. The presented algorithm follows the guidelines suggested in [9] and consists of an iterative approach that can be divided into the following steps:

1. Estimation of an initial synchronization reference for each site, which can be obtained by assuming that the initial location of the UE is the position of the sector with strongest signal.
2. Selection of a set of very well-conditioned events (reporting a high number of different sites, without multipath propagation, etc.) that are expected to speed up the convergence to the actual synchronization reference.
3. Geo-location of those selected events using Equation A.5 with the previously estimated synchronization reference.
4. Update of the synchronization reference according to the latest geo-location.
5. Repetition of the procedure until the synchronization reference converges.

A.3.3. Filtering of Events

There are multiple sources that can distort the OTD measurements of certain events such as strong multipath propagation. Those events should be clearly detected beforehand and not geo-located because they can worsen the estimated location. On the other hand, this kind of analysis could be also beneficial for the operator to identify certain configuration problems in the network. The main filtering criteria are listed below:

- Wrong Scrambling Code (SC) assignment: decoding the proper site based on the reported SC, so that the site coordinates are correct, is essential for an accurate geo-location. The probability of failing on the SC assignment depends on the SC planning and the area in which the MRs are collected. For instance, this probability increases in very dense areas.
- Multipath propagation: due to building and terrain reflections the signal can arrive with different delays, degrading the reported OTD measurements.
- Handover effects: according to 3GPP specifications [6], OTDs reporting the serving sector must be a multiple of 256 chips if there is good synchronization. However, after a handover process some time is needed for recovering such synchronization to the new serving sector. Events collected during this period may not be reliable.

A.4.   Scenario and Working Assumptions

Both topology and traffic conditions may have an impact on the frequency and the characteristics (potential multipath propagation, number of reported sites, etc.) of the collected MRs, and hence, on the achievable accuracy. This Appendix studies two different RNCs corresponding to a sub-urban area in Europe and an urban area in US. Forty-five minutes of call traces have been collected in each scenario containing enough data, that is, more than 200 000 events, for a detailed analysis. Accuracy is measured by comparing the estimated geo-location with a drive-test performed during the period in which call traces were collected in the case of Europe, and with two static locations in the US scenario. Figure A.1 shows where MRs were collected during the drive test route followed in the European scenario. As commented in Section A.1, there are long periods of time with almost total absence of MRs, which makes proper traffic estimation in those areas more difficult.

A.5.   Results

The presented analysis shows the performance of the proposed geo-location algorithm based on OTDs under two different scenarios from real networks. All results consider only events that can be directly located, i.e. those reporting at least three different sites, except in Section A.5.4, where the impact of using PD measurements is studied.

A.5.1.   Reported Sites per Event

The amount of different reported sites is a key factor for an event to be properly geo-located. Figure A.2 shows the Probability Density Function (PDF) of the number of reported sites per event in both scenarios. As expected, events reporting fewer sites are more common, but it is important to focus on the weight that two-site events, which cannot be directly located, have in the PDF distribution. For the scenario in US, this percentage reaches around 60%, which may indicate that this network has better dominance, i.e. lower number of interferers as compared to the network from Europe.

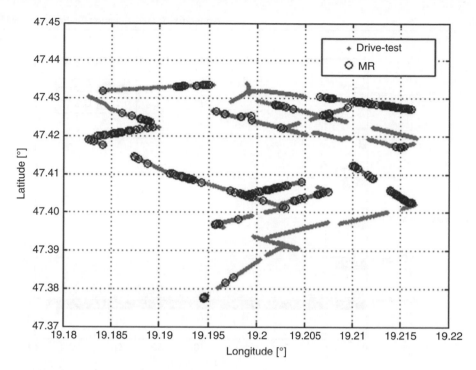

Figure A.1 Location of MRs during the drive test route followed in the European scenario.

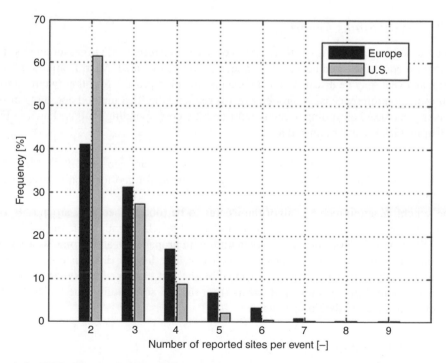

Figure A.2 PDF of the number of different reported sites per event for scenarios in US and Europe.

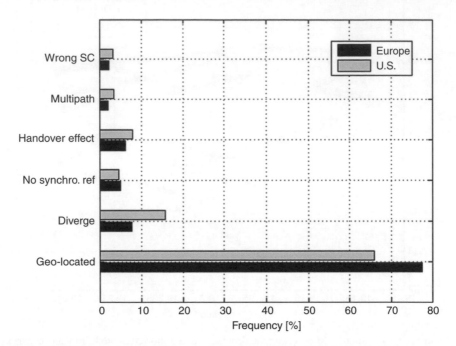

Figure A.3   PDF of the event status for scenarios in US and Europe, including only events reporting at least three different sites.

A.5.2.   Event Status Report

As mentioned in Section A.3.3, there are events with distorted OTD measurements due to propagation conditions, and hence, they should not be geo-located. Apart from those filtering criteria, an event may be discarded during the geo-location process for other two reasons: the synchronization reference is not properly recovered, and the RLS method does not converge. Events are classified according to their status and the corresponding PDF is shown in Figure A.3, from which it can be drawn that:

- Most of the events reporting at least three different sites are geo-located, that is between 65–80%, which indicates that using OTD measurements for geo-location in a real network is a feasible approach.
- The synchronization recovery algorithm seems to be robust since only around 5% of the events in both scenarios are discarded due to lack of synchronization reference.
- The US scenario shows worse numbers in multipath propagation and wrong SC assignment as compared to the selected European scenario. The largest difference occurs for the divergence criterion, which can be explained by the fact that an event is more robust when it reports more different sites. And events from the US scenario include noticeably fewer reported sites than in the European case.

A.5.3.   Geo-Location Accuracy

Accuracy numbers are shown in Figure A.4 for both scenarios in Europe and US. As expected, results are worse in the US scenario due to the higher number of distorted OTD measurements

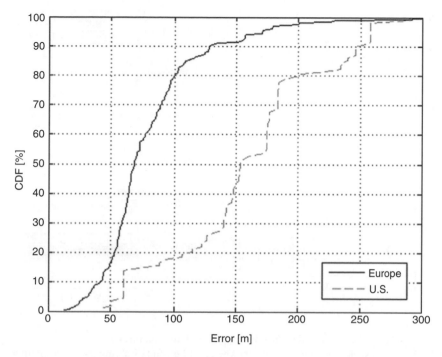

Figure A.4 CDF of the geo-location error for scenarios in US and Europe, including only events reporting at least three different sites.

(see Figure A.3). However, the median error for the scenario in US is less than 150 m, which is very promising considering that OTD measurements have an intrinsic uncertainty of a chip.

When the event filtering is disabled, firstly the upper tail of the Cumulative Distribution Function (CDF) gets worse and, secondly, the number of events that diverge increases, which proves that bad-conditioned events are more likely to be erroneously geo-located. Figure A.5 shows the CDF of the geo-location error (a) and the event status summary (b) for the European scenario comparing results when enabling and disabling the event filtering.

A.5.4. Impact of Using PD Measurements

Previous results are focused only on events reporting at least three different sites. However, two-site events can be geo-located by combining PD and OTD measurements. The use of PD is reliable when it is close enough in time to the call establishment. In this case a time-window of 20 s is considered. The CDF of the geo-location error (a) and the event status summary (b) for the European scenario are depicted in Figure A.6. As expected, only a small percentage of events (less than 40%) can make use of the reported PD, which also shows poorer accuracy than events reporting at least three different sites. This means that the benefit of PD measurements is limited.

A.6. Concluding Remarks

This Appendix evaluates a time-delay based geo-location algorithm providing promising results in both accuracy and percentage of located events in two different real UMTS networks

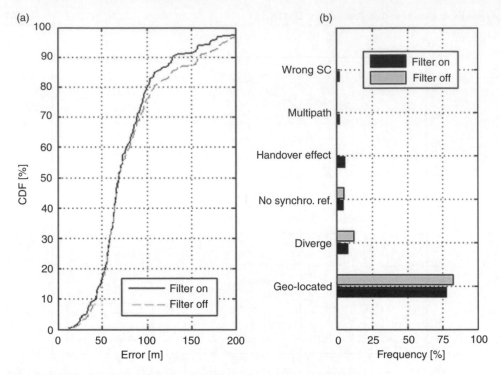

Figure A.5 CDF of the geo-location error (a) and event status summary
(b) for the European scenario comparing results when enabling and disabling the event filtering. This
analysis only includes events reporting at least three different sites.

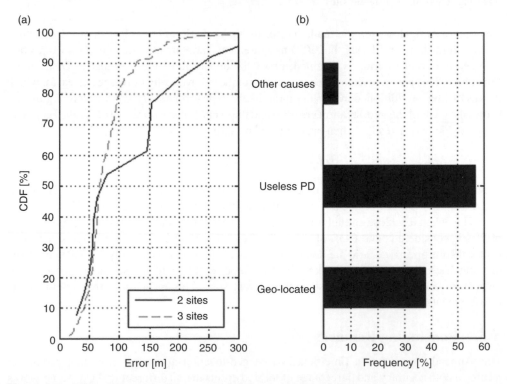

Figure A.6 CDF of the geo-location error (a) and event status summary
(b) for the European scenario considering only two-site events.

[10]. Between 65–80% of the events are geo-located with an error lower than 150 m at the fiftieth percentile. The algorithm performs slightly worse in the selected US scenario than in the European one that was analyzed, which can be explained by the fact that the former includes noticeably less reported sites per event, and such events are also worse conditioned. The use of filtering prior to geo-location is proven to be beneficial since it accurately discards events that are more likely to be erroneously geo-located. It is also shown that there are long periods of time with almost total absence of MRs, which makes proper traffic estimation in those areas more difficult. Finally, the study illustrates that there is a large percentage of two-site events, which cannot be directly geo-located. However, the use of PD measurements has a limited benefit, which leaves room for improvement using, for instance, a combination of the aforementioned techniques with RSL measurements.

A.7.   References

[1] Deng, P. and Fan, P.Z. (2000) An AOA Assisted TOA Positioning System, *Proc. World Computer Congress – International Conference Communication-Technology*, August 2000.

[2] Cong, L. and Zhuang, W. (2002) Hybrid TDOA/AOA Mobile User Location for Wideband CDMA Cellular Systems, *IEEE Transaction on Wireless Communications*.

[3] McGuire, M., Plataniotis, K.N. and Venetsanopoulos, A.N. (2005) Data Fusion of Power and Time Measurements for Mobile Terminal Location, *IEEE Transactions on Mobile Computing*.

[4] Nájar, M., Cabrera, M. and Játiva, R. (2001) Kalman Tracking for UMTS Mobile Location, *IEEE 53rd Vehicular Technology Conference*, May 2001.

[5] Zaidi, Z.R. and Mark, B.L. (2005) Real-Time Mobility Tracking Algorithms for Cellular Networks Based on Kalman Filtering, *IEEE Transactions on Mobile Computing*.

[6] 3GPP, Technical Specification, Technical Specification Group Radio Access Network (2010), Physical Layer; Measurements (FDD), 3GPP TS 25.215 Version 9.2.0, Release 9, 30 March 2010, http://www.3gpp.org/ftp/ Specs/archive/25_series/25.215/25215-920.zip (accessed 3 June 2011).

[7] Selçuk Bayin, S. (2006) *Mathematical Methods in Science and Engineering*, John Wiley & Sons, Inc., New York.

[8] Xu, G. (2007) *GPS: Theory, Algorithms, and Applications*, Springer, Heidelberg.

[9] Brunner, C. and Flore, D. (2009) Generation of Pathloss and Interference Maps as SON Enabler in Deployed UMTS Networks, *IEEE 69th Vehicular Technology Conference*, April 2009.

[10] Úbeda, C., Romero, J. and Ramiro, J. (2010) Evaluation of a Time-Delay based Geolocation Algorithm in Real UMTS Networks, *Fifth International Conference on Broadband and Biomedical Communications*, December 2010.

Appendix B

X-Map Estimation for LTE*

Michaela Neuland, Mehdi Amirijoo and Thomas Kürner

B.1. Introduction

Information about the radio coverage is essential for network planning, RF tuning, and Radio Resource Management (RRM) parameters optimization. Nowadays, network operators use prediction tools to produce this information. These tools are based on maps providing topographic and land use information (e.g. buildings, natural areas and roads), as well as on tuned propagation models. However, this approach is not fully accurate. Reasons for the inaccuracies are imperfections in the used geographic data, simplifications or approximations in the applied propagation models, and changes in the environment caused by, for example, constructions/demolitions or seasonal effects (foliage changes). Furthermore, variations in the traffic distribution and user profiles are a source for inaccurate results provided by prediction tools. For this reason, rigorous drive tests are performed in order to get more information about the real situation in the network and to gather measurement data for calibrating the used propagation models. Drive tests provide a picture of the end user perception in the field and enable the operator to identify locations causing poor performance and their corresponding cause (e.g. incorrect tilt or handover settings). Drive tests are, however, not ideal since they are expensive, time-consuming and cover only a limited (outdoor) part of the network due to access restrictions. Another disadvantage is that only a snapshot in time of the conditions in the field is captured. These difficulties with drive tests could be overcome if the User Equipments (UEs) in the network could be used for reporting the observed service quality along with the positions where the measurements are taken. The standardization of such UE reports is currently being carried out in the Third Generation Partnership Project (3GPP) [1]. These UE reports can be used to create a geographic map with overlay performance information, referred to as an X-map where X can stand for different types of performance

*The work presented in Appendix B was carried out within the FP7 SOCRATES project, which is partially funded by the Commission of the European Union. See http://www.fp7-socrates.eu/.

Self-Organizing Networks: Self-Planning, Self-Optimization and Self-Healing for GSM, UMTS and LTE,
First Edition. Edited by Juan Ramiro and Khalid Hamied.
© 2012 John Wiley & Sons, Ltd. Published 2012 by John Wiley & Sons, Ltd.

information. A coverage map, for instance, would be a specific kind of X-map. The UE report processing is done by an X-map estimation function that continuously monitors the network, estimates its spatial performance characteristics, e.g. the coverage and throughput, and maps the UE measurements to an estimated geographic position. In Long Term Evolution (LTE), three different positioning techniques are foreseen, namely, the network-assisted version of Global Navigation Satellite Systems (GNSS) such as Global Positioning System (GPS) or Galileo, Observed Time Difference Of Arrival (OTDOA) and an enhanced Cell-ID positioning method [2]. In addition to the measurement data, the X-map estimation function may also use other sources of information, such as prediction data. The advantage of the X-map estimation function is that rigorous drive tests can be reduced. This will significantly reduce network maintenance costs for operators, ensure faster optimization cycles resulting in higher customer satisfaction and provide measurement data from areas that are not accessible for drive tests, e.g. narrow roads, forests, private land, houses or offices [1]. Apart from assisting the network operator in observing the network performance, the information embedded in an X-map may be used as an integral part of Self-Organizing Networks (SON) [3, 4], especially in functionalities addressing the optimization of coverage, capacity and quality.

The accuracy of an X-map depends on many factors such as the applied UE positioning technique, the UE measurement accuracy, the number of measurements taken and the network architecture. In particular, the positioning technique has a significant impact on the overall X-map estimation accuracy.

B.2.   X-Map Estimation Approach

An overview of the X-map estimation approach is given in Figure B.1. The UE Location and Measurement Unit (LMU) gathers measurement data such as the Reference Signal Received Power (RSRP) or the Reference Signal Received Quality (RSRQ), location data, and the corresponding time from the UEs in the network. In order to manage the positioning of UEs in LTE, the Evolved Serving Mobile Location Center (E-SMLC) is introduced [2]. Depending on the conditions in the network and on the UE/enhanced NodeB (eNodeB) positioning capabilities, this entity decides which positioning method should be used and combines all the received results to determine one single position estimate for the UE. Furthermore, it can provide additional information such as the accuracy of the position estimate or the velocity of the UE. Thus, the E-SMLC provides all important positioning features that have to be fulfilled by the LMU and may be utilized for providing the necessary data for an X-map estimation function. In addition to the LMU, a RAN Measurement Unit (RMU) is introduced which collects the measurement data from the RAN, e.g. the interference and load, and utilizes this data in order to better estimate the network state. The measurement data of both measurement units together with the positions of the UEs are then delivered to the X-map estimation function, which uses this data directly for updating the corresponding bins in the X-map (approach 1) or for updating a propagation model (approach 2).

If new measurement data together with a UE position estimate become available, approach 1 uses this information for updating the corresponding bin in the X-map where the UE is located, similar to the approaches in [5, 6]. Various update mechanisms are possible, e.g. the latest measurement value can replace the current value in the bin or it can be combined with the current value in the bin, such as by applying a filter approach. In the presented simulations, the average of the value in the X-map and the UE reports gathered over a time interval is

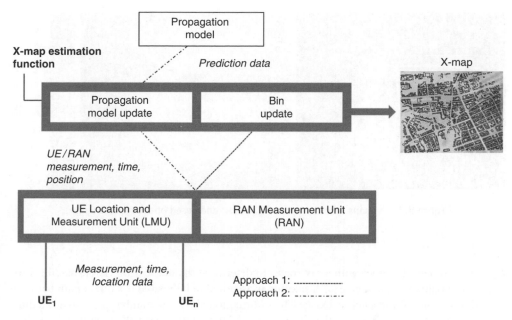

Figure B.1   Overview of the X-map estimation approaches.

calculated. In areas where no measurement data can be taken, e.g. due to coverage holes, no information is available in the corresponding part of the X-map.

In approach 2, the X-map is based on prediction data created by means of a propagation model which is adapted to the environment of the corresponding eNodeB with the help of the provided measurement data. In the presented simulations, the Okumura-Hata model [7] is used and correction factors are calculated for the considered land use classes based on the collected measurement data. This is part of the calibration method described in [8] and similar to the method introduced in [9]. By calibrating the correction factors per land use class, the accuracy of the Okumura-Hata model can be improved.

B.3.   Simulation Results

For simulating the two approaches, a scenario of 1.5 km×1.5 km in an urban area is used [10]. In this scenario, 20 users move along the streets and send measurement reports every 200 ms to their serving eNodeBs during a simulation time of 1000 s. For the LTE network, the site locations, sector orientations and antenna tilts are taken from an existing UMTS layout. For these sites, realistic path loss predictions at 2.6 GHz are available. These predictions are used to determine the 30 strongest cells for each user position. Furthermore, the predictions serve as a reference for determining the accuracy of the resulting X-maps (in terms of mean error and standard deviation) by comparing every pixel with a valid RSRP value in the X-map with the corresponding pixel of the prediction. For the simulation area, land use data is derived from the freely available information provided by the OpenStreetMap project [http://www. openstreetmap.org]. Five different land use classes are distinguished, namely, buildings, natural areas, waterways, streets and railways. Furthermore, satellite orbits for a specific date

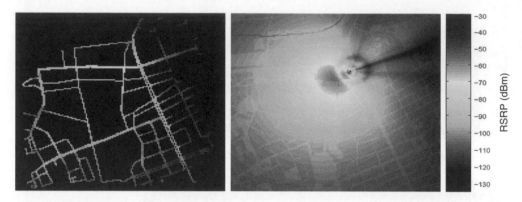

Figure B.2   X-maps based on approach 1 (left) and based on approach 2 (right).

and time are used together with a ray tracer to determine the number of visible satellites for each user position in the scenario. A satellite is assumed to be visible if a direct path between the UE position and the satellite exists. The information about the number of visible satellites and about the geometry between the UE position and the satellite, as well as about the standard deviation of the measurement error, is utilized to model the position error based on the approach described in [11]. The calculated position errors are then added to the real positions provided by the scenario in order to obtain the estimated user positions. If less than four satellites are visible, e.g. in dense urban canyons, it is assumed that positioning with GPS is not possible. In such cases, other positioning methods should be used.

In Figure B.2, two examples of X-maps with a resolution of $10\,m \times 10\,m$ showing RSRP in dBm for a specific site of the scenario can be seen. On the left side, approach 1 is used to create the X-map, whereas the X-map on the right side is based on approach 2. For both X-maps, the real positions as provided by the scenario are applied.

The advantage of approach 1 is that the resulting X-map reflects the real situation in the network since the measurement data is used directly for creating the X-map. However, the disadvantage of this approach is that the X-map provides RSRP values only for those pixels that are covered by the UEs as can be seen in Figure B.2 (the dark area means that there is no RSRP value available). In contrast, every pixel in the area has an RSRP value when using approach 2, i.e. the coverage of the X-map is greater compared to approach 1. The accuracy, however, is not as high as for the X-map based on approach 1 with a mean error of 2.1 dB and a standard deviation of 6.6 dB as shown in Table B.1. Note that the accuracy is determined by comparing every pixel with an available RSRP value with the reference values, i.e. pixels that are not included in the calibration process are also considered when determining the accuracy of the X-map based on approach 2.

The accuracy for approach 2 is comparable to what can be achieved with today's calibrated propagation models for small macro- and microcells. For the models presented in [7], for example, about 7 dB up to 9 dB is mentioned as the average standard deviation of the prediction error. The absolute values of the mean errors typically range from about 0 dB up to 6 dB. For the corresponding analyses, building data and measurement data at 947 MHz were available for an area in downtown Munich [7].

Table B.1 Mean error (mean) and standard deviation (std.) for different X-maps using the real positions and GPS as positioning method

	Approach 1		Approach 2	
	Mean	Std.	Mean	Std.
Real position	0.0	0.2	2.1	6.6
Position estimate	0.1	2.3	2.6	6.6

Approach 1 strongly depends on the applied positioning method as can be seen in Table B.1. The standard deviation of the X-map based on approach 1 is increased by 2 dB using GPS as positioning method instead of using the real positions. In contrast, approach 2 is not sensitive to the applied positioning method regarding the standard deviation, which is the same for using the real position or a position estimate applying GPS.

B.4. References

[1] 3GPP, Technical Report, Technical Specification Group Radio Access Network (2010) *Study on Minimization of Drive-Tests in Next Generation Networks*, 3GPP TR 36.805 Version 9.0.0, Release 9, 5 January 2010, http://www.3gpp.org/ftp/Specs/archive/36_series/36.805/36805-900.zip (accessed 3 June 2011).

[2] 3GPP, Technical Specification, Technical Specification Group Radio Access Network (2010) *Stage 2 Functional Specification of User Equipment (UE) positioning in E-UTRAN*", 3GPP TS 36.305 Version 9.2.0, Release 9, 28 April 2010, http://www.3gpp.org/ftp/Specs/archive/36_series/36.305/36305-920.zip (accessed 3 June 2011).

[3] 3GPP, Technical Report, Technical Specification Group Radio Access Network (2009) *Self-Configuring and Self-Optimizing Network Use Cases and Solutions*, 3GPP TR 36.902 Version 9.0.0, Release 9, 9 October 2009, http://www.3gpp.org/ftp/Specs/archive/36_series/36.902/36902-900.zip (accessed 3 June 2011).

[4] van den Berg, J.L., Litjens, R., Eisenblätter, A., Amirijoo, M., Linnell, O., Blondia, C., Kürner, T., Scully, N., Oszmianski, J. and Schmelz, L.C. (2008) Self-Organisation in Future Mobile Communication Networks, *Proc. ICT - Mobile Summit 2008*, Stockholm, Sweden, June 10–12 2008.

[5] Catovic, A. and Dills, J.F.M.M.K. (2009) *Apparatus and method for generating performance measurements in wireless networks*, Patent US 2009/0310501, issued December 2009.

[6] Brunner, C. and D. Flore, D. (2009) Generation of Pathloss and Interference Maps as SON Enabler in Deployed UMTS Networks, *Proc. IEEE 69th Vehicular Technology Conference (VTC 2009-Spring)*, pp. 1–5.

[7] COST 231 (1999) *Final Report* http://www.lx.it.pt/cost231/ (accessed 3 June 2011).

[8] Erik, A. and S. Holm, S. (2004) Tuning of Empirical Radio Propagation Models Effect of Location Accuracy, *Wireless Personal Communications*, **30**, pp. 267–281, Kluwer Academic Publishers, Dordrecht.

[9] Clancy, J.G. (2003) *Clutter database enhancement methodology*, Patent US 6,580,911 B1, issued June 2003.

[10] SOCRATES Project (2009) Deliverable: *Review of use cases and framework II*, December 2009, http://www.fp7-socrates.eu (accessed 3 June 2011).

[11] Fritsche, C. and Klein, A. (2008) Cramér-Rao Lower Bounds for Hybrid Localization of Mobile Terminals, *Proc. 5th Workshop on Positioning, Navigation and Communication (WPNC '08)*, March 2008.

Index

Self-Organizing Networks: Self-Planning, Self-Optimization and Self-Healing for GSM, UMTS and LTE,
First Edition. Edited by Juan Ramiro and Khalid Hamied.
© 2012 John Wiley & Sons, Ltd. Published 2012 by John Wiley & Sons, Ltd.